Recommendations for Design and Analysis of Earth Structures using Geosynthetic Reinforcements – EBGEO

Ernst & Sohn
A Wiley Company

Recommendations for Design and Analysis of Earth Structures using Geosynthetic Reinforcements – EBGEO

Translation of the
2nd German Edition

Published by
the German Geotechnical Society
(Deutsche Gesellschaft für Geotechnik e. V., DGGT)

Deutsche Gesellschaft
für Geotechnik e. V.
German Geotechnical Society

Working Group 5.2 ‚Analysis and Dimensioning of Soil Structures using Geosynthetic Reinforcements' Technical Group ‚Synthetics in Geotechnical Engineering' of the German Geotechnical Society
Chairman: AOR Dipl.-Ing. Gerhard Bräu
Technische Universität München
Zentrum Geotechnik
Lehrstuhl und Prüfamt für Grundbau, Bodenmechanik,
Felsmechanik und Tunnelbau
Baumbachstraße 7
81245 München
Germany

Translator: Alan Johnson, Nordstemmen/Adensen, Germany

Cover photographs:
Top left: Geotextile-encased sand columns on the Mühlenberger Loch project, Hamburg (photo: Huesker Synthetic GmbH); Top right: Steep embankment for heavy goods vehicle transport in Aalen, reinforced using geogrids (photo: Tensar International GmbH); Bottom left: Landslide rehabilitation using geogrids along the B115 road near Altenmarkt in Austria (photo: TenCate Geosynthetics Deutschland GmbH); Bottom right: Geogrids as base course reinforcement in redevelopment of a former mining spoil tip along the A38 autobahn near Leipzig (photo: Naue GmbH & Co. KG)

Library of Congress Card No.:
applied for

British Library Cataloguing-in-Publication Data
A catalogue record for this book is available from the British Library.

Bibliographic information published by the Deutsche Nationalbibliothek
The Deutsche Nationalbibliothek lists this publication in the Deutsche Nationalbibliografie; detailed bibliographic data are available on the Internet at <http://dnb.d-nb.de>.

Coverdesign: Designpur, Berlin
Production Management: pp030 – Produktionsbüro Heike Praetor, Berlin
Typesetting: Manuela Treindl, Fürth
Printing and Binding: Strauss GmbH, Mörlenbach

Printed in the Federal Republic of Germany.
Printed on acid-free paper.

ISBN 978-3-433-02983-1
Electronic version available. O-Book ISBN 978-3-433-60093-1

German Geotechnical Society

Working Group 5.2
Analysis and Dimensioning of Earth structures using
Geosynthetic Reinforcements

Chairman: Dipl.-Ing. Bräu, Munich
Deputy Chairman: Dipl.-Ing. Herold, Weimar

Members of the Working Group:

Dr.-Ing. Alexiew, Gescher
Dr.-Ing. Bauer, Olching
Dipl.-Ing. Blume, Overath
Dipl.-Geol. Blume, Bergisch-Gladbach
Dipl.-Ing. Dollowski, Bonn
Prof. Dr.-Ing. Göbel, Dresden
Dipl.-Ing. Hubal, Munich (formerly)
Dipl.-Ing. Jas, AM Oostvoorne (formerly)
Prof. Dr.-Ing. Kempfert, Kassel
Prof. Dr.-Ing. Klapperich, Freiberg
Dr.-Ing. Köhler, Erfurt
Dr.-Ing. Magnus, Leipzig
Dipl.-Ing. Mannsbart, Linz (formerly)
Prof. Dr.-Ing. Meyer, Clausthal-Zellerfeld
Prof. Dr.-Ing. Müller-Rochholz, Münster
Dipl.-Ing. Murray, Dietzenbach
Dipl.-Ing. Naciri, Bonn
Prof. Dr.-Ing. Nimmesgern, Wurzburg
Dipl.-Ing. Pachomow, Cottbus
Prof. Dipl.-Ing. Paul, Lindau
Dr.-Ing. Raithel, Würzburg
Dr.-Ing. Retzlaff, Steinfurt
Dr.-Ing. Reuter, Minden
Prof. Dr.-Ing. Riße, Rostock (formerly)
Dipl.-Ing. Scheu, Lübbecke (formerly)
Dipl.-Ing. Schön, Erlangen
Dr.-Ing. Schwerdt, Köthen
Dr.-Ing. Trunk, Offenbach
Dipl.-Ing. Vogel, Munich
Dipl.-Ing. Vollmert, Espelkamp
Dr. Wilmers, Wetzlar
Prof. Dr.-Ing. Ziegler, Aachen

Recommendations for Design and Analysis of Earth Structures using
Geosynthetic Reinforcements (EBGEO). German Geotechnical Society.   V
© 2011 Ernst & Sohn GmbH & Co. KG.
Published by Ernst & Sohn GmbH & Co. KG.

Preface

The 1997 edition of EBGEO presented the profession with recommendations for designing and analysing earth structures using geosynthetic reinforcements. It adopted the partial safety factor concept used in geotechnical standards, which was then still being developed, more or less in its entirety. The introduction of the 2005 edition of DIN 1054 as part of the body of legally binding building regulations and the associated European regulations made it necessary to revise EBGEO. In addition, unification of the various analysis approaches was necessary to keep pace with fundamental product developments and new applications. These were implemented exhaustively by the members and guests of the German Geotechnical Society's *(Deutsche Gesellschaft für Geotechnik e. V. (DGGT))* Working Group AK 5.2 'Analysis and Dimensioning of Soil Structures using Geosynthetic Reinforcements' in innumerable meetings comprising both small and large groups. We would like to take this opportunity to say many thanks to all involved!

In addition to a thorough revision of the existing sections, where both practical construction experience and the most recent national and international research results have been incorporated, new sections covering:

- Reinforced Earth Structures over Point or Linear Bearing Elements,
- Foundation Systems Using Geotextile-encased Columns,
- Bridging Subsidence and
- Dynamic Actions on Geosynthetic-reinforced Systems

were included.

Positive experience was gathered on a number of construction projects during the Recommendations' compilation phase and their applicability confirmed – in cluding on international projects. The Working Group also regards this edition of EBGEO as an intermediate stage, because in many cases it is still only possible to design in terms of individual components, but not in terms of the actual 'soil/ geosynthetic' composite construction material. However, the latter represents the primary objective, which will be pursued by way of more research and monitoring measures on active construction projects.

EBGEO follows the tradition of similar DGGT recommendations such as the EAB (Recommendations on Excavations) or the Recommendations of the Working Group on Piles, which now represent established best practice. The user is referred to the Notes for the User with regard to the compulsory nature of these Recommendations (see Page XXI, taken from EAB (2006), 4th edition, Ernst & Sohn).

The German Geotechnical Society's *(Deutsche Gesellschaft für Geotechnik e. V. (DGGT))* Working Group AK 5.2 'Analysis and Dimensioning of Soil Structures using Geosynthetic Reinforcements' asks you to send any suggestions and correspondence concerning further development of the Recommendations to the Chairman of AK 5.2 (see Page IV for address).

Munich, 2010

G. Bräu

Recommendations for Design and Analysis of Earth Structures using Geosynthetic Reinforcements (EBGEO). German Geotechnical Society.
© 2011 Ernst & Sohn GmbH & Co. KG.
Published by Ernst & Sohn GmbH & Co. KG.

Preface to the English edition

This edition is a translation of the 2nd edition of EBGEO published in April 2010. To improve understanding among the international readership the German limit state designations were translated using the terms employed in EN 1997 (EC7):

GZ 1A   EQU
GZ 1B   STR
GZ 1C   GEO
GZ 2   SLS

However, this does not mean that EBGEO is now based technically on EN 1997 – it is still based on the 2005 edition of the German DIN 1054. If any confusion arises as a consequence of translation, the German original is the authoritative text.

Working Group 5.2 would like to thank everybody involved in publishing the English-language edition, in particular Mr Alan Johnson, who did an excellent job of translating the German original.

Munich, March 2011   *G. Bräu*

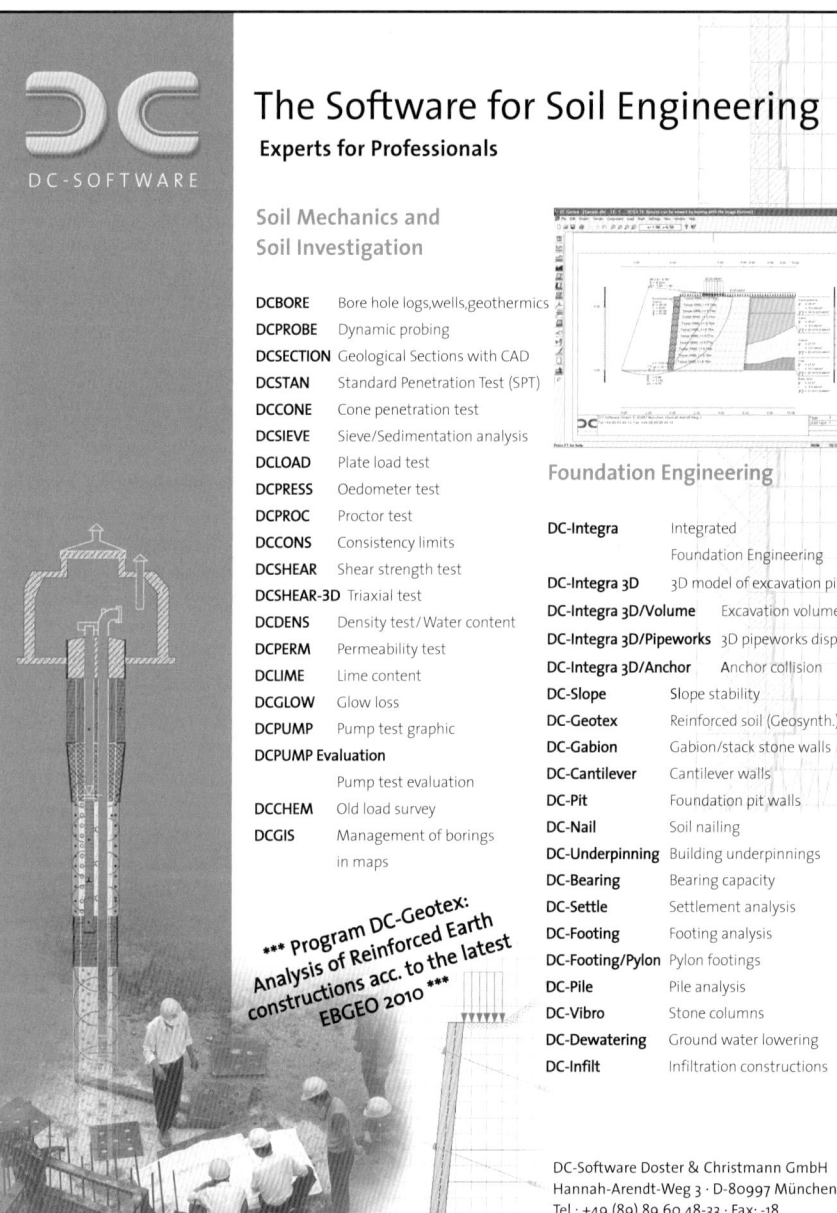

Contents

Recommendations for Design and Analysis of Earth Structures using
Geosynthetic Reinforcements (EBGEO). German Geotechnical Society.
© 2011 Ernst & Sohn GmbH & Co. KG.
Published by Ernst & Sohn GmbH & Co. KG.

XV

XIX

Notes for the User

1. The Recommendations of the Working Group on 'Analysis and Dimensioning of Soil Structures using Geosynthetic Reinforcements' represent technical regulations. They are the result of voluntary efforts within the technical-scientific community, are based on valid and current professional principles, and have been tried and tested as 'general best practice'.

2. The Recommendations of the Working Group on 'Analysis and Dimensioning of Soil Structures using Geosynthetic Reinforcements' may be freely applied by anyone. They represent a yardstick for flawless technical performance; this yardstick is also of legal relevance. A duty to apply the recommendations may result from legislative or administrative provisions, contractual obligations or other legal requirements.

3. The Recommendations of the Working Group on 'Analysis and Dimensioning of Soil Structures using Geosynthetic Reinforcements' represent an important source of information for professional conduct in normal design cases. They cannot reproduce all possible special cases in which advanced or more restrictive measures may be required. Note also that they can only reflect best practice at the time of publication of the respective edition.

4. Deviations from the suggested analysis approaches may prove necessary in individual cases, if founded on appropriate analyses, measurements or on empirical data.

5. Use of the Recommendations of the Working Group on 'Analysis and Dimensioning of Soil Structures using Geosynthetic Reinforcements' does not release anybody from their own professional responsibility. In this respect, everybody works at their own risk.

Recommendations for Design and Analysis of Earth Structures using Geosynthetic Reinforcements (EBGEO). German Geotechnical Society.
© 2011 Ernst & Sohn GmbH & Co. KG.
Published by Ernst & Sohn GmbH & Co. KG.

1 Introduction to the Recommendations and their Application Principles

Note: The following paragraphs are taken in part from the EAB (2006) or are based on them.

1.1 National and International Regulations

In Germany the analysis and design of reinforced fill structures, as well as the required safety stipulations, are controlled by DIN 1054 and other relevant standards. These Recommendations are based on DIN 1054:2005-01 'Subsoil – Verification of the Safety of Earthworks and Foundations' and analyses are performed using the partial safety factor approach. In addition, the European design standard EN 1997-1 (EC 7-1) 'Eurocode 7: Draft, Geotechnical Design' is also referenced; it too deals with reinforced structures. See Section 1.2 for details of the formal and planning control use of these two standards.

The following manufacturing standard is used for the individual reinforcement systems:

– DIN EN 14475: 'Execution of Special Geotechnical Work – Reinforced Fill'.

The following standards and regulations apply to quality assurance:

– DIN EN 13251: 'Geotextiles and Geotextile-related Products – Required Characteristics for use in Earthworks, Foundations and Retaining Structures',
– DIN EN 13249: 'Geotextiles and Geotextile-related Products – Required Characteristics for use in the Construction of Roads and other Trafficked Areas',
– *Merkblatt über die Anwendung von Geokunststoffen im Erdbau des Straßenbaus*, M-Geok E 05, FGSV 535, Forschungsgesellschaft für Straßen- und Verkehrswesen,
– *Technische Lieferbedingungen für Geokunststoffe im Erdbau des Straßenbaus*, TL Geok E-StB 05, FGSV 549, Forschungsgesellschaft für Straßen- und Verkehrswesen,
– Guidelines for Determining the Long-term Strength of Geosynthetics for Soil Reinforcement, English Edition ISO/TR 20432.

Inasmuch as no information to the contrary is given in these Recommendations, the respective current editions of the relevant technical regulations (e.g. standards, guidelines, codes of practice and recommendations) shall be observed. They are named in the appropriate sections.

A summary can be found at: http://www.gb.bv.tum.de/fachsektion/index.htm.

Hereinafter, references to standards are given without the publication date. If a certain paragraph is referred to directly the edition is also given.

Recommendations for Design and Analysis of Earth Structures using Geosynthetic Reinforcements (EBGEO). German Geotechnical Society.
© 2011 Ernst & Sohn GmbH & Co. KG.
Published by Ernst & Sohn GmbH & Co. KG.

Details of reference literature are given at the end of each respective section of these Recommendations.

1.2 Types of Analysis and Limit States using the Partial Safety Factor Approach

1.2.1 New Standards Generation and Transitional Regulations

A European Commission decision aims to replace the governing national building design and execution standards by European standards. Numerous European design and execution standards now exist for special geotechnical engineering.

The governing European execution standard for manufacturing reinforced fill structures is given in Section 1.1.

Analysis and design of reinforced fill structures in Europe are dealt with in EN 1997-1: 'Draft, Geotechnical Design' (Eurocode EC 7-1 (EC 7)). The German edition is published with the title DIN EN 1997-1:2005-10 and triggers a transition period within which a National Annex to Eurocode EC 7-1 shall be compiled to comply with European agreements. The National Annex (NA DIN EN 1997-1) will contain national specifications on those sections defined for this purpose in Eurocode EC 7-1. Simultaneously, another transition period begins, by the end of which Eurocode EC 7-1 will be introduced into building regulations in conjunction with the National Annex and all contradictory national regulations are withdrawn. A collateral DIN 1054:2010 standard to be compiled by 2009 may then only include non-contradictory supplements to Eurocode EC 7-1 in conjunction with the National Annex. The National Annex and the DIN 1054:2010 collateral standard have now been compiled in NA 005-05-01-01 and will be published in draft form in 2009. To simplify use of the three parallel standards they will be published together in a standards manual accompanying DIN EN 1997-1:2005 and DIN 1054:2009 'Draft, Geotechnical Design'. The regulations in the National Annex and the collateral standard have been adopted in the text of EC 7-1, and are specially marked.

Until the introduction of the Eurocodes a temporary generation of national standards using the partial safety factor approach meets the needs of all fields of structural engineering.

The following regulations, in particular, govern the construction of geosynthetic-reinforced structures:

- DIN 1055: 'Actions on Structures', in conjunction with DIN Fachbericht (*Technical Report*) 101,
- DIN 1054:2005-01: 'Verification of the Safety of Earthworks and Foundations'.

2

1.2.2 Effects and Resistances

The foundation for stability analyses is represented by the characteristic values for actions and resistances. The characteristic value, characterised by the index 'k', is a value with an assumed probability neither exceeded nor fallen short of during the reference period, taking the design working life or the corresponding design situation of the civil engineering structure into consideration. Characteristic values are generally specified on the basis of test results, measurements, analyses and/or empiricism.

The characteristic values of effects are multiplied by partial safety factors, those of resistances are divided. The variables acquired in this way are known as the design values of effects or resistances respectively and are characterised by the index 'd'. Different limit states are differentiated for stability analyses.

1.2.3 Limit States

The following limit states are differentiated in the partial safety factor approach:

- The ultimate limit state is a condition of the structure which, if exceeded, immediately leads to a numerical collapse or another form of failure. It is known as the ultimate limit state (ULS) in DIN 1054. Three cases of ultimate limit state are differentiated.
- The serviceability limit state (SLS) is a condition of the structure which, if exceeded, no longer fulfils the conditions specified for its use. It is known as the serviceability limit state (SLS) in DIN 1054.

The EQU limit state describes the loss of static equilibrium. It includes:

- analysis of safety against overturning,
- analysis of heave or uplift safety,
- analysis of hydraulic heave safety.

The EQU limit state incorporates favourable and unfavourable actions only, but no resistances.

The governing limit state condition is:

$$F_d = F_k \cdot \gamma_{dst} \leq G_k \cdot \gamma_{stb} = G_d , \qquad \text{Eq. (1.1)}$$

i.e. the destabilising actions F_k, multiplied by the partial safety factor $\gamma_{dst} > 1.0$, may only be as large as the stabilising action G_k, multiplied by the partial safety factor $\gamma_{stb} < 1.0$.

The STR limit state describes the failure of structures and structural elements or failure of the ground. It includes:

- analysis of the bearing capacity of structures and structural elements subjected to ground loads or supported by the ground,

3

- analysis of the bearing capacity of the ground, e.g. provided by passive earth pressure or bearing resistance, to ensure it is not exceeded.

Analysis of the bearing capacity of the ground to ensure it is not exceeded is performed in exactly the same way as for any other construction material. The limit state condition is always the governing condition:

$$E_d = E_k \cdot \gamma_F \le R_d , \qquad\qquad\qquad \text{Eq. (1.2)}$$

$$R_d = \frac{R_k}{\gamma_R} , \qquad\qquad\qquad\qquad \text{Eq. (1.3)}$$

i.e. the characteristic action or effect E_k, multiplied by the partial safety factor γ_F, may only be as large as the characteristic resistance R_k, divided by the partial safety factor γ_R. A characteristic of the STR limit state is that the effects and resistances are determined using characteristic values. The partial safety factors do not come into play until applying the limit state equation.

The GEO limit state is peculiar to geotechnical engineering. It describes the loss of overall stability. It includes:

- analysis of slope stability,
- analysis of global stability.

The governing limit state condition is:

$$E_d \le R_d , \qquad\qquad\qquad\qquad \text{Eq. (1.4)}$$

i.e. the design value E_d of the effects may only be as large as the design value of the resistances R_d. The geotechnical actions and resistances are determined using the design values for shear strength:

$$\tan \varphi_d' = \frac{\tan \varphi_k'}{\gamma_\varphi} \quad \text{and} \quad c_d' = \frac{c_k'}{\gamma_c} , \qquad\qquad \text{Eq. (1.5)}$$

and

$$\tan \varphi_{u,d} = \frac{\tan \varphi_{u,k}}{\gamma_{\varphi u}} \quad \text{and} \quad c_{u,d} = \frac{c_{u,k}}{\gamma_{cu}} \qquad\qquad \text{Eq. (1.6)}$$

i.e. the friction $\tan \varphi$ and cohesion c values adopted in the calculations are reduced from the outset using the partial safety factors γ_φ, $\gamma_{\varphi u}$, γ_c and γ_{cu}. An analogous procedure applies to the interface friction angle and adhesion.

The serviceability limit state describes the state of a structure or structural element at which the conditions specified for its use are no longer met, but without loss of bearing capacity. It is based on a serviceability analysis, i.e. that the antici-pated displacements and deformations are compatible with the purpose of the structure. Analysis uses characteristic values, where all partial safety factors are generally 1.0.

4

1.2.4   Applying EBGEO in Conjunction with DIN EN 1997-1

This edition of EBGEO is based on the stipulations made in DIN 1054. This in turn was closely harmonised with DIN EN 1997-1, Eurocode EC 7-1. DIN 1054 is not identical to Eurocode 7-1 in all details. At the transition to Eurocode 7-1/NA EC 7-1 (see 1.2.1) DIN 1054:2005-01 will be replaced by the collateral standard DIN 1054:2010. The consequences associated with this for applying the present edition of the Recommendations are related below, as well as a preview will allow.

Legally binding rules in terms of the applicability of the individual regulations are specified by the respective controlling authorities. The controlling agencies are deemed to be:

- the building regulations control authorities of the federal German states for building measures subject to the respective state building code; at regular intervals the upper building regulations control authorities of the respective federal states publish a list of technical building regulations applicable to that state.
- the departments of the Federal Ministry of Transport, Building and Urban Affairs responsible for inland waterways, federal roads and road bridges, and the Federal Railway Authority responsible for rail traffic.

Stability analyses as described in Section 1.2.3, Eurocode EC 7-1, provide three options in terms of the STR limit state. DIN 1054 is based on analysis procedure 2 to Eurocode EC 7-1, inasmuch as the partial safety factors are applied to both the effects and the resistances. To differentiate between this and the other permitted scenario, in which the partial safety factors are not applied to the effects but to the actions, this procedure is known as analysis method 2* in the Commentary to Eurocode EC 7-1.

The National Annex represents the link between Eurocode EC 7-1 and national standards. It states which of the possible analysis methods and partial safety factors are applicable in the respective national domains. Remarks, clarifications or supplements to Eurocode EC 7-1 are not permitted. However, the applicable, complementary national codes may be given. The complementary national codes may not, however, contradict Eurocode EC 7-1. Moreover, the National Annex may not repeat information already given in Eurocode EC 7-1.

The revised DIN 1054 will be paramount in the complementary national code; it has the working title 'DIN 1054:2010' and is the application rule to Eurocode EC 7-1.

The supplements, improvements and modifications included shall be adhered to inasmuch as they affect the regulations of the EBGEO, if the respective geosynthetic-reinforced structure is designed to Eurocode EC 7-1. However, they may also be utilised accordingly if design is based on DIN 1054.

In the current edition Eurocode EC 7-1 defines the following limit states instead of the limit states GZ 1A, GZ 1B and GZ 1C to DIN 1054:

- EQU: loss of equilibrium of the structure or the ground, which is regarded as rigid. The designation is derived from 'equilibrium'.
- STR: internal failure or very large deformation of the structure or its components, where the strength of the materials governs resistance. The designation is derived from 'structural failure'.
- GEO: failure or very large deformation of the structure or the ground, where the strength of the soil or rock governs resistance. The designation is derived from 'geotechnical failure'.
- UPL: loss of equilibrium of the structure or ground due to buoyancy or water pressure. The designation is derived from 'uplift'.
- HYD: hydraulic failure, internal erosion or piping in the ground, caused by a flow gradient. The designation is derived from 'hydraulic failure'.

In order to convey the GZ 1B und GZ 1C (STR and GEO) limit states from DIN 1054 to the terminology used in Eurocode EC 7-1 the GEO limit state is divided into GEO-2 and GEO-3:

- GEO-2: failure or very large deformation of the ground in conjunction with determining the action effects and dimensions; i.e. when utilising the shear strength for passive earth pressure or bearing resistance. The GEO-2 limit state comprises analysis method 2* to Eurocode EC 7-1.
- GEO-3: failure or very large deformation of the ground in conjunction with analysis of overall stability, i.e. when utilising the shear strength for analysis of slope stability and global stability and, generally, when analysing the stability of engineered slope stabilisation measures, including that of structural elements. The GEO-3 limit state comprises analysis method 3 to Eurocode EC 7-1.

The previous limit states are replaced as follows:

- The previous limit state GZ 1A to DIN 1054 now corresponds without restrictions to the EQU, UPL and HYD limit states to Eurocode EC 7-1.
- The previous GZ 1B limit state to DIN 1054 now corresponds in all facets to the Eurocode EC 7-1 STR limit state. The GEO-2 limit state to Eurocode EC 7-1 is also used in conjunction with the design dimensions for foundation elements.
- The previous GZ 1C limit state to DIN 1054 corresponds to the GEO-3 limit state to Eurocode EC 7-1 in conjunction with analysis of overall stability.

Analyses of the stability of engineered slope stabilisation measures are always allocated to the GEO limit state. Depending on the engineering design and function (see DIN 1054) they may be dealt with either according to the previous GZ 1B limit state or the GEO-2 limit state, or according to the previous GZ 1C limit state or the GEO-3 limit state. The geosynthetic material is designed for the STR limit state.

1.3   Examples of Reinforced Earth Structures

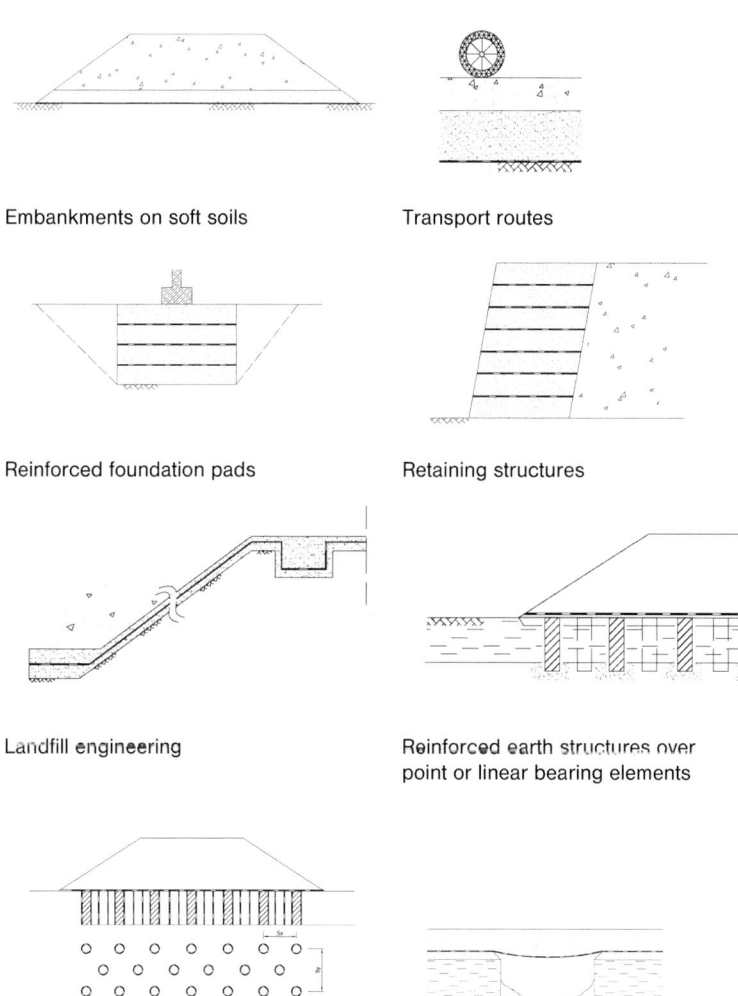

Embankments on soft soils

Transport routes

Reinforced foundation pads

Retaining structures

Landfill engineering

Reinforced earth structures over point or linear bearing elements

Geosynthetic-encased columns

Overbridging systems in areas prone to subsidence

Figure 1.1 Examples of reinforced earth structures

1.4   General Definitions

Reinforced fill or reinforced earth structures are engineered earthworks where the bearing capacity is increased by introducing geosynthetics.

Reinforcement in earth structures in the terms of these Recommendations comprises oriented geosynthetics installed in layers, which may form either continuous surfaces or grids. The stiffness, limiting strain and tensile strength of isotropic geosynthetics are the same in both directions (machine and cross-machine directions); in anisotropic geosynthetics they are different.

Fill soil is the soil within the reinforced earth structure.

Facing is the frontage on the visible surface of a reinforced earth structure; it retains the fill material between the reinforcing layers and protects against erosion.

Backfill area is the ground outside the reinforced earth structure extending to the top of the structure.

Cover fill zone is the ground above the reinforced earth structure.

2   Demands on Materials

2.1   Soil

2.1.1   Ground Investigation

Before building a reinforced earth structure geotechnical ground investigations shall be carried out according to DIN 4020 as they would be for similar, conventional structures.

2.1.2   Fill Soil

2.1.2.1   Soil Mechanics Demands

The soil mechanics demands on fill soil depend on the demands on the structure, where bearing capacity, deformation behaviour, frost hazard and drainage behaviour in particular are important, as well as the actions. If water is present locally or percolates in from outside and is not collected by other means the fill soil shall be sufficiently permeable, filter stable and resistant to weathering.

The demands on the fill soil are differentiated for structures subjected to predominantly static loads and those subjected to predominantly dynamic loads (see DIN 1054, 6.1.4 or Section 12).

2.1.2.1.1 Predominantly Statically Loaded Structures

For predominantly statically loaded structures generally only the necessary soil mechanics analyses in terms of the friction angle of the soil and any possible cohesion are required, in addition to compactability. Depending on the application and the soil type (in particular for mixed- and fine-grained soils) it may be necessary to determine the coefficient of permeability. Supplementary investigations of the composite action with the reinforcement are also necessary (also see Section 2.2.4.11).

In principle the following soil types classified to DIN 18196 may be used for predominantly static loading, inasmuch as application-specific suitability can be demonstrated or the soil properties are taken into consideration for the specific application:

- coarse-grained soil types of groups SW, SI, SE, GW, GI and GE,
- mixed-grain soil types of groups SU, ST, GU, GT, SU*, GT*, GU*, ST*,
- fine-grained soil types of groups UL, UM, TL, TM,
- maximum grain size $\leq 2/3$ of fill layer thickness (ZTV E-StB).

The suitability of other soils and materials, e.g. industrial by-products and recycled materials, shall be demonstrated separately.

The fill soils shall be adequately compactable. The demands of *ZTV E-StB* in terms of compactability shall be adhered to, inasmuch as more stringent or other deviating demands are not specified in the respective sections.

Note: Individual solutions such as reinforced noise abatement walls or embankments built using a sandwich method, for example, where geosynthetic drainage layers accelerate the consolidation of saturated fill soils and thus increase their shear strength, may be exempt. Separate analysis is required in these cases.

2.1.2.1.2 Predominantly Dynamically Loaded Structures

The dynamic actions on predominantly dynamically loaded structures can be taken into consideration by adopting quasi-static actions to DIN 1054, 6.1.4. If these demands are met the demands of Section 2.1.2.1.1 also apply to the soils.

In cases in which the dynamic actions shall be explicitly taken into consideration, the demands on the materials are given in Section 12.

2.1.2.2 Soil Chemistry Demands

Fill soils shall be of uniform quality and free from harmful constituents. They may not attack or damage the reinforcement, the frontage of retaining structures and any connecting elements used. Chemicals contained in the soil can shorten the design working life of polymer reinforcements (also see Section 2.2.4.8). The soil chemistry properties may not be altered, even temporarily, e.g. due to groundwater fluctuations or the discharge of harmful substances (also see Section 2.1.3), unless this unfavourable state is incorporated in the design.

The fill soils discussed in Section 2.1.2.1 can be used for permanent structures without additional analyses as described in Section 2.2.4.8 if the soil pH is $4 < pH < 9$. The pH of fill soils and groundwater is determined to DIN 19684.

If soils with deviating pH values or other soils (e.g. industrial by-products, recycled materials) are used, or if aggressive groundwater or gases are anticipated, additional suitable investigations of the compatibility of the fill soil and the reinforcement shall be carried out (also see Section 2.2.4.8).

2.1.2.3 Execution

The *ZTV E-StB* regulations and the DIN EN 14475 execution standard for reinforced earth structures apply to the execution of earthworks, in addition to the notes in the respective sections of these Recommendations.

2.1.3 Back-fill and Cover-fill Soils

The demands of *ZTV E-StB* apply to back-fill and cover-fill soils.

The soil mechanics demands described in Section 2.1.2.2 also apply to back-fill and cover-fill soils, if the reinforced earth structure is not reliably separated from

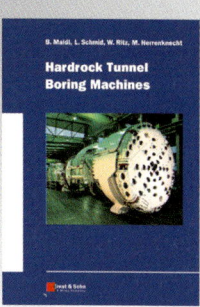

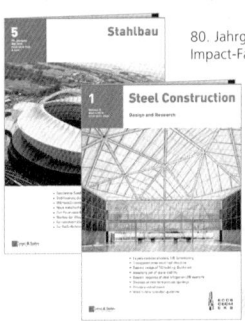

the back-fill and cover-fill zone by engineered means (e.g. liner, separating layer, drainage).

2.2 Geosynthetics

2.2.1 General Recommendations

Geosynthetics are harmonised European building products; their conformity is documented by the CE mark. In Germany they are controlled by the Construction Products Act (*Bauproduktengesetz – BauPG*).

Geosynthetic designations are controlled by DIN EN ISO 10318. They can be differentiated according to structure as follows:

- geotextiles (GTX), e.g. woven, nonwoven and knitted products,
- geotextile-related products (GTP)
- geogrids (GGR, extruded, woven, Raschel-knit, bonded),
- geocomposites (GCO).

Note: In DIN 1054 the term 'geotextiles' is incorrectly used as a generic term in the same sense as the term 'geosynthetics' or, in places, 'product' here. To simplify matters only the term 'geosynthetic' or 'product' is used here in place of the complete formulation of 'reinforcing geosynthetics', 'reinforcing products', 'geosynthetics with reinforcing function', etc.

2.2.2 Raw Materials

The following polymers are regarded as parent materials for geosynthetics in the scope of these Recommendations (in alphabetical order):

- aramid   (AR),
- polyamide   (PA),
- polyester (polyethylene terephtalate) (PET),
- polyolefines
  - · polyethylene   (PE, PEHD),
  - · polypropylene   (PP),
- polyvinyl alcohol   (PVA).

Additional designations, acronyms and symbols are included in DIN EN ISO 10318, DIN 60001 and DIN ISO 2076.

2.2.3 Product Properties and Demands

Raw materials, manufacturing method and structure have a governing impact on the properties of geosynthetics (also see [1]).

Geosynthetics are selected for reinforcement tasks such that the forces allocated can be transferred throughout the planned design working life, taking the allowed deformations of the overall system into consideration.

The following demands on the geosynthetics and the corresponding properties are therefore relevant:

- accept tensile forces, taking deformations into consideration, see Sections 2.2.4.4 and 2.2.4.5,
- transfer forces between reinforcement and the soil (composite action, anchoring), see Section 2.2.4.11,
- durability against mechanical damage during transportation and installation (robustness), see Section 2.2.4.6,
- adequate permeability to prevent water ponding,
- chemical and microbiological durability, see Sections 2.2.4.7 and 2.2.4.9,
- weather-resistant (UV-resistant), see Section 2.2.4.9.3.

The characteristic value for the short-term strength of geosynthetics ($R_{B,k0}$) shall be verified as the 5% minimum quantile, see Section 2.2.4.

The required geosynthetic material resistance depends on the planned design working life of the structure or the duration of load application in cases where the geosynthetics are only temporarily loaded, e.g. for construction stages or in structures used for short periods only (see Section 1).

To allow any changes in the mechanical properties of the geosynthetics to be taken into consideration in design, the long-term chemical and microbiological durability of the geosynthetics in the ground shall be demonstrated, see Section 2.2.4.7. Additional demands and notes are given in [1] and [2].

Plastics are not permanently UV-stable without special measures being taken or modifications made. UV stability shall be demonstrated for the design working life in the installation phase and for geosynthetics exposed to weather conditions during use (e.g. unprotected reinforcement in the facing, wrap-around) (see Section 2.2.4.9.3).

Note: Investigations of chemical durability [14] and more than 30 years of practical building experience show that geosynthetics currently used for reinforcement display high durability in the ground.

2.2.4 Testing and Reduction Factors

2.2.4.1 General Recommendations

The product properties discussed in Section 2.2.3 shall be demonstrated by testing.

Note: The product properties have previously been predominantly determined using 'index tests' without ground contact, because tests with ground contact (performance tests), which better model real conditions, are generally very complex.

Design values X_d are determined from the characteristic values X_k by applying reduction factors and a partial safety factor (see Section 3.3).

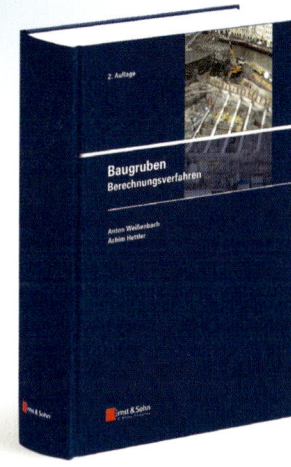

The tests are described in the appropriate standards and in other codes of practice and recommendations (see Section 2.3). The principal tests and special boundary conditions are therefore only discussed briefly below.

In permanent applications in which the product plays a governing role in the stability of the reinforced structure, specimens may be installed in the geosynthetic/soil system such that they can be removed for examination at extended intervals in order to verify parameters and adopted reduction factors (also see [1]). The specimens shall be exposed to the same conditions as the reinforcement, i.e. where possible they should also be subjected to the same tensile stresses. Both a factory-fresh specimen and a specimen that has been subjected to installation conditions shall be used as references for identifying changes (Section 2.2.4.6).

2.2.4.2 Product Identification (DIN EN ISO 10320)

Geosynthetics shall be clearly marked compliant with DIN EN ISO 10320 and DIN EN 13249 pp (CE marking). Every product shall carry the name of the manufacturer and the product type designation.

2.2.4.3 Mass Per Unit Area (DIN EN ISO 9864)

The mass per unit area influences the properties of the geosynthetics. However, numerical data do not have a direct impact on the reinforcement effect.

2.2.4.4 Short-term Load-Extenson Behaviour

2.2.4.4.1 Tensile Strength and Strain (DIN EN ISO 10319)

The tensile strain behaviour and the ultimate tensile force (short-term strength) are determined in tensile tests on 200 mm wide specimens.

The stresses determined from these tests at 2%, 3%, 5% and, if possible, 10% strain are given as mean values and not as 5% quantile values. In polymers with low yield strengths (e.g. AR, PVA) the test stresses are determined at three typical strains.

Note: Mean test stress values can be used for analyses of the serviceability limit state (SLS), for example.

A representative load-extension curve shall be provided.

The reinforcing action of nonwovens is tested in tensile tests with soil contact [7].

Zones representing the various short-term load-extension curves for a selection of geosynthetic products are shown in Figure 2.1; however, the whole spectrum

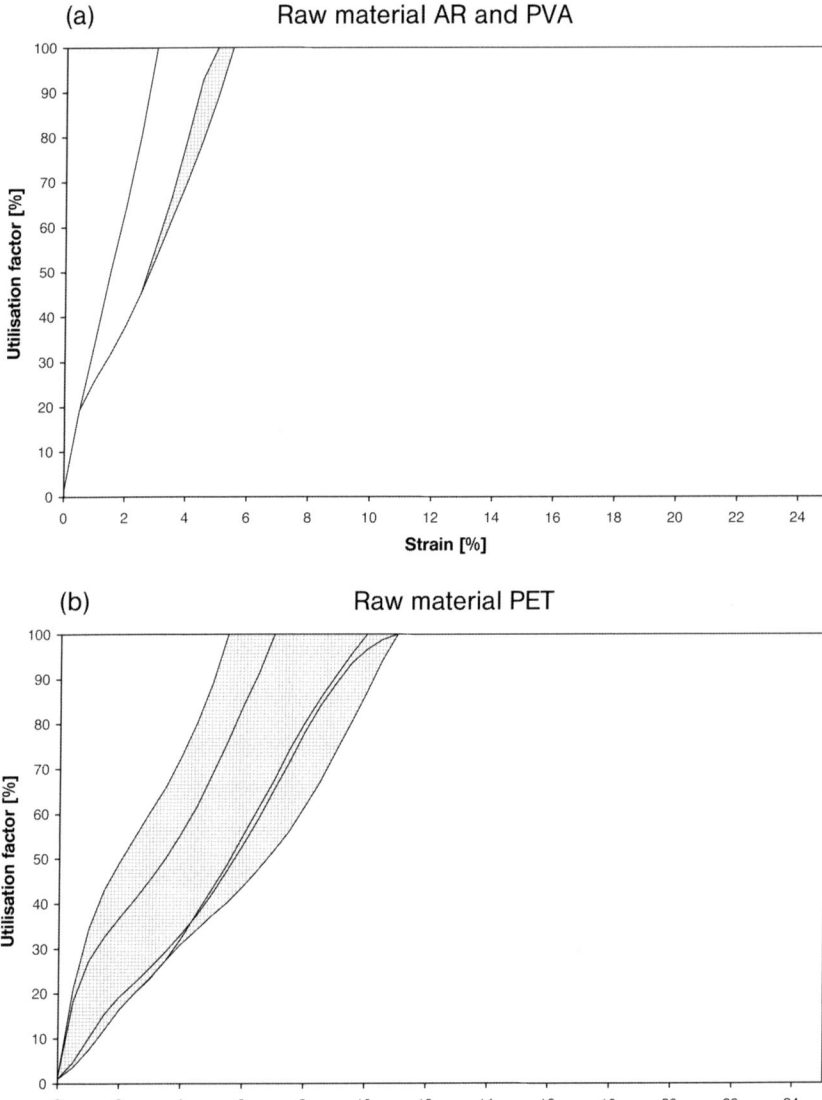

Figure 2.1 Typical geosynthetic load-extension zones [6]

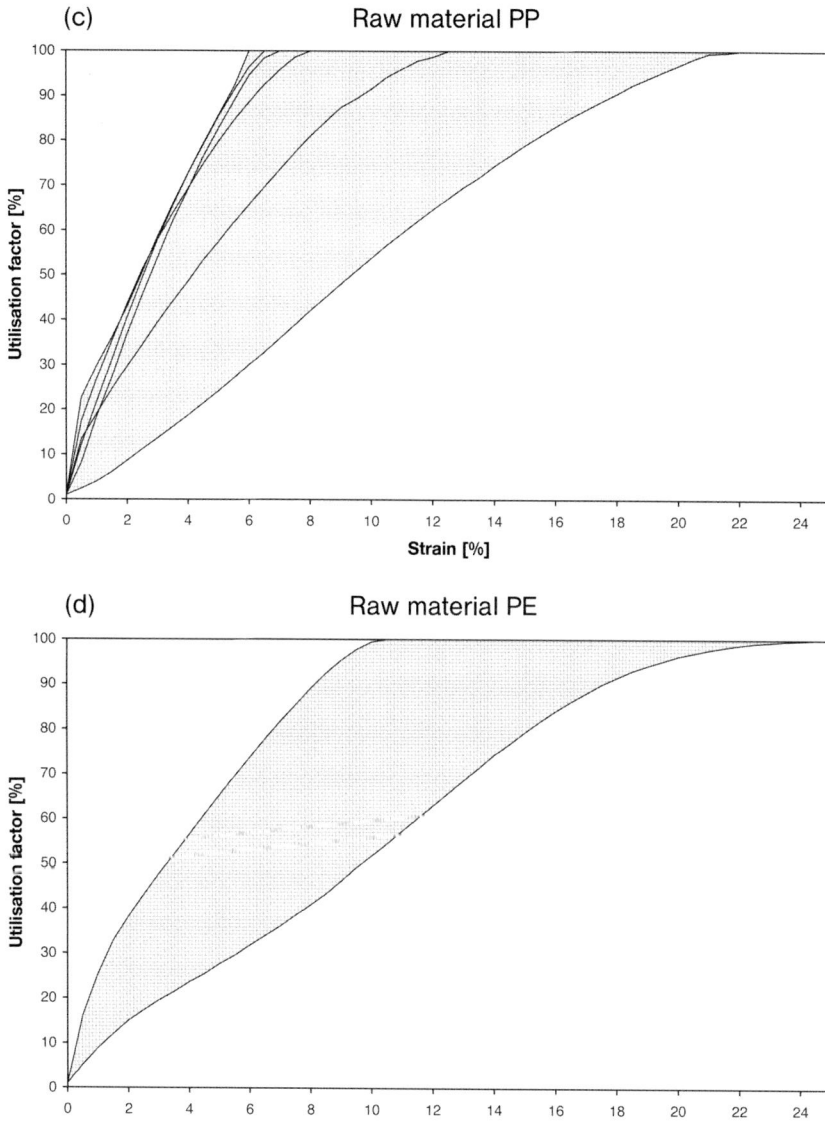

Figure 2.1 (continued)

of products commonly available on the market cannot be covered. The short-term stresses are determined from tests to DIN EN ISO 10319 (in air) and are normalised, i.e., the respective short-term strength is related to 100% ('utilisation factor', also see Section 2.2.4.5.1). Allocation to design strengths is not possible using this method and can only be determined directly from product information and the corresponding reduction factor.

Table 2.1 Typical short-term strengths of geosynthetics

Raw material	Product type	Typical short-term strengths [kN/m]			Typical elongations at failure [%]	
		from	to	max.	from	to
AR	Woven and Raschel-knit geogrids	40	1,200	2,200	2	4
	Woven Geotextile	100	1,400	2,400	2	4
PE	Woven and Raschel-knit geogrids	20	150	300	15	20
	Extruded geogrids	40	150	200	10	15
	Woven Geotextile	30	200	400	15	20
PET	Woven and Raschel-knit geogrids	20	800	1,200	8	15
	Bonded geogrids	20	400	500	6	10
	Woven Geotextile	100	1,000	1,600	8	15
PP	Woven and Raschel-knit geogrids	20	200	500	8	15
	Bonded geogrids	20	200	400	8	15
	Extruded geogrids	20	50		8	20
	Woven Geotextile	20	200	600	8	20
PVA	Woven and Raschel-knit geogrids	30	1,000	1,600	4	5
	Woven Geotextile	30	900	1,800	4	5

Note: The values given in the table are for initial orientation only. Products may be manufactured with substantially deviating parameters. Composites comprising combinations of the products with nonwovens, for example, have similar reinforcing properties.

2.2.4.4.2 Axial Stiffness

The axial stiffness of a geosynthetic reinforcement may be approximately given as the secant modulus J and is a measure of the load-extension behaviour of a reinforcement. The short-term axial stiffness is determined on the basis of the DIN EN ISO 10319 test procedure and the representative short-term load-extension curve.

The following equation applies:

$$J_{a-b,k,0} = \frac{F_b - F_a}{\varepsilon_b - \varepsilon_a}$$

Eq. (2.1)

where:

$J_{a-b,\,k0}$ characteristic short-term axial stiffness for the range ε_a to ε_b [kN/m],
F stress at a given strain ε [kN/m],
ε given strain [–].

Isochrones are used to determine long-term axial stiffnesses $J_{a-b,k,t}$ in analogy to the above procedure (Section 2.2.4.5.4).

Note: Long-term axial stiffnesses are lower than short-term axial stiffnesses due to creep (see Section 2.2.4.5).

In a simplified approach the axial stiffness J_k passing through the origin ($\varepsilon_a = 0$) and with a selected maximum allowable reference strain ε_b may be determined. If required, $J_{a-b,k}$ can be determined for a specified strain range (Figure 2.2).

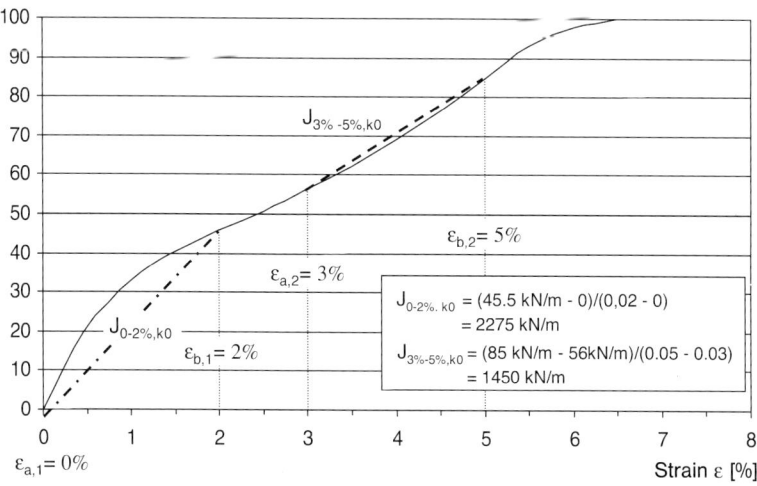

Figure 2.2 Example of the evaluation of short-term axial stiffnesses J_k for two strain ranges based on a short-term load-extension curve

17

Note: It may be more appropriate to use the tangent modulus for FEM analyses.

2.2.4.4.3 Uniaxial and Biaxial Reinforcement

The use of planar models for analysis purposes is common for slopes, retaining structures and embankments. The reinforcement is primarily loaded uniaxially in the direction of the principal tensile force.

Biaxial stresses may also occur in some areas, in particular in the applications discussed in Sections 6, 9, 10, and 11. The tensile strength parameters in Section 2.2.4.4.1 are then adopted.

Note: The extent to which separately determined material parameters are adopted in this case remains the subject of research.

2.2.4.4.4 Serviceability Limit State/Strain Behaviour

The load-extension behaviour of the geocomposite material (geosynthetics/fill soil) or, simply, the individual fill soil and geosynthetic reinforcement components, shall be estimated and incorporated in the analysis to limit the deformations in reinforced earth structures.

The strain suffered by the reinforcement in the serviceability limit state can be determined for the anticipated utilisation factor using isochrones as described in Section 2.2.4.5.3.

Note: Notes on the allowable or anticipated structural deformations and the demands on the geosynthetics derived from them can be found in the respective application sections of these Recommendations and in [15]

2.2.4.5 Long-term Load-Extension Behaviour (Creep Rupture, Creep)

2.2.4.5.1 General Recommendations

Geosynthetics consist of polymer raw materials (Section 2.2.2) displaying elasto-plastic behaviour. Under load, not only elastic (short-term) deformations occur, but also viscose, time-dependent creep processes.

They have structurally relevant consequences:

– reduced strength and/or
– greater strain in the reinforcement compared to short-term behaviour (Section 2.2.4.4).

The reduction in strength may lead to failure (creep rupture) and an increase in strain (creep strain) may lead to unacceptable deformations in the structure. When designing to these Recommendations creep rupture is taken into consideration by applying a reduction factor A_1 (Section 2.2.4.5.2) and the creep strain is taken into consideration by isochrones (Section 2.2.4.5.4).

Creep rupture ist relevant to analysis of the ultimate limit state (ULS) (DIN 1054) and creep strain for analysis of SLS (DIN 1054).

Creep and creep rupture of geosynthetics are tested to DIN EN ISO 13431 (also see *TL Geok E-StB*). The results of testing are creep and creep rupture curves.

Creep processes are a function of:

- the polymer, the type of polymer, the raw material,
- their processing,
- the level of tensile stress,
- the duration of the load and
- the temperature.

Note: These impacts are incorporated in the corresponding EBGEO reduction factors/design approaches. The impact of other factors, such as embedment, multiaxial stress states, loading speed and history, etc., is the subject of research. Any transfer of such test results to the respective practical construction application shall be considered on a case-by-case basis.

Creep curves ('strains plotted over time') are determined for a variety of utilisation factors at room temperature and in atmosphere. The utilisation factor β is defined as follows:

$$\beta = \frac{\text{tensile force F}}{\text{short term tensile strength } R_{B,k0}} \qquad \text{Eq. (2.2)}$$

Note: The tensile force F is provided by the results of long-term testing (Section 2.2.4.5.2).

Isochrones are produced from the creep curves for practical applications (see Section 2.2.4.5.4).

Creep rupture curves ('utilisation factor drawn over time until failure') are determined for a variety of utilisation factors at room temperature and in atmosphere.

The utilisation factors adopted for testing are reassigned to the stresses in the structure (DIN 1054).

2.2.4.5.2 Determining Reduction Factor A_1 from Creep Testing

The creep rupture strength of geosynthetics is tested to DIN EN ISO 13431. The test results are given as creep rupture curves (graph of utilisation factor over time until failure, see Figure 2.3). From the curves the acceptable tensile forces can be determined for a given duration and from these in turn a reduction factor A_1.

Note: More detailed notes on determining long-term durability can also be found in ISO/TR 20432 [11].

As can be seen in Figure 2.3 the time to failure is measured for geosynthetic specimens subjected to a variety of utilisations factors β. DIN EN ISO 13431 envisages

19

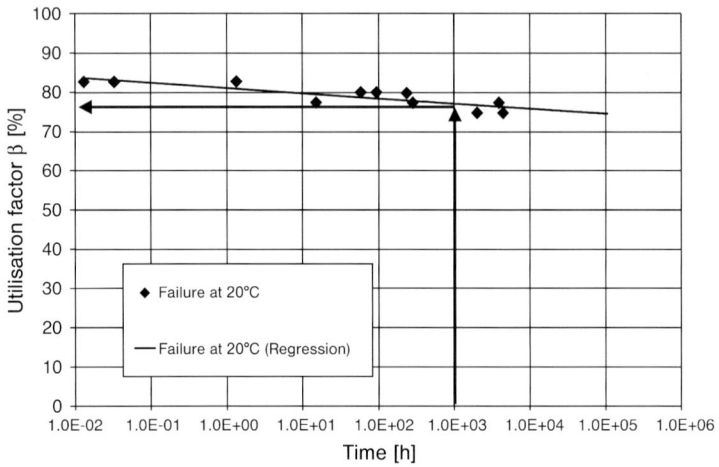

Figure 2.3 Reading off the allowable utilisation factor β to determine the reduction factor A_1 from creep rupture results ([6], [11], [13])

load durations of at least one year (10^4 h). Using these test results product- and material-independent extrapolation to the times given in DIN EN 13249 pp. for permanent structures (up to ≥ 100 years) is then possible. The procedure is explained in [1], [2] and [10].

Note: The following procedures are also possible:
1) Short duration tests at higher temperatures (time/temperature shift method, stepped isothermal method SIM, [12]). The tests shall be verified by a standard test to DIN EN ISO 13431 with a duration of at least 1,000 hours.
2) Tests performed on specimens taken from an at least 20 year old structure.

The reduction factor A_1 used in ULS analyses (see Section 1.2.3) is determined from creep rupture results to DIN EN ISO 13431 using a regression curve. The allowable utilisation factor β for the planned design working life is given by the intersection with the regression curve. The reduction factor A_1 is the reciprocal of the allowable utilisation factor.

Note: The example shown in Figure 2.3 gives β = 0.77 for a design working life of 6 weeks (1,000 h). A_1 = $1/β$ = $1/0.77$ = 1.30.

2.2.4.5.3 Reduction Factor A_1 for Creep Failure Behaviour

Product-specific values of A_1 are determined in tests as described in Section 2.2.4.5.2. If no product-specific test results are available the reduction factors A_1 given in Table 2.2 are adopted as a function of the raw material employed:

Table 2.2 Reduction factor A_1

Material	Acronym	Common values for A_1 from product-specific analyses		Minimum A_1 if analysis unavailable
		from	to	
Aramid	AR	1.5	2.0	3.5
Polyamide	PA	1.5	2.0	3.5
Polyethylene	PE	2.0	3.5	6.0
Polyester	PET	1.5	2.5	3.5
Polypropylene	PP	2.5	4.0	6.0
Polyvinyl alcohol	PVA	1.5	2.5	3.5

Note: *Where analysis results are unavailable the values of A_1 given in the table above have been increased compared to the previous edition of these Recommendations, because the values could not be confirmed as 'conservative' due to the great diversity of products and manufacturer's data available (see additional notes in [5]).*

2.2.4.5.4 Identifying Long-term Strain Behaviour by Evaluating Isochrones

Isochrone evaluation may be used to take the long-term behaviour of geosyn thetic reinforcements in serviceability limit state analyses into consideration (see Section 3). It is possible to restrict the results to a defined deformation state.

Utilisation of the isochrone values for safety against failure states (ultimate limit states, ULS) is not permitted (see Sections 9 and 11 for exceptions).

Creep tests to determine isochrones are carried out at a variety of load conditions. The strain and load condition value pairs are taken from the curves attained (Figure 2.4, left) for a given time and drawn in a graph (Figure 2.4, right). The 'utilisation factor' is commonly used on the ordinate for this type of representation (see Section 2.2.4.5.1).

Note: *Isochrones as shown in Figure 2.5 can be interpreted as follows: For a very short design working life (in this case approx. 1 minute) and a utilisation factor of 100% (corresponds to short-term strength) an elongation at failure of approx. 10% is anticipated. This roughly corresponds to the elongation at failure in a short tensile test. For a specified/allowable strain resulting from an estimate of the serviceability limit state, e.g. 6% in this case, a utilisation factor of approx. 51% may be adopted for a design working life of approx. 120 years.*

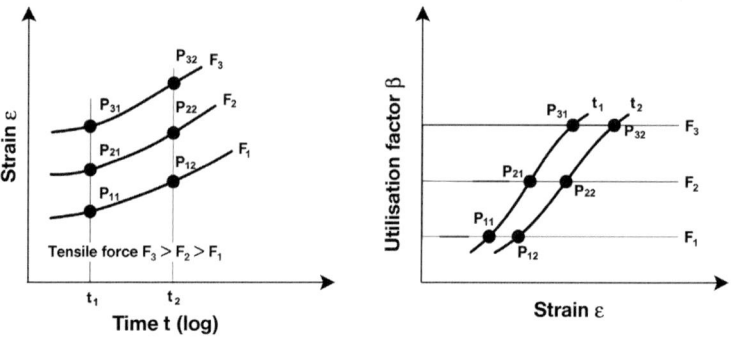

Figure 2.4 Determining isochrones (right) from creep curves (left)

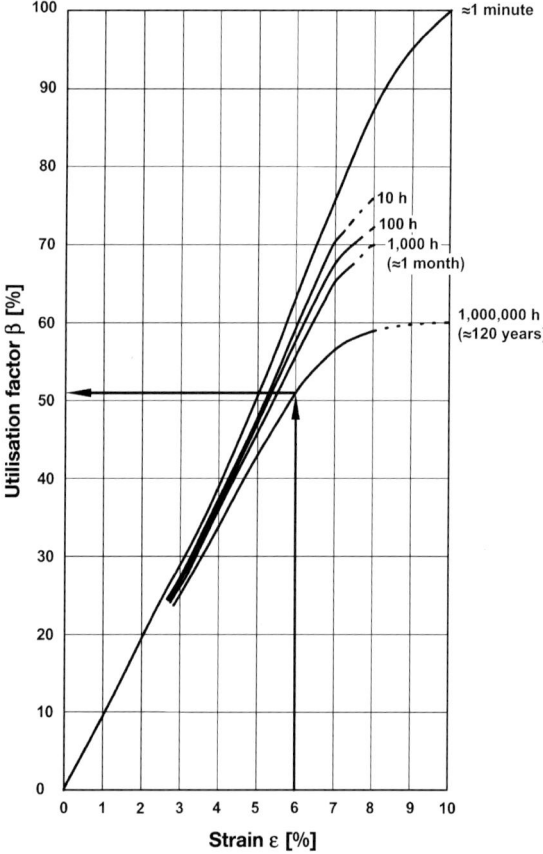

Figure 2.5 Principle of isochrones and reading example (see note)

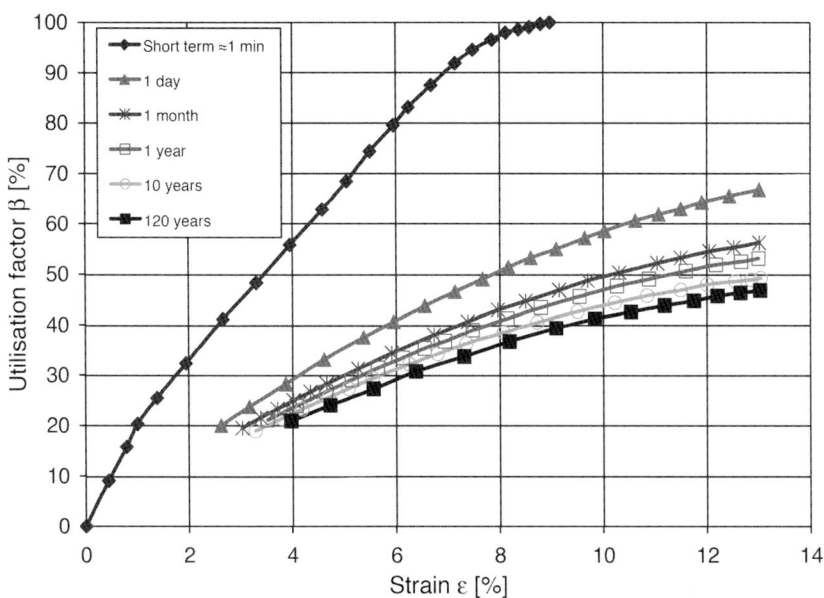

Figure 2.6 Example isochrones for a PEHD extruded geogrid

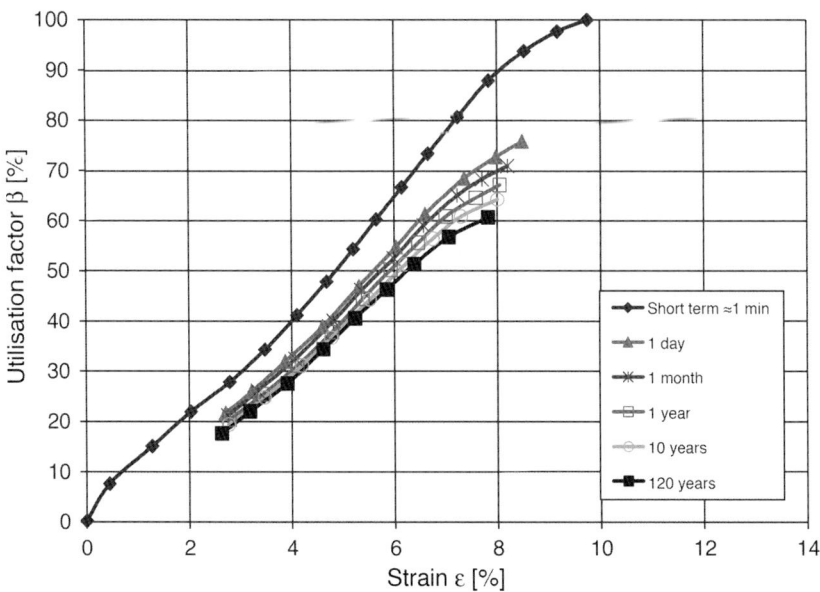

Figure 2.7 Example isochrones for a PET product

23

Figure 2.6 and Figure 2.7 show example isochrones for individual products. The product-specific curves are used for design purposes.

If these isochrone evaluations are utilised for analysis of the serviceability limit state, the reduction factor A_1 is already incorporated and need not be additionally adopted. The remaining reduction factors A_2 to A_5 remain untouched by this evaluation and are adopted as usual.

2.2.4.6 Resistance to Mechanical Damage During Installation

2.2.4.6.1 General Recommendations

The anticipated mechanical damage to geosynthetic reinforcements during installation is taken into consideration by a reduction factor based on installation damage testing. *The geotextile robustness classes in TL-Geok E-StB are not adopted for this purpose.*

At the beginning of a construction project installation tests under real site conditions are recommended (geosynthetic, underlay, fill material, layer thickness, compaction equipment and compacting runs), together with inspection of any changes after removing the specimen [16].

Note: The result also serves as a reference for assessing specimens removed at intervals of several years, in order to identify any changes resulting from chemical attack or to determine residual strength (Section 2.2.4.7).

Inasmuch as reduction factors obtained under directly comparable site conditions are available from earlier projects, they may be used.

2.2.4.6.2 Reduction Factor A_2 for Damage to Geosynthetics During Transportation, Installation and Compaction

If fine-grained soils are used for permanent structures an A_2 **value of at least 1.5** shall be adopted.

If mixed-grain and coarse-grained soils with rounded grains are used for permanent structures an A_2 **value of at least 2.0** shall be adopted.

Site testing must always be carried out where broken and angular grains, stone fill or recycled materials (RC materials) are used if mechanical damage cannot be minimised by engineering measures, e.g. protective layers of fine- and coarse-grained soils with rounded grains or protective geotextiles.

Note: The composite properties of fill soil/geosynthetics may be influenced by the use of protective layers. This shall be taken into consideration in design.

Lower values of A_2 for a specific product shall be demonstrated by suitable field or site tests. Tests are also necessary if special conditions are anticipated while installing the geosynthetics (see Section 1, for example).

2.2.4.6.3 In-situ Testing

During in-situ testing ([1], [2], [16]) the installation conditions are adapted as nearly as possible to the boundary conditions of the actual installation. Planar geosynthetic installations and special installations (e.g. encased columns) are differentiated.

Testing procedures for determining the reduction factor A_2 are described below. The boundary conditions shall be examined for transferability to the respective construction project and adapted correspondingly.

- Installation test for planar geosynthetics
  A few square metres of the product are removed and kept as a reserve speci-men. The tested product is then laid out on a defined surface, cover filled and the cover fill compacted. It is then excavated and removed, and care is taken that no additional damage is caused during removal.
  Reinforcements for slopes and retaining structures have an underlay consist-ing of a compacted layer of the fill material or the planned in-situ underlay. This consists of a fine-grained soil of soft consistency or the natural ground for separating layers and reinforcements installed in the fill base.
  The cover fill consists of either the planned fill material or broken natural stone with 0/45 mm grading for crushed stone base courses compliant with *ZTV SoB-StB*. The layer is applied to a load-bearing, compacted, 25 cm thick base, or the thickness planned for installation in-situ. In soft ground the fill thickness is selected such that it can be traversed by the compacting equip-ment. Compaction is carried out using a vibratory roller with approx. 10 t to 12 t gross weight, vibrating at high amplitude (approx. 1.5 to 2.0 mm) or using the compaction equipment planned for use at the site until the planned relative compaction is achieved (inasmuch as no other requirements are specified: $D_{pr} = 100\%$ of normal Proctor density).

- Installation testing for geosynthetic-encased columns
  Where geosynthetic-encased columns are employed the specified standard procedure for defining the value of A_2 by in-situ testing cannot be adopted. Appropriate specimens shall be taken from the geotextile tubes of at least three test columns manufactured under similar conditions.

The area of the product to be tested is specified before installation and examined thoroughly after removal. The minimum size of the specimen is 1 m · 1 m. It shall be removed immediately following the in-situ test. Care should be taken during removal that no additional damage is caused.

The damage caused is described, including the number of holes per m^2, where necessary classified according to hole size, and the shape and type of damage.

The tensile strength and the strain are determined for the reserve specimens as described in Section 2.2.4.4.1. The reduction factor A_2 is the quotient of the mean values from the reserve specimens and the removed in-situ specimens.

Note: *If tensile strength testing is not possible on a specimen due to damage, the specimen is entered into the evaluation with a tensile strength of 'zero'.*

The quotient of the elongations at failure is determined in an analogous manner, but does not represent a reduction factor. Manufacturer's data may not be used in place of testing for reserve specimens. The reserve specimens may not display any previous damage (e.g. transportation or similar damage).

Note: *The procedure described only applies for the short-term impact of damage. However, research reveals no additional impacts on the creep failure behaviour of the products tested [9].*

2.2.4.6.4 Laboratory Testing (DIN EN ISO 10722)

The DIN EN ISO 10772 index test does not provide useful A_2 design values.

Note: *This laboratory test cannot realistically reproduce the in-situ load conditions and cannot completely replace in-situ testing. However, a modified test using the actual site material and a grain size with a defined upper limit provides useful guide values.*

2.2.4.7 Joins and Connections

2.2.4.7.1 General Recommendations

DIN EN ISO 10321 is adopted for testing the tensile strength and strains of seams and other joining techniques.

2.2.4.7.2 Reduction Factor A_3 for Junctions, Joins, Seams and Connections to Other Structural Elements

A_3 **equals 1.0** if there are no junctions, joins or seams in the force direction and no connections to other structural elements are required.

2.2.4.7.3 Determining the Reduction Factor A_3 by Testing

Where geosynthetics are joined to each other (connectors, seams, adhesive bonding, etc.) an appropriate analysis of the transfer of tensile forces is required. The tests are carried out as specified in DIN EN ISO 10321. The factor A_3 can be determined from the test results by comparing to the geosynthetic's characteristic short-term strength. Any strain in the join deviating from the anticipated product strain shall be taken into consideration for design.

Special investigations based on DIN EN ISO 10321 shall be performed where geosynthetics connect to other structural elements.

Note: *Connections to concrete revetments can be investigated compliant with ASTM D 6638.*

2.2.4.8 Chemical Resistance

2.2.4.8.1 Reduction Factor A_4 for Environmental Chemical Impacts

For operational lives up to 5 and up to 25 years respectively the manufacturer's specifications on the applicability of the geosynthetics apply in compliance with DIN EN 13249 pp. Annex B. DIN EN 13249 pp. does not require information on the reduction factor A_4 to be given. The manufacturer must provide a reasoned reduction factor A_4 for use as reinforcement in the terms of EBGEO.

The reduction factor A_4 is adopted without recourse to analysis for operational lives greater than 25 years and up to 100 years (permanent structures), depending on the polymer used as given in Table 2.3 (also see [1]). Lesser values or values for applications outside of the $4 \leq pH \leq 9$ pH range shall be determined by additional, product-specific investigations or by proof of appropriate long-term experience and measurements.

Table 2.3 Reduction factor A_4 (without analysis, $4 \leq pH \leq 9$)

Material	Acronym	A_4
Aramid	AR	3.3
Polyamide	PA	3.3
Polyester	PET	2.0
Polyethylene	PE	3.3
Polypropylene	PP	3.3
Polyvinyl alcohol	PVA	2.0

Due to their alkali sensitivity, polyester and aramid products may not be used in building applications for extended periods when in contact with soils improved or strengthened using cement or lime, or in direct contact with cement concrete (including crushed concrete), regardless of the operational life. An exception is made if their suitability is determined by additional, product-specific investigations or by proof of appropriate long-term experience and measurements.

2.2.4.8.2 Determining Chemical Resistance by Testing

The basis for determining the chemical resistance of reinforcement products by the manufacturer for operational lives up to 5 years and up to 25 years respectively, based on DIN EN 13249 pp., is specified in [2]. In addition, a procedure for identifying the durability and deriving the residual strength, or an appropriate reduction factor, is given for an operational life up to 100 years. The *Leitfaden zur Beurteilung der Beständigkeit von Geokunststoffen* (Guidelines for Assessing the Resistance of Geosynthetics) [11], DIN Fachbericht 86 [10] and recent developments shall be taken into consideration.

27

2.2.4.9 Additional Environmental Impacts

2.2.4.9.1 Microbiological Resistance

DIN EN 12225, 'Geotextiles and Geotextile-related Products – Method for Determining the Microbiological Resistance by a Soil Burial Test' applies for the resistance to microbiological attack.

Empiricism indicates that common polymers are not impaired by microbiological attack.

2.2.4.9.2 Biological Resistance and Vandalism

Geosynthetic-reinforced structures can be damaged by biological attack (e.g. rodents) or by vandalism.

Empiricism indicates that this does not present a problem, but it should be investigated on a case-by-case basis and engineering measures implemented where necessary (e.g. facings on slopes using wrap-around methods, vole mesh).

2.2.4.9.3 Weathering Resistance (UV Resistance)

If products are not immediately cover-filled their weathering resistance shall be taken into consideration.

To estimate the weathering resistance to DIN EN 13249 pp. the residual strength for a defined degree of weathering is first determined in a laboratory test to DIN EN 12224. The geosynthetic is then classified according to residual strength, resulting in the greatest allowable exposure duration until covering (Table 2.4).

Table 2.4 Weathering and greatest allowable exposure duration classes (to DIN EN 13249, Annex B.1 normative, Table B.1)

	Reinforcement		
Residual strength to DIN EN 12224	> 80%	60% to 80%	< 60%
Greatest allowable exposure duration	1 month	2 weeks	1 day

The greatest allowable exposure duration for reinforcement products is given in the CE accompanying documentation ('Cover after xx days/xx weeks/xx months').

Not only is the strength of the product impaired by weathering, but also the resistance to chemical attack. The time until protection is applied (cover-fill, facing) should therefore always be kept as short as possible. Where possible, consequently, the exposure duration given in Table 2.4 should not be completely utilised.

Note: Additional information can be found in the literature: [10] and [14].

2.2.4.10 Effects of Predominantly Dynamic Actions

2.2.4.10.1 Reduction Factor A_5 for Predominantly Dynamic Actions

In most applications dealt with by these Recommendations geosynthetics are subjected to predominantly static actions. A reduction factor A_5 of 1.0 can therefore generally be adopted.

In Section 12 a variety of load cases are discussed which require that predominantly dynamic actions (dynamic/cyclic) impacting the strength of the reinforcement are considered. Information on how to determine an appropriate reduction factor is given.

Note: Applications that may require individual analysis include reinforcements immediately beneath machine footings, on railway lines at shallow depths below the track, the uppermost reinforcement layer in the bridge abutment/earthworks transition zone and reinforced structures outside seismic zone 0.

2.2.4.10.2 Determining the Reduction Factor A_5 for Predominantly Dynamic Actions by Testing

The procedure for determining A_5 by testing is given in Section 12 for load cases where the impact of dynamic actions on the design strength of the geosynthetics needs to be determined.

Note: In addition, Section 12 includes information on determining damage to the reinforcement by testing and the impacts from predominantly dynamic actions on the composite behaviour of geosynthetics/fill soil.

2.2.4.11 Friction and Composite Behaviour

2.2.4.11.1 General Recommendations

Soil reinforcement depends on the transfer of forces (stresses) from the soil to the geosynthetics and *vice versa*. This requires adequate composite action between the reinforcement and the soil.

Simply, composite action is described by the friction coefficient $f_{sg,k}$, defined as follows:

$$f_{sg,k} = \lambda \cdot \tan \varphi_k \qquad \text{Eq. (2.3)}$$

where:

λ composite coefficient of friction $\lambda = \dfrac{\tan \delta}{\tan \varphi}$,

$\tan \delta$ composite coefficient of the geosynthetics/soil (measured),
$\tan \varphi$ composite coefficient of the soil (measured),
$\tan \varphi_k$ characteristic friction coefficient of the soil.

Inasmuch as cohesion is adopted for calculating composite action, the composite action can be described by the shear coefficient $f_{scg,k}$, defined as follows:

$$f_{scg,k} = \lambda_c \cdot c_k \qquad\qquad \text{Eq. (2.4)}$$

where:

λ_c composite coefficient of cohesion $\lambda_c = \dfrac{a}{c}$,

a adhesion of the geosynthetics/soil (measured),

c cohesion of the soil (measured),

c_k characteristic cohesion of the soil.

When adopting adhesion components in particular the effect shall be guaranteed for the entire design working life (also see [8] and [17]).

Note: When determining the composite coefficients the 'geosynthetic/soil' and 'soil only' shear coefficients shall be determined in the same boxes as far as possible (e.g. 300 mm × 300 mm, see Section 2.2.4.11.2). The ratio is formed on the basis of the real measured data, not the characteristic data.

The composite coefficients are generally important in two cases:

– when analysing sliding/shear in a geosynthetic/soil (or geosynthetic/geosynthetic) contact plane and
– when analysing reinforcement pull-out stability.

These situations are illustrated as examples for a steep reinforced slope in details 'A' and 'B' in Figure 2.8.

The composite coefficients may be different for the two situations. Accordingly, separate shear and pull-out tests are carried out with the geosynthetics and the ground.

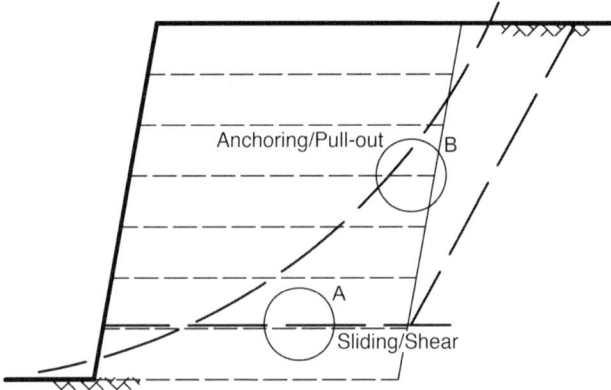

Figure 2.8 Relevance of composite behaviour for two typical situations (sliding/shear 'A' and anchoring/pull-out 'B'), oversteep slope example

2.2.4.11.2 Determining Composite Coefficients by Testing

Direct shear tests are carried out in shear boxes (minimum dimensions 300 mm × 300 mm) based on DIN EN ISO 12957-1 for the failure mechanism as shown in Figure 2.8, Detail A, using the planned geosynthetic and fill soil. The entire contact surface (including any apertures, e.g. as in geogrids) between the geosynthetic and the soil is adopted to determine the friction coefficients. The result of the test is the friction coefficient in the 'shear' mode.

Pull-out tests in pull-out boxes are carried out based on DIN 60009 and DIN EN 13738 for the failure mechanism as shown in Figure 2.8, Detail B, using the planned geosynthetic and fill soil. Although the investigations to DIN EN 13738 differentiate between the composite coefficients of friction and cohesion given in 2.2.4.11.1, DIN 60009 is oriented around the ratio of the various measured normal stresses to the shear stress.

The box dimensions (shear and pull-out tests) are dependent on the maximum soil grain size and the geometry of the geosynthetics.

Note: *The box dimensions must allow load transfer in geogrids across at least three consecutive elements in both directions.*

The same friction and pull-out coefficients may be used for draft design.

Inasmuch as no test results are available, the following minimum friction coefficients may be adopted for draft design:

- geosynthetic/fill soil: $f_{sg,k} = 0.50 \tan \varphi'_k$,
  $f_{cg,k} = 0.50\, c'_k$ or $0.50\, c_u$,
- geosynthetic/geosynthetic: $f_{gg,k} = 0.20$.

Note: *To date, empiricism indicates that the geosynthetic/fill soil friction coefficient lies between $0.5 \tan \varphi'_k$ and $1.0 \tan \varphi'_k$.*
Reinforcement is increasingly used in conjunction with soils with natural or 'artificial' cohesion (e.g. following stabilisation). The composite action then consists of both friction ('friction coefficient', see above) and adhesion components. To determine the required parameters it is necessary to carry out shear tests (including fill soil/fill soil) based on DIN EN ISO 12957-1 and DIN EN 13738 or DIN 60009 in 300 mm × 300 mm to 500 mm × 500 mm boxes under boundary conditions corresponding to the practical application.

2.3　　　Bibliography

[1]　Merkblatt über die Anwendung von Geokunststoffen im Erdbau des Straßenbaus, Ausgabe 2005, Forschungsgesellschaft für Straßen- und Verkehrswesen, Arbeitsgruppe Erd- und Grundbau, Cologne, FGSV Heft Nr. 535.

[2]　Technische Lieferbedingungen für Geokunststoffe im Erdbau des Straßenbaus, TL Geok E-StB 05, FGSV 549, Forschungsgesellschaft für Straßen- und Verkehrswesen, Arbeitsgruppe Erd- und Grundbau, Cologne.

[3]　Anwendung von Geotextilien im Wasserbau, DVWK Merkblatt No. 221/1992 of the Deutsche Verband für Wasserwirtschaft und Kulturbau e. V., Verlag Paul Parey, Hamburg and Berlin.

[4]　Anwendung und Prüfung von Kunststoffen im Erdbau und Wasserbau, DVWK Merkblatt No. 76/1986 of the Deutsche Verband für Wasserwirtschaft und Kulturbau e. V., Verlag Paul Parey, Hamburg and Berlin.

[5]　Schweizerischer Verband für Geokunststoffe (2003): Handbuch Bauen mit Geokunststoffen.

[6]　Müller-Rochholz, J. (2007): Geokunststoffe im Erd- und Straßenbau. Werner Verlag.

[7]　Bauer A., Bräu G. (2002): Entwicklung eines Bemessungsverfahrens für die Bodenbewehrung mit Vliesstoffen basierend auf Zugversuchen im Bodenkontakt. Bundesministerium für Verkehr, Bau- und Wohnungswesen, Forschung Straßenbau und Straßenverkehrstechnik, Heft 831.

[8]　Alexiew, D. (2005): Zur Berechnung und Ausführung geokunststoffbewehrter 'Böschungen' und 'Wände': aktuelle Kommentare und Projektbeispiele. Conference Proceedings 5, Österreichische Geotechniktagung Vienna, February 2005, pp. 87–105.

[9]　Müller-Rochholz, J., Mannsbart, G. (2004): Installation stress testing – results and interpretation. Third European Geosynthetics Conference – EuroGeo3, Munich, 2004.

[10]　DIN-Fachbericht 86 (2000): Geotextilien und geotextilverwandte Produkte – Leitfaden zur Beständigkeit (CEN Report 13434).

[11]　Guidelines for Determining the Long-term Strength of Geosynthetics for Soil Reinforcement, English Edition ISO/TR 20432.

[12]　Standard Test Method for Accelerated Tensile Creep and Creep-Rupture of Geosynthetic Materials Based on Time-Temperature Superposition Using the Stepped Isothermal Method ASTM D 6992.

[13]　Brown, R. P., Greenwood, J. H. (2002): Practical Guide to the Assessment of the Useful Life of Plastics, Rapra.

[14]　FE 05.122: Chemische Veränderungen von Geotextilien unter Bodenkontakt (Chemical changes in geotextiles making contact with soil). Bundesanstalt für Materialforschung und -prüfung (BAM), Berlin, Berichte der BASt, Heft S 41, 2005.

[15] Verwendung von Geotextilien im Erd- und Grundbau für den Bau und die Erhaltung von Straßen: Bericht einer Arbeitsgruppe des Straßenforschungsprogramms der OECD. Paris 1991/published by Federal Ministry of Transport, Highways Engineering Department, 1993.

[16] Bräu, G., Floss, R., Bauer, A., Vogt, N. (2004): Verhalten von Geotextilien und Geokunststoffen im Boden bei Beanspruchung während des Einbaus und unter dynamischen Belastungen. Federal Ministry of Transport, Building and Urban Affairs, Forschung Straßenbau und Straßenverkehrstechnik, Heft 893.

[17] Aydogmus, T., Alexiew, D., Klapperich, H. (2005): Über das Verbundverhalten von zementstabilisiertem bindigem Boden mit PVA Geogittern im Schermodus. FS-KGeo 2005.

[18] Nimmesgern, M., Lieberenz, K. (2001): Geogitterbewehrter Bahndamm – Ausgrabung nach 10-jähriger Beanspruchung. FS-KGeo 2001.

3   Analysis Principles

3.1   General Principles

The correct application of geosynthetics in the subsurface leads to increased bearing capacity and improved serviceability. These are based on the transfer of stresses between the ground and the geosynthetic reinforcements under tension. Stresses are transferred via friction, interlocking and/or adhesion between the reinforcement and the ground. The governing properties for analysing the interaction of earth structures and geosynthetic reinforcements are:

- the effective shear resistance between the geosynthetics and the fill soil,
- the resistance of the geosynthetics (tensile strength),
- the axial stiffness of the geosynthetics in the ground/geosynthetics composite system.

The safety of earth structures with geosynthetic reinforcement layers is analysed. The analysis investigates the ultimate limit state and the serviceability limit state. Table 3.1 shows the allocation of analyses to the limit states defined in DIN 1054.

Note: Supplementary regulations to DIN 1054 for geosynthetics are introduced as described in Section 6.1.3. They are obligatory for designing geosynthetic-reinforced structures.

a) Ultimate Limit State (ULS)

The ultimate limit state is generally determined using classical limit equilibrium methods.

For STR limit state analyses the effects and resistances of the structural elements are first determined using characteristic values. They are only converted and compared using the partial factors for effects or resistances following structural analysis.

For GEO limit state analyses the design values of the actions are formed from the characteristic values by multiplication before structural analysis is carried out. The design values of the shear parameters friction and cohesion are determined by dividing by the partial safety factors. The design effects acquired from the analysis are compared to the design resistances.

b) Serviceability Limit State (SLS)

Serviceability limit state analyses are carried out using characteristic values for actions and resistances to DIN 1054.

Explicit analysis can be generally dispensed with in the following cases, inasmuch as the ultimate limit state is satisfied:

Table 3.1 Allocation of analyses to the limit states defined in DIN 1054

Analysis group	Analysis	Limit state	Notes/standards Remarks
Ultimate limit state	Bearing capacity failure	STR	Regarded as quasi-mono-lithic, *cf.* DIN 1054:2005-01, Para. 12.4.4 (2), 7.5.2
	Sliding	STR	Regarded as quasi-mono-lithic, *cf.* DIN 1054:2005-01, Para. 12.4.4 (2), 7.5.3
	General failure/slope failure	GEO	Regarded as quasi-mono-lithic, *cf.* DIN 1054:2005-01, Para. 12.3
	'Overturning' alternatively via position of bearing pressure resultant	EQU[1]	Regarded as quasi-mono-lithic, *cf.* DIN 1054:2005-01, Para. 7.5.1 (1)
	General failure/slope failure	GEO	Reinforcement layer inter-sected, *cf.* DIN 1054:2005-01, Para. 12.3, 12.4.3, *cf.* EBGEO, Section 3.4
	Design strength of reinforcement	STR	*cf.* EBGEO, Section 3.4
	Pull-out resistance of reinforcement	GEO/STR[2]	*cf.* DIN 1054:2005-01, Para. 12.4.3 *cf.* EBGEO, Section 3.4
	Analysis of connection of the outer skin	STR	*cf.* EBGEO, Section 3.4
	Analysis of reinforcement overlapping/joining (reinforcement junctions)	STR	*cf.* DIN 1054:2005-01, Para. 7.5.3 *cf.* EBGEO, Section 3.4
Service-ability limit state	Deformation of the structure	SLS	*cf.* EBGEO, Section 3.1
	Settlement in the contact area	SLS	*cf.* EBGEO, Section 3.1
	Analysis of bearing pressure resultant	SLS	Regarded as quasi-mono-lithic, *cf.* DIN 1054:2005-01, Para. 12.4.4 (2), 7.6.1

Notes/explanations:

[1] Because these are not rigid structures the stability of supporting structures against 'overturning' is regarded as adequate if:
   a) The load resultant defined in DIN 1054, Paragraph 7.5.1 (3) does not leave the second kernel width (calculated using characteristic values).
   b) The position of the bearing pressure resultant (first kernel width) SLS is analysed.

[2] The pull-out resistance is analysed for the limit state in which the design effects where determined (deficit forces).

- structures in Geotechnical Category GC 1,
- retaining structures with wide-area live loads $p_k \leq 10$ kN/m^2, allocated to GC 2 and displaying a utilisation factor $\mu \leq 0.75$ compliant with GEO,
- structures for which well-founded empirical or measured data are available for similar structures using the same construction method and in similar ground conditions (*cf.* DIN 1054).

If, for Geotechnical Category 3 structures, no well-founded empirical or measured data are available for similar structures using the same construction method (*cf.* DIN 1054), against which the deformation analyses can be calibrated, instrumented monitoring of the structure and/or the observational method shall be adopted.

3.2 Allocation of Geosynthetic-reinforced Structures to Geotechnical Categories

The structures discussed in these Recommendations can be allocated to geotechnical categories as shown in Table 3.2, inasmuch as more detailed information is not provided in the appropriate sections.

If DIN 1054 specifies a higher category, the higher category is always adopted.

Table 3.2 Recommendation for allocating geosynthetic-reinforced structures to geotechnical categories

Structure	Geotechnical Category 1	Geotechnical Category 2	Geotechnical Category 3
Retaining structures	$H < 3$ m	3 m $\leq$ H < 9 m	$H \geq 9$ m
Bridge abutments	–	$H < 2$ m	$H \geq 2$ m
Embankments	$H < 3$ m	3 m $\leq$ H < 9 m	$H \geq 9$ m
Doline stabilisation	–	–	Doline stabilisation
Foundation pads	5 cm $\leq s_{allow.} \leq 10$ cm	2 cm $\leq s_{allow.} \leq 5$ cm	1 cm $\leq s_{allow.} \leq 2$ cm
Special geosynthetic-reinforced structures	–	Foundation system using geosynthetic-encased columns	
	–	Reinforced earth structures over point or linear bearing elements	

3.3 Design Resistances

3.3.1 Structural Resistance of Geosynthetics

The structural resistance of a geosynthetic refers to its tensile strength $R_{B,d}$. It is based on the load-extension curve determined in a tensile test for the respective geosynthetic. The short-term strength $R_{B,k0}$ determined from the maximum tensile force identified in the test is given relative to a width of 1 m. To take production tolerances into consideration the characteristic value of the short-term strength $R_{B,k0}$ is given as the 5% quantile. The characteristic values of the short-term strength $R_{B,k0}$ are determined as described in Section 2.2.4.4. The long-term strength of the geosynthetic $R_{B,k}$ is calculated from the short-term strength $R_{B,k0}$ by dividing by the reduction factors A_1 to A_5. The reduction factors take into consideration the impacts of creep (A_1), damage to the geosynthetics during transportation, installation and compaction (A_2), the impacts of junctions, seams and connections (A_3), environmental impacts such as weathering, chemicals and microorganisms (A_4), and impacts from predominantly dynamic actions (A_5).

$$R_{B,k} = R_{B,k0} / (A_1 \cdot A_2 \cdot A_3 \cdot A_4 \cdot A_5) \qquad \text{Eq. (3.1)}$$

where:

$R_{B,k0}$ characteristic value of the short-term strength of the geosynthetics (5% quantile),

$R_{B,k}$ characteristic value of the long-term strength of the geosynthetics,

A_1 reduction factor for considering creep strain or creep rupture behaviour,

A_2 reduction factor for considering any damage caused during installation, transportation and compaction,

A_3 reduction factor for considering processing (seams, connections, joins),

A_4 reduction factor for considering environmental impacts (weathering resistance, resistance against chemicals, microorganisms, animals),

A_5 reduction factor for considering the impact of dynamic actions (*cf.* Section 12).

The design resistance of the geosynthetics $R_{B,d}$ is calculated by dividing the characteristic long-term strength $R_{B,k}$ by the partial safety factor γ_M for the structural resistance of the reinforcement (*cf.* Section 3.4). Among other things it takes into consideration any deviations in the geometry of the structure and in the characteristic values of the geosynthetics compared to those identified in the laboratory. The design resistance $R_{B,d}$ of the geosynthetics is calculated as follows:

$$R_{B,d} = R_{B,k} / \gamma_M \qquad \text{Eq. (3.2)}$$

where:

$R_{B,d}$ design resistance of the geosynthetic reinforcement,

γ_m partial safety factor for the structural resistance of flexible reinforcement elements as described in Section 3.4.

3.3.2 Determining Reduction Factors

The required product parameters and the reduction factors shall be provided by the manufacturer, for example by:

- test reports provided by independent institutes with appropriate personnel and equipment or
- by data from technical approvals.

Alternatively, the reduction factors used in EBGEO, Section 2.2.4 shall be adopted. If, due to a combination of the reduction factors given above, $R_{B,d} \leq 0.10 \cdot R_{B,k0}$ for permanent structures, a variety of options are available to avoid casting doubt on the execution of the structures as a whole. Using appropriate engineering measures, and by the selection of geosynthetics and fill soils it is possible to achieve a technically correct and economical solution. The following points may be mentioned here as examples:

- selection of raw materials (for A_1),
- configuration of protective layers (for A_2),
- use of geocomposites (for A_1 and A_2),
- changing the fill soil (for A_2 and A_4).

The respective solution shall be adopted to suit the actual application.

3.3.3 Pull-out Resistance of Geosynthetics

3.3.3.1 Characteristic Pull-out Resistance of Geosynthetics

The characteristic **pull-out resistance** is the integral of the shear stresses mobilised in the reinforcement plane. In the limit state the characteristic pull-out resistance is:

$$R_{A,k} = \sigma_{v,k} \cdot L_A \cdot f_{sg,k} \cdot n \qquad \text{Eq. (3.3)}$$

where:

$R_{A,k}$ characteristic pull-out resistance of the reinforcement relative to 1 m width,

$\sigma_{v,k}$ characteristic value of the normal stress in the reinforcement plane,

L_A anchorage length of reinforcement behind the failure plane under consideration,

$f_{sg,k}$ characteristic value of the mean friction coefficient between the fill soil and the plane formed by the geosynthetics and the intermediate ground as described in Section 2.2.4.11,

n number of adoptable friction surfaces.

3.3.3.2 GEO Pull-out Resistance Design Values

The pull-out resistance design value for GEO limit state stability analyses of intersected reinforcements is determined from the characteristic value of the pull-out resistance by dividing by the partial safety factor for the pull-out resistance of flexible reinforcement elements γ_B to DIN 1054, Table 3. The design value of the pull-out resistance $R_{A,d}$ is:

$$R_{A,d} = R_{A,k} / \gamma_B \qquad \text{Eq. (3.4)}$$

where:

$R_{A,d}$ design value of the pull-out resistance of the reinforcement,

γ_B partial safety factor for the pull-out resistance of the reinforcement.

3.3.3.3 STR Pull-out Resistance Design Value

The design value of the pull-out resistance for analysis of the required overlapping of the reinforcement (reinforcement junctions) in the STR limit state is determined from the characteristic pull-out resistance by dividing by the partial safety factor γ_{GL} to DIN 1054, Table 3, based on the sliding resistance to DIN 1054. The design value of the pull-out resistance $R_{A,d}$ is:

$$R_{A,d} = R_{A,d} / \gamma_{Gl} \qquad \text{Eq. (3.5)}$$

where:

$R_{A,d}$ design value of the pull-out resistance of the reinforcement,

γ_{Gl} partial safety factor for sliding resistance.

3.3.4   Axial Stiffness of Geosynthetics in the Serviceability Limit State

The axial stiffness of the geosynthetics is determined as a conservative characteristic value from the load-extension curve of the geosynthetic or from its isochrones (without considering ground contact).

3.4   Partial Safety Factors – Supplementary Regulations to DIN 1054

The partial safety factors in Table 3.3 are defined as supplementary factors to those in DIN 1054 for flexible reinforcement elements (geosynthetics). They are adopted for design.

Table 3.3 Partial safety factors for resistances supplementary to DIN 1054

Resistance	Notation	Load Case		
		LC 1	LC 2	LC 3
STR: Limit state of failure of the structure and of structural elements:				
Resistances of flexible reinforcement elements				
Structural resistance of reinforcement	γ_m	1.40	1.30	1.20

Wide equivalent loads where $p_k \leq 10$ kN/m^2 are always regarded as *permanent actions* for designing geosynthetic-reinforced structures in analogy to DIN 1054.

4   Embankments on Soft Soils

4.1   General Recommendations

The stability of embankments on soft soils can be enhanced by the use of geosynthetic reinforcements installed in the embankment contact zone. This does not prevent settlement, but generally makes it more uniform.

[1], [2], [6] and [7] report on the modus operandi, instrumented monitoring and experiences in terms of the long-term behaviour of geosynthetic reinforcements in embankment contact zones.

Only stability analyses are dealt with below. The separating function of a geosynthetic layer in the embankment contact zone is analysed separately [4].

When analysing and designing the reinforcement the initial stability, any construction stages and the final stability are investigated. If the calculated stability is inadequate, reinforcement provides an additional resisting force in the equilibrium conditions.

If the stability in the final state is adequate without reinforcement, the operational life of the reinforcement corresponds to the consolidation period. If the calculated stability of the unreinforced embankment cannot be guaranteed in the final state, the reinforcement shall be designed for the operational life of the embankment.

In the following equations and figures the friction angle and cohesion are designated by the generalised symbols φ and c. The actual application determines whether these refer to the shear parameters φ' and c' in the drained or the undrained conditions (φ_u and c_u).

The embankment structure is designated by the index '1', the softer subsoil by the index '2'.

All possible failure mechanisms shall be considered when investigating the stability of embankments on soft soils. They are based on the global failure described in DIN 4084, where both circular and linearly bounded failure bodies are investigated. The transition zones between the ground and the upper and lower surfaces of the geosynthetic reinforcement respectively, on which the embankment structure may slip, represent failure planes requiring special investigations.

Note: Because these failure mechanisms merely represent special variations of the global failure mode, they are analysed here in the GEO limit state, while according to DIN 1054 the 'sliding' analysis in the STR limit state may be the governing mode.

Recommendations for Design and Analysis of Earth Structures using
Geosynthetic Reinforcements (EBGEO). German Geotechnical Society.
© 2011 Ernst & Sohn GmbH & Co. KG.
Published by Ernst & Sohn GmbH & Co. KG.

4.2 Analysing Global Failure

4.2.1 General Recommendations

When building an embankment on soft soil it is necessary to investigate the global stability of slip planes which:

- remain within the embankment structure and do not intersect reinforcement layers,
- remain within the embankment structure and intersect reinforcement layers,
- are within the embankment structure and the ground and intersect reinforcement layers (Figure 4.1).

The resistance of the reinforcement is adopted as a restraining force, where the smaller of the following values is adopted:

- the design stiffness $R_{B,d}$ of the reinforcement layer (STR),
- the design value (GEO) of the pull-out resistance force of the reinforcement layer from the surrounding fill soil at the 'left' ($R_{AL,d}$) or 'right' ($R_{AR,d}$) of the respective slip line,
- the design value of the frictional resistance on the top of the geosynthetics $R_{O,d}$ (STR) at the 'right' of the respective slip line.

The line of action of this resisting force is adopted conservatively in the undeformed state.

The initial state, any construction stages and the final state are differentiated (see Section 4.1).

The safety against global failure is adequate if the following condition is met for all possible failure mechanisms by the values for the actions and resistances:

$$E_d \leq R_d + \min (R_{B,d}; R_{AL,d}; R_{AR,d}; R_{O,d}) .$$

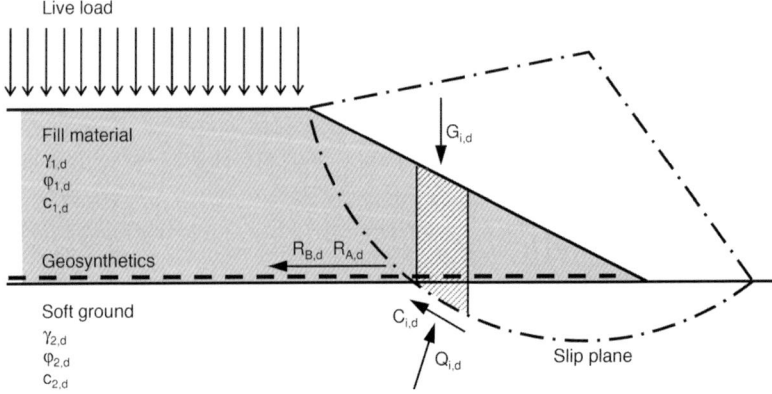

Figure 4.1 Analysis of global stability with intersected reinforcement

4.2.2 Failure Mechanisms

4.2.2.1 Failure on Circular Slip Planes

Analysis of global stability for circular slip planes is carried out to DIN 4084 for the GEO limit state.

4.2.2.2 Defined Slip Plane in Soft Soil

For structurally and/or geologically defined slip planes (e.g. flexible membranes with very low adhesion or friction coefficients below the embankment and/or for deeper, thin layers with very low shear strength) global stability should be analysed using composite failure mechanisms (e.g. as shown in Figure 4.2) (also see DIN 4084).

4.2.2.3 Slip Plane Between Geosynthetics and Fill Soil or Between Geosynthetics and Soft Soil

The boundary planes between embankment fill material/geosynthetics and geosynthetics/ground represent preferential slip planes (Figure 4.3). Adequate sliding stability exists if the following conditions are met:

$$E_{ah,d} \leq R_{O,d} \qquad\qquad \text{Eq. (4.1)}$$

$$E_{ah,d} \leq R_{U,d} + \min (R_{B,d}; R_{A,d}) \qquad\qquad \text{Eq. (4.2)}$$

where:

$E_{ah,d}$ design value of the horizontal component of the active earth pressure,

$R_{O,d}$ design value of the friction resistance between the embankment fill material and the top of the geosynthetics,

$R_{U,d}$ design value of the friction resistance between the bottom of the geosynthetics and the ground,

$R_{B,d}$ design resistance of the reinforcement layer (STR),

$R_{A,d}$ design value (GEO) of the pull-out resistance force from the surrounding ground.

The smaller of the values of $R_{B,d}$ and $R_{A,d}$ is the governing value.

4.2.2.4 Adopting Reinforcement Wrap-around

Where friction resistance $R_{O,d}$ is inadequate empiricism [1] indicates that sliding stability can be substantially increased by wrapping the geosynthetic reinforcement back at the embankment toe. Structures configured as in Figure 4.4 are analysed by testing various sections.

Analysis of the sliding stability on the bottom of the reinforcement is also carried out using Equation (4.2) for structures with reinforcement wrap-around.

Sections must also be investigated for sliding of the embankment flank above the wrap-around (Figure 4.4 a) and for sliding of the embankment above the reinforcement layer (Figure 4.4 b). The following conditions shall be met:

46

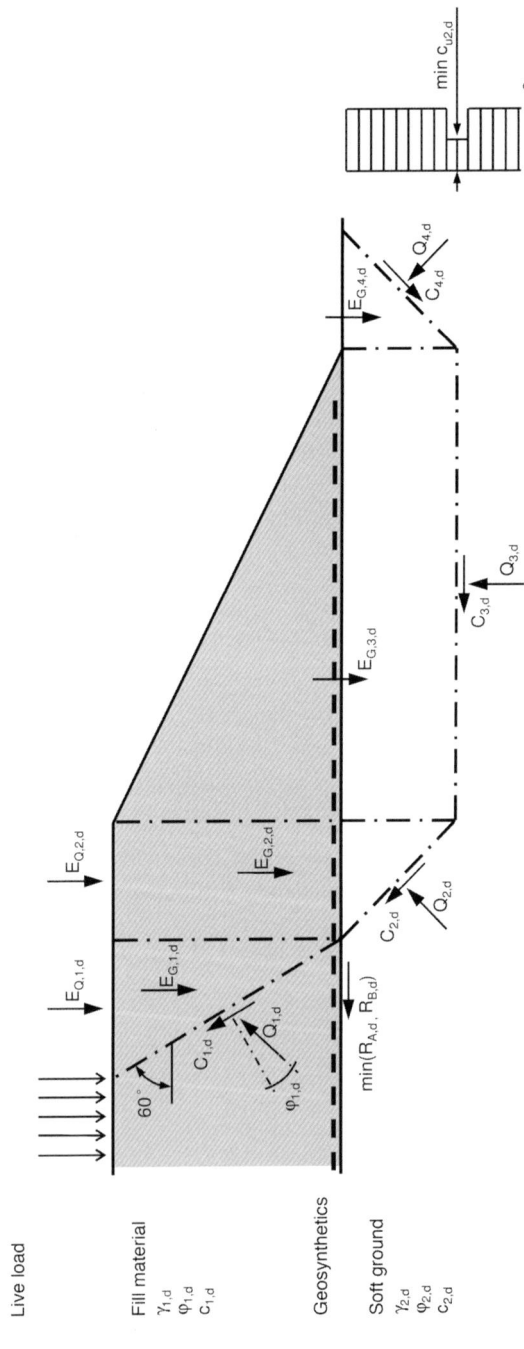

Figure 4.2 Analysis of global stability for defined slip planes

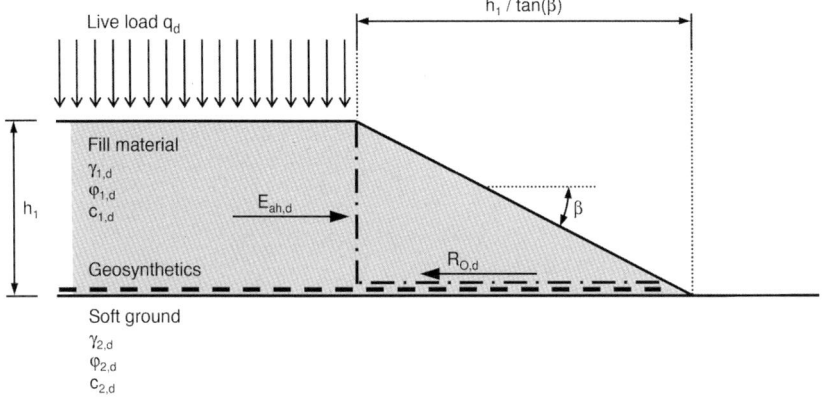

a) Above the geosynthetics

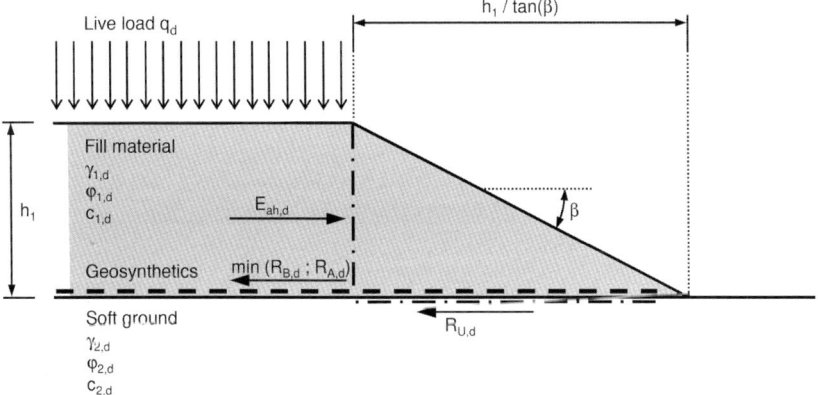

b) Below the geosynthetics

Figure 4.3 Analysis of the sliding stability of an embankment without wrap-around of the geotextile reinforcement at the embankment toe

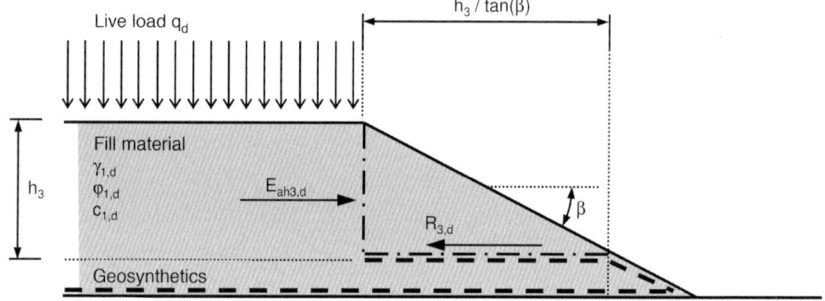

a) Embankment sliding above the wrap-around

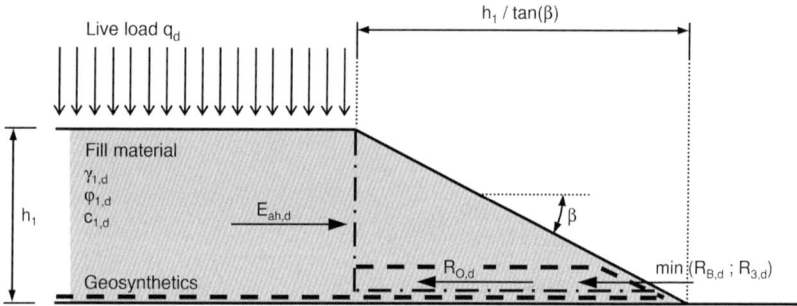

b) Embankment sliding above the reinforcement layer

Figure 4.4 Analysis of the sliding stability of an embankment
with wrap-around of the geotextile reinforcement at the embankment toe

$$E_{ah3,d} \leq R_{3,d} \qquad \text{Eq. (4.3)}$$

$$E_{ah,d} \leq R_{O,d} + \min (R_{3,d} ; R_{B,d}) \qquad \text{Eq. (4.4)}$$

where:

$E_{ah,d}$ design value of the horizontal component of the active earth pressure
(relative to embankment height h),

$E_{ah3,d}$ design value of the horizontal component of the active earth pressure
(relative to height h_3),

$R_{O,d}$ design value of the friction resistance between the embankment fill
material and the top of the geosynthetics,

$R_{3,d}$ design value of the friction resistance between the embankment fill
material and the top of the geosynthetics
(relative to the length $h_3 / \tan \beta$),

$R_{B,d}$ design resistance of the reinforcement layer (STR).
The smaller of the values of $R_{B,d}$ and $R_{3,d}$ is the governing value.

4.2.3 Actions

The earth pressures from soil dead weight and live loads on the embankment crest are adopted as actions:

$$E_{ah,d} = \gamma_G \cdot (\gamma_{1,k} \cdot 0,5 \cdot h_1 \cdot h_1 \cdot K_{agh}) + \gamma_Q \cdot (p_k \cdot h_1 \cdot K_{aph}) \qquad \text{Eq. (4.5)}$$

where:

$E_{ah,d}$ design value of the earth pressure for the total height,
h_1 total height of the embankment,
K_{agh} horizontal earth pressure coefficient from dead load (DIN 4085),
K_{aph} horizontal earth pressure coefficient from live loads (DIN 4085),
p_k characteristic live load,
$\gamma_{1,k}$ characteristic value of the unit weight of the embankment fill material,
γ_G partial safety factor for permanent actions in the GEO limit state,
γ_Q partial safety factor for variable actions in the GEO limit state.

Note: Actions from earth pressures are determined analogously for other heights.

4.2.4 Resistances

4.2.4.1 Design Value of the Friction Resistance on Top of the Geosynthetics $R_{O,d}$

The friction resistance between the embankment fill material and the geosynthetics is:

$$R_{O,d} = 1/2 \cdot \gamma_{1,d} \cdot (h_1 / \tan \beta) \cdot h_1 \cdot f_{1g,d} \qquad \text{Eq. (4.6)}$$

where:

β slope angle ($\tan \beta = 1 : n$),
h_1 total height of the embankment,
$\gamma_{1,d}$ design value of the unit weight of the embankment fill material ($\gamma_{1,d} = \gamma_{1,k}$),
$f_{1g,d}$ characteristic value of the friction coefficient between the embankment fill material and the geosynthetics (see Section 2.2.4.11.2 where $\tan \varphi_d = \tan \varphi_k / \gamma_\varphi$).

4.2.4.2 Design Value of the Shear Resistance on the Bottom of the Geosynthetics $R_{U,d}$

The initial state and the final state are differentiated when determining the shear resistance between the geosynthetics and the ground.

In the initial state:

$$R_{U,d} = c_{u2,d} \cdot h_1 / \tan \beta . \qquad \text{Eq. (4.7)}$$

In the final state:

$$R_{U,d} = c'_{2,d} \cdot h_1 / \tan\beta + 1/2 \cdot \gamma_{1,d} \cdot (h_1 / \tan\beta) \cdot h_1 \cdot f_{2g,d} .$$ Eq. (4.8)

where:

β slope angle (tan β = 1 : n),

h_1 total height of the embankment,

$\gamma_{1,d}$ design value of the unit weight of the embankment fill material ($\gamma_{1,d} = \gamma_{1,k}$),

$c_{u2,d}$ design value of the shear strength of the undrained soil ($c_{u2,d} = c_{u2,k} / \gamma_{cu}$),

$c'_{2,d}$ design value of the shear strength of the drained soil ($c'_{2,d} = c'_{2,k} / \gamma_c$),

$f_{2g,d}$ design value of the friction coefficient between the ground and the geosynthetics (see Section 2.2.4.11.2 where $\varphi_d = \varphi_k / \gamma_\varphi$).

4.2.4.3   Design Value of the Pull-out Resistance $R_{A,d}$

The design value of the pull-out resistance $R_{A,d}$ is determined as described in Sections 2.2.4.11 and 3.

4.2.4.4   Design Resistance of the Geosynthetic Reinforcement $R_{B,d}$

Safety against failure of the reinforcement is analysed as described in Sections 2 and 3.

4.2.4.5   Design Value of the Friction Resistance on Top of the Geosynthetic $R_{3,d}$

The friction resistance between the embankment fill material and the geosynthetic is:

$$R_{3,d} = 1/2 \cdot \gamma_{1,d} \cdot (h_3 / \tan\beta) \cdot h_3 \cdot f_{1g,d}$$ Eq. (4.9)

where:

β slope angle (tan β = 1 : n),

h_3 height as in Figure 4.4,

$\gamma_{1,d}$ design value of the unit weight of the embankment fill material ($\gamma_{1,d} = \gamma_{1,k}$),

$f_{1g,d}$ design value of the friction coefficient between the embankment fill material and the geosynthetics (see Section 2.2.4.11.2 where $\varphi_d = \varphi_k / \gamma_\varphi$).

4.3 Analysing the Stability of the Ground against 'Squeezing Out'

Ground 'squeezing' is particularly prevalent when installing the embankment structure, especially in the initial state when the soil is very weak and the thickness of the soft soil is limited (also see [3]) (Figure 4.5). The GEO limit state is analysed.

When adopting the undrained shear strength of the ground the following action on the soil monolith in question resulting from tipping the embankment shall be considered:

$$E_{ah4,d} = \gamma_G \cdot (\gamma_{1,k} \cdot h_1 \cdot h_4 + 0.5 \cdot \gamma_{2,k} \cdot h_4^2 - 2 \cdot c_{u2,k} \cdot h_4) \qquad \text{Eq. (4.10)}$$

where:

$E_{ah4,d}$ design value of the earth pressure, determined using the undrained shear strength of the ground,

Note: *Live loads may need to be taken into consideration when determining the earth pressure.*
In the equation above the earth pressure coefficient K_{agh} = 1.0 and is not explicitly given because φ_u = 0 for the ground.

h_1 total height of the embankment,
h_4 height of the soil monolith ($h_4 \leq h_2$),
$\gamma_{1,k}$ characteristic value of the unit weight of the embankment fill material,
$\gamma_{2,k}$ characteristic value of the unit weight of the ground,
$c_{u2,k}$ characteristic value of the undrained shear strength of the ground,
γ_G partial safety factor for permanent actions in the GEO limit state.

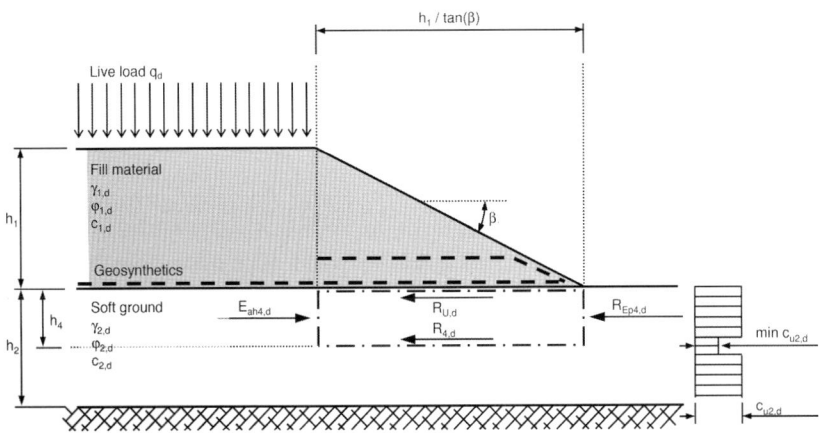

Figure 4.5 Ground 'squeezing'

The following resistances are adopted:

- the passive earth pressure in front of the soil monolith (passive earth pressure in the undrained state)

$$R_{Ep4,d} = 0,5 \cdot \gamma_{2,d} \cdot h_4^2 + 2 \cdot c_{u2,d} \cdot h_4 ,$$
Eq. (4.11)

- the characteristic value of the friction resistance on the bottom of the geosynthetics $R_{U,d}$ (see 4.2.4.2),
- the characteristic value of the friction resistance in the ground on the bottom of the soil monolith

$$R_{4,d} = c_{u2,d} \cdot L = c_{u2,d} \cdot h_1 / \tan\beta ,$$
Eq. (4.12)

where:

β slope angle (tan β = 1 : n),
h_1 total height of the embankment,
h_4 height of the soil monolith ($h_4 \leq h_2$),
$\gamma_{1,d}$ design value of the unit weight of the embankment fill material ($\gamma_{1,d} = \gamma_{1,k}$),
$\gamma_{2,d}$ design value of the unit weight of the ground ($\gamma_{2,d} = \gamma_{2,k}$),
$c_{u2,d}$ design value of the undrained shear strength of the ground ($c_{u2,d} = c_{u2,k} / \gamma_{cu}$).

In homogeneous ground the height h_4 corresponds to the height h_2. The height of any special weak zones in the ground (also see Figure 4.2) is adopted for h_4.

Adequate squeezing stability of the ground exists if the following conditions are met:

$$E_{ah4,d} \leq R_{Ep4,d} + R_{U,d} + R_{4,d}$$
Eq. (4.13)

and

$$R_{U,d} \leq \min (R_{B,d} ; R_{A,d}) .$$
Eq. (4.14)

4.4   Analysing Bearing Capacity

Bearing capacity to **DIN 4017** is always analysed in accordance with **DIN 1054** including for embankments and fill on ground with low shear strength. The embankment is regarded as quasi-monolithic. Reinforcement layers are not intersected. The STR limit state is analysed.An examination of whether the failure model based on the DIN 4017 analysis can actually form shall be carried out, in particular when the soft soil is not very thick. Usually the global failure analysis is the governing one, not the bearing capacity.

4.5   Engineering Notes

Draft sketches for embankments on fine-grained, soft soils and engineering notes on the selection and processing of reinforcements are given in [4] and [5].

4.6   Bibliography

[1]   Blume, K.-H. (1995): Großversuch zum Tragverhalten textiler Bewehrung unter einer Dammaufstandsfläche. FS-KGEO 1995.

[2]   Bürger, M., Blosfeld, J., Blume, K.-H., Hillmann, R. (2005): 30 Jahre Erfahrungen mit Straßen auf wenig tragfähigem Untergrund. Bundesanstalt für Straßenwesen.

[3]   Code of practice for strengthened/reinforced soils and other fills (British Standard, BS8006).

[4]   Forschungsgesellschaft für Straßen- und Verkehrswesen, Merkblatt über die Anwendung von Geokunststoffen im Erdbau des Straßenbaus.

[5]   Forschungsgesellschaft für Straßen- und Verkehrswesen, Merkblatt über Straßenbau auf wenig tragfähigem Untergrund.

[6]   Blume, K.-H., Alexiew, D. (1998): Long-Term Experience with Reinforced Embankments on Soft Subsoils: Mechanical Behavior and Durability. 6th Int. Conf. on Geosythetics, Atlanta.

[7]   Blume, K., Alexiew, D., Glötzl, F. (2006): The new federal highway (Autobahn) A26 in Germany with high geosynthetic reinforced embankments on soft soils. 8th Int. Conf. on Geosynthetics, Yokohama, Millpress, Rotterdam.

4.7   Example Embankment on Soft Soil

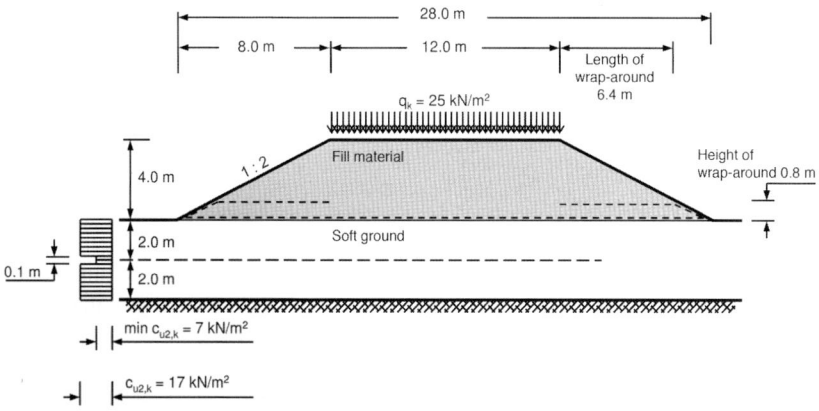

Figure 4.6 Embankment dimensions

Soil parameter	Unit weight	Drained shear parameters		Undrained shear parameters	
		φ'	c'	φ_u	c_u
Fill material	$\gamma_{1,k} = 20\ \text{kN/m}^3$	$\varphi'_{1,k} = 35°$	$c'_{1,k} = 0\ \text{kN/m}^2$	–	–
Soft ground	$\gamma_{2,k} = 15\ \text{kN/m}^3$	$\varphi'_{2,k} = 20°$	$c'_{2,k} = 0\ \text{kN/m}^2$	$\varphi_{u,2,k} = 0°$	$c_{u,2,k} = 20\ \text{kN/m}^2$ min $c_{u,2,k} = 10\ \text{kN/m}^2$

4.7.1 Failure on Circular Slip Planes

An embankment on soft soil can fail along a slip plane passing through the embankment fill material and the ground. Global failure is analysed to **DIN 4084** in the GEO limit state and the reinforcement resistance adopted as a restraining force.

The resistances are calculated using the design values of the shear strengths of the individual soil strata. The initial and the final states are differentiated. The design value of the permanent actions from horizontal earth pressure is given by the quotient of the characteristic friction angle and the characteristic unit weight of the soil (GEO limit state), and the partial safety factor for permanent actions for the GEO limit state for Load Case 1 $\gamma_{Gl} = 1.00$.

Adequate safety against failure is given if the general condition for the ultimate limit state

$$E_d \leq R_d \qquad \text{Eq. (4.15)}$$

is met. When investigating by varying slip circles no failure mechanism may violate the condition for the ultimate limit state, where the equation for circular failure planes (index 'M') in the notation

$$E_M / R_M = \mu \leq 1 \qquad \text{Eq. (4.16)}$$

is used with μ as the utilisation factor.

The design value of the pull-out resistance may be adopted to DIN 1054 as an additional resistance with the partial safety factor γ_B for favourably acting reinforcement layers. A software application was used to vary the circular slip planes using the method of slices. The tensile force in the intersected reinforcement layer was varied until Equation (4.16) was fulfilled for $\mu = 1.00$.

4.7.1.1 Initial Stability

Design values of shear parameters:

– Ground

$$c_{u,2,d} = c_{u,2,k}/\gamma_{cu}$$
$$c_{u,2,d} = 20/1.25 = 16.0 \text{ kN/m}^2$$
$$\min c_{u,2,d} = 10/1.25 = 8.0 \text{ kN/m}^2.$$

– Fill material

$$\varphi'_{1,d} = \text{arc tan } [\tan (\varphi'_{1,k})/\gamma_\varphi]$$
$$\varphi'_{1,d} = \text{arc tan } [\tan (35°)/1.25] = 29.3°.$$

The actions are determined from the dimensions and characteristic soil unit weights as shown in Figure 4.11. The partial safety factor for permanent actions is $\gamma_G = 1.00$ for the GEO limit state LC1. The characteristic live load of $q_k = 25.0 \text{ kN/m}^2$ is multiplied by the partial safety factor $\gamma_Q = 1.30$ to acquire the design value $q_d = 32.5 \text{ kN/m}^2$.

Analysis results (initial stability):

Equation (4.16) ($\mu = 1.0$) is fulfilled for a tensile force in the geosynthetics layer of **42 kN/m**. The governing slip circle results in a maximum possible effective bond length of $\mathbf{L_{Ai} = 12.5}$ **m** within the circular failure mechanism as far as the toe of the slope (see Figure 4.7). Outside of this zone the geosynthetic layer has a possible anchorage length of around 15 m.

Initial state:

$$\min (\text{H}_{A,d} ; \text{R}_{B,d}) > 42 \text{ kN/m}$$

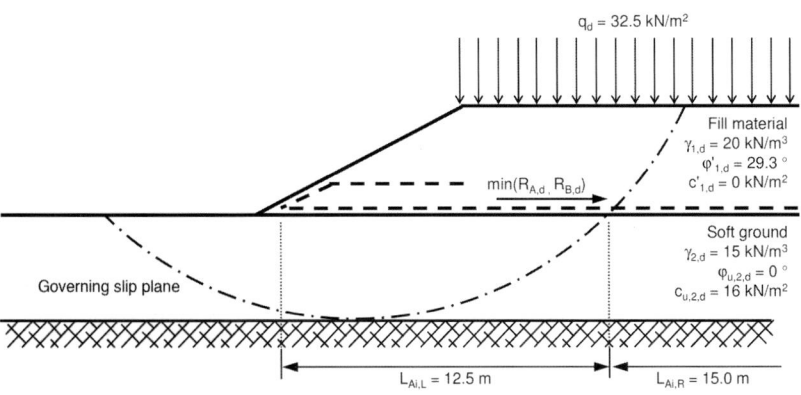

Figure 4.7 Global stability – initial state

4.7.1.2   Final Stability

Design values of shear parameters:

– Ground

$$\varphi'_{2,d} = \text{arc tan } [\tan{(\varphi'_{2,k})}/\gamma_\varphi]$$
$$\varphi'_{2,d} = \text{arc tan } [\tan{(20°)}/1.25] = 16.2°.$$

– Fill material

$$\varphi'_{1,d} = \text{arc tan } [\tan{(\varphi'_{1,k})}/\gamma_\varphi]$$
$$\varphi'_{1,d} = \text{arc tan } [\tan{(35°)}/1.25] = 29.3°.$$

Dimensions and soil unit weights correspond to Figure 4.11, the partial safety factor for permanent actions (horizontal earth pressure) $\gamma_G = 1.00$ for LC1 and the GEO limit state. The characteristic live load of $q_k = 25.0$ kN/m^2 is multiplied by the partial safety factor $\gamma_Q = 1.30$ to get the design value $q_d = 32.5$ kN/m^2. The analysis is carried out as described above with a search for the tensile force in the geosynthetics which fulfils Equation (4.16) ($\mu = 1.00$).

Analysis results (final stability):

Equation (4.16) is fulfilled for a design force of **5 kN/m** transferred by the reinforcement layer. The governing slip circle results in a maximum possible effective bond length of $\mathbf{L_{Ai}} = \mathbf{6.5}$ **m** within the circular failure mechanism as far as the toe of the slope (see Figure 4.8). Outside of this zone the geosynthetic layer has a possible anchorage length of around 20 m.

Final state:

min $(R_{A,d} ; R_{B,d}) > 5$ kN/m

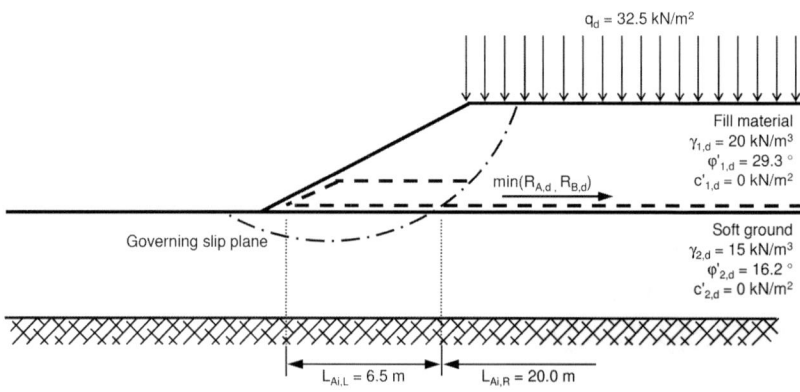

Figure 4.8 Global stability – final state

4.7.2 Defined Slip Plane in Soft Soil

4.7.2.1 Initial Stability

A separate analysis shall be carried out for the initial state of a geotechnically predefined slip plane in ground with very low undrained shear strength (see Figure 4.11), for example using general wedge mechanisms with planar slip planes. A graphical structural analysis is carried out here as an example for a possible mechanism (see Figure 4.9). The GEO limit state, Load Case 1, is analysed.

The governing design values for the shear parameters $\varphi'_{1,d}$, $c_{u,2,d}$ and min $c_{u,2,d}$ for the initial state are determined analogous to the analyses for circular slip planes.

The geometry of the failure mechanism is dictated by the following boundary conditions:

– the base of the failure mass lies within the soft soil strata –2.00 m below ground level,
– the failure angle in the undrained soil strata is always 45°,
– the failure angle through the embankment fill material is:
$45° + (\varphi'_{1,d}/2) = 45° + (29.3°/2) = 60°$.

The given boundary conditions produce the failure model shown in Figure 4.9.

Depending on the geometry, it can be useful to introduce the following auxiliary variables as a function of wedges 1, 2, 3 and 4:

Slice number	Width of slice	Length of shear plane
1	$b_1 = 2.3$ m	$l_1 = 4.6$ m
2	$b_2 = 2.0$ m	$l_2 = 2.8$ m
3	$b_3 = 8.0$ m	$l_3 = 8.0$ m
4	$b_4 = 2.0$ m	$l_4 = 2.8$ m

First, the adopted actions from soil dead weight and live loads are identified (cf. Figure 4.9). The partial safety factors for permanent and unfavourable, variable actions for the GEO limit state, LC1 are required:

$\gamma_G = 1.00$

$\gamma_Q = 1.30$

The actions from soil dead weight are then given by:

$$E_{G,1,d} = 1/2 \cdot b_1 \cdot h_1 \cdot \gamma_{1,k} \cdot \gamma_G$$
$$= 1/2 \cdot 2.3 \cdot 4.0 \cdot 20 \cdot 1.00 = 92.0 \text{ kN/m}$$

$$E_{G,2,d} = b_2 \cdot h_1 \cdot \gamma_{1,k} \cdot \gamma_G + 1/2 \cdot b_2 \cdot h_4 \cdot \gamma_{2,k} \cdot \gamma_G$$
$$= 2.0 \cdot 4.0 \cdot 20 \cdot 1.00 + 1/2 \cdot 2.0 \cdot 2.0 \cdot 15 \cdot 1.00 = 190.0 \text{ kN/m}$$

58

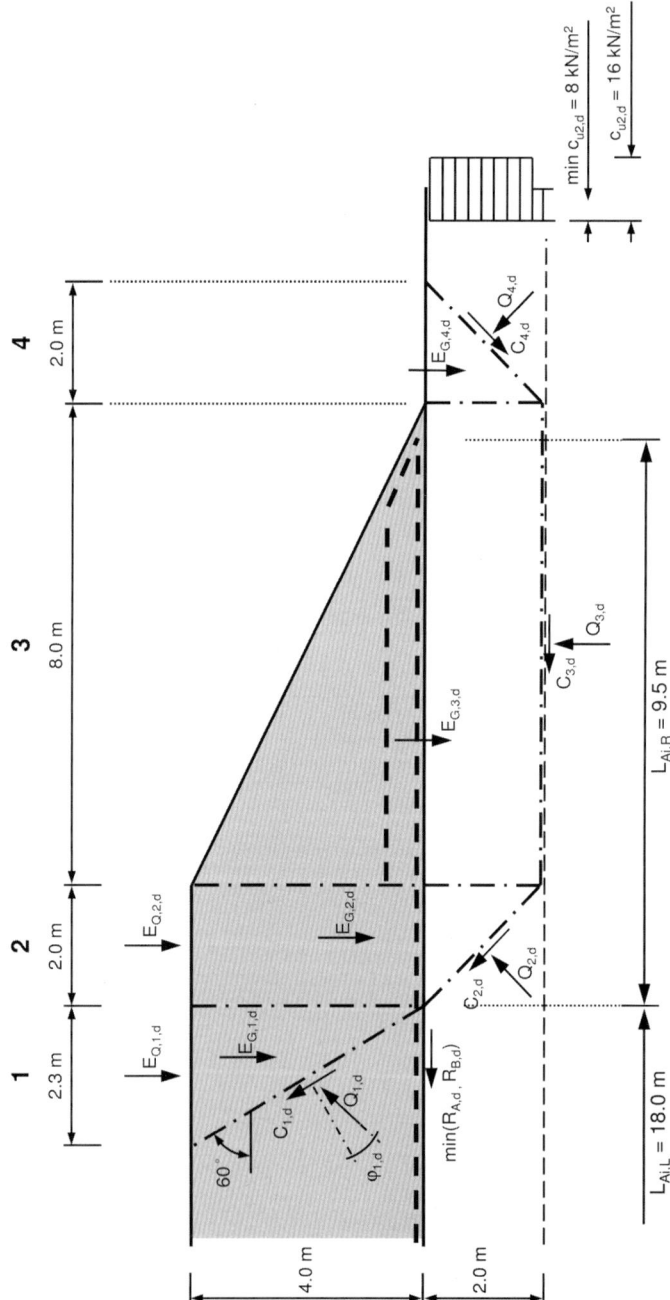

Figure 4.9 Geometry and forces for analysing global stability on a predefined slip plane

$$E_{G,3,d} = 1/2 \cdot b_3 \cdot h_1 \cdot \gamma_{1,k} \cdot \gamma_G + b_3 \cdot h_4 \cdot \gamma_{2,k} \cdot \gamma_G$$
$$= 1/2 \cdot 8.0 \cdot 4.0 \cdot 20 \cdot 1.00 + 8.0 \cdot 2.0 \cdot 15 \cdot 1.00 = 560.0 \text{ kN/m}$$

$$E_{G,4,d} = 1/2 \cdot b_4 \cdot h_4 \cdot \gamma_{2,k} \cdot \gamma_G$$
$$= 1/2 \cdot 2.0 \cdot 2.0 \cdot 15 \cdot 1.00 = 30.0 \text{ kN/m.} \qquad \text{Eq. (4.17)}$$

The action from the live load is given by:

$$E_{Q,1,d} = b_1 \cdot q_k \cdot \gamma_Q = 2.3 \cdot 25 \cdot 1.30 = 74.8 \text{ kN/m}$$

$$E_{Q,2,d} = b_2 \cdot q_k \cdot \gamma_Q = 2.0 \cdot 25 \cdot 1.30 = 65.0 \text{ kN/m.}$$

The resistances to wedge sliding of the embankment section on the geological weak zone are given by the undrained cohesion of the soft ground and the friction angle of the embankment fill material.

The design values of the cohesive forces are given by:

$$C_{1,d} = 0.0 \text{ kN/m}$$

$$C_{2,d} = l_2 \cdot c_{u,2,d} = 2.8 \cdot 16 = 44.8 \text{ kN/m}$$

$$C_{3,d} = l_3 \cdot \min c_{u,2,d} = 8.00 \cdot 8 = 64.0 \text{ kN/m}$$

$$C_{4,d} = l_4 \cdot c_{u,2,d} = 2.8 \cdot 16 = 44.8 \text{ kN/m}$$

Using the forces of the actions $E_{G,i,d}$, $E_{Q,i,d}$ and the forces of $C_{i,d}$ (i = 1 to 4) a resultant horizontal force can be calculated for each slice. Calculation of this horizontal force requires that the directions of the forces $Q_{i,d}$ (i = 1 to 4) be defined. The directions of the forces $Q_{i,d}$ are given by $\varphi_{u,2,k} = 0°$ for slip planes in undrained, soft soil, because they are normal to the slip plane. The slip plane in slice 1 passes through the gravel fill, which has a friction angle $\varphi'_{1,d}$. The force $Q_{1,d}$ is at an angle of $90° - \varphi'_{1,d}$ to the slip plane.

When the directions of all forces acting on the soil wedge are known the horizontal force can be calculated for each slice. This can also be visualised using a graphical equilibrium diagram based on a scaled drawing of the force vectors.

$$H_{1,d} = (E_{G,1,d} + E_{Q,1,d}) \sin (45° - \varphi'_{1,d}/2) / \sin (45° + \varphi'_{1,d}/2)$$
$$= (92.0 + 74.8) \sin (45° - 29.3°/2) / \sin (45° + 29.3°/2) = 97.6 \text{ kN/m}$$

$$H_{2,d} = E_{G,2,d} + E_{Q,2,d} - 2\,C_{2,d}\,2^{-0.5} = 190.0 + 65.0 - 2\,44.8\,2^{-0.5} = 191.6 \text{ kN/m}$$

$$H_{3,d} = -C_{3,d} = -64.0 \text{ kN/m}$$

$$H_{4,d} = -E_{G,4,d} - 2\,C_{4,d}\,2^{-0.5} = -30.0 - 2\,44.8\,2^{-0.5} = -93.4 \text{ kN/m}$$

The resultant mobilising horizontal force is:

$$\text{res } H = H_{1,d} + H_{2,d} + H_{3,d} + H_{4,d} = 97.6 + 191.6 - 64.0 - 93.4 = 131.9 \text{ kN/m}$$

In order for the soil wedge to remain in equilibrium a geosynthetic design strength of $R_{B,d} = 132$ kN/m is required for the slip mechanism shown in LC 1 and the GEO limit state. Analysis of the pull-out resistance must show that $R_{A,d} = 132$ kN/m.

A governing anchorage length L_{Ai} of approx. 9.5 m shall be demonstrated, without consideration of the action of the reinforcement wrap-around.

4.7.2.2 Final Stability

There are no geologically predefined slip planes in the soft ground in the final state. The final stability of this state therefore does not require investigation.

4.7.3 Slip Plane Between Geosynthetics and Fill Soil or Between Geosynthetics and Soft Soil taking the Reinforcement Wrap-around into Consideration

4.7.3.1 General Recommendations

If the reinforcement is wrapped-around, the slip planes in Figure 4.10 shall be investigated for sliding of soil wedges on the preferential slip planes between the ground and the geosynthetics in the initial and the final states.

The following three unequations shall be fulfilled for the GEO limit state:

$$E_{ah3,d} \leq R_{3,d}$$

$$E_{ah,d} \leq R_{O,d} + \min (R_{3,d}; R_{B,d})$$

$$E_{ah,d} \leq R_{U,d} + \min (R_{B,d}; R_{A,d})$$

In this example the actions from soil dead weight and live loads are calculated independently of the initial and final stability and are given by:

$$
\begin{aligned}
E_{ah3,d} &= \gamma_G \cdot (\gamma_{1,k} \cdot 0.5 \cdot h_3 \cdot h_3 \cdot K_{agh}) + \gamma_Q \cdot (p_k \cdot h_3 \cdot K_{aph}) \\
&= 1.00 \cdot (20 \cdot 0.5 \cdot 3.2 \cdot 3.2 \cdot 0.27) + 1.30 \cdot (25 \cdot 3.2 \cdot 0.27) \\
&= 55.7 \text{ kN/m}
\end{aligned}
$$

$$
\begin{aligned}
E_{ah,d} &= \gamma_G \cdot (\gamma_{1,k} \cdot 0.5 \cdot h_1 \cdot h_1 \cdot K_{agh}) + \gamma_Q \cdot (p_k \cdot h_1 \cdot K_{aph}) \\
&= 1.00 \cdot (20 \cdot 0.5 \cdot 4.0 \cdot 4.0 \cdot 0.27) + 1.30 \cdot (25 \cdot 4.0 \cdot 0.27) \\
&= 78.3 \text{ kN/m}
\end{aligned}
$$

with the active horizontal earth pressure coefficients:

$$K_{agh} = \tan^2 (45° - \varphi'_{1,k}/2) = \tan^2 (45° - 35°/2) = 0.27$$

$$K_{aph} = K_{agh} = 0.27.$$

4.7.3.2 Initial Stability

The resistances required for the analyses are given by:

$$
\begin{aligned}
R_{3,d} &= 1/2 \cdot \gamma_{1,d} \cdot (h_3/\tan \beta) \cdot h_3 \cdot f_{1g,d} \\
&= 1/2 \cdot 20 \cdot (3.2/0.5) \cdot 3.2 \cdot 0.28 = 57.3 \text{ kN/m}
\end{aligned}
$$

where:

$$f_{1g,d} = 0.5 \cdot \tan \varphi'_{1,d} = 0.5 \cdot \tan (29.3°) = 0.28$$

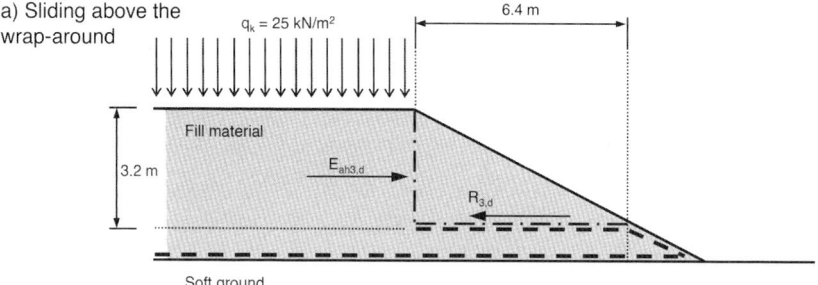

a) Sliding above the wrap-around

$q_k = 25$ kN/m²

6.4 m

Fill material

3.2 m

$E_{ah3,d}$

$R_{3,d}$

Soft ground

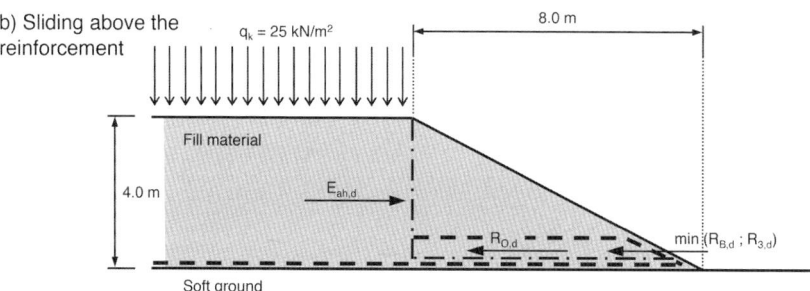

b) Sliding above the reinforcement

$q_k = 25$ kN/m²

8.0 m

Fill material

4.0 m

$E_{ah,d}$

$R_{O,d}$

min $(R_{B,d} ; R_{3,d})$

Soft ground

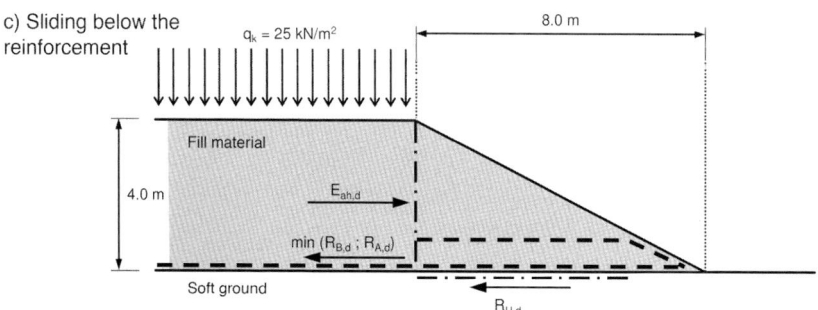

c) Sliding below the reinforcement

$q_k = 25$ kN/m²

8.0 m

Fill material

4.0 m

$E_{ah,d}$

min $(R_{B,d} ; R_{A,d})$

Soft ground

$R_{U,d}$

Figure 4.10 Possible slip planes between geosynthetics and fill soil or between geosynthetics and soft soil taking the reinforcement wrap-around into consideration

$$R_{O,d} = 1/2 \cdot \gamma_{1,d} \cdot (h_1 / \tan \beta) \cdot h_1 \cdot f_{1g,d}$$
$$= 1/2 \cdot 20 \cdot (4.0 / 0.5) \cdot 4.0 \cdot 0.28 = 89.6 \text{ kN/m}$$

$$R_{U,d} = c_{u2,d} \cdot h_1 / \tan \beta = 16 \cdot 4.0 / 0.5 = 128.0 \text{ kN/m}.$$

61

Analysis of the sliding of a soil wedge above the wrap-around:

$E_{ah3,d} \leq R_{3,d}$

$55.7 \leq 57.3$

is thus complete.

Sliding above and below the reinforcement layer is also analysed. The earth pressure $E_{ah,d}$ is adopted as an action. The forces $R_{O,d}$, $R_{3,d}$ and $R_{B,d}$ act as resistances against sliding above the reinforcement layer:

$E_{ah,d} \leq R_{O,d} + \min (R_{3,d}; R_{B,d})$

$78.3 \leq 89.6 + \min (R_{3,d}; R_{B,d})$.

Because the force $R_{O,d}$ is greater than the action $E_{ah,d}$, no additional geosynthetic reinforcement resistance is required for this analysis.

The forces $R_{U,d}$, $R_{B,d}$ und $R_{A,d}$ are required for analysis of sliding below the reinforcement layer (geosynthetics intersected):

$E_{ah,d} \leq R_{U,d} + \min (R_{B,d}; R_{A,d})$

$78.3 \leq 128 + \min (R_{B,d}; R_{A,d})$.

The force $R_{O,d}$ is also greater than the action $E_{ah,d}$ for this failure mechanism. No additional geosynthetic reinforcement resistance is therefore required for this analysis.

4.7.3.3   Final Stability

The soft soil is characterised by its drained shear parameters for analysis of the final stability. This alters the analysis of sliding of a soil wedge below the reinforcement layer. The resistance $R_{U,d}$ is:

$$R_{U,d} = c'_{2,d} \cdot h_1 / \tan \beta + 1/2 \cdot \gamma_{1,d} \cdot (h_1 / \tan \beta) \cdot h_1 \cdot f_{2g,d}$$
$$= 0 \cdot (4.0/0.5) + 1/2 \cdot 20 \cdot (4.0/0.5) \cdot 4 \cdot 0.15 = 48.0 \text{ kN/m}$$

where:

$f_{2g,d} = 0.5 \cdot \tan \varphi'_{2,d} = 0.5 \cdot \tan (16.2°) = 0.15$.

Analysis can now be carried out giving:

$E_{ah,d} \leq R_{U,d} + \min (R_{B,d}; R_{A,d})$

$78.3 \leq 48 + \min (R_{B,d}; R_{A,d})$

$\min (R_{B,d}; R_{A,d}) > 78.3 - 48 = 30.3 \text{ kN/m}$.

This means that a geosynthetics design strength of $R_{B,d} = 30.3$ kN/m is required in LC 1 and the GEO limit state. The pull-out resistance $R_{A,d}$ of at least 30.3 kN/m must also be demonstrated. The governing anchorage length without reinforcement wrap-around is $L_{Ai} = 7.5$ m.

4.7.4 Analysing the Stability of the Ground against 'Squeezing Out'

Section 4.3 notes how ground 'squeezing' is particularly prevalent in the initial state, in particular in very weak soil and where the soft soil is of only limited thickness. The conditions of Equation (4.13) shall be adhered to for the GEO limit state:

$$E_{ah4,d} \leq R_{Ep4,d} + R_{U,d} + R_{4,d}$$

and

$$R_{U,d} \leq \min (R_{B,d}; R_{A,d}).$$

If particularly weak zones are present in the ground the base of the soil wedge being considered shall be located within the weak zone.

The action is given by:

$$E_{ah4,d} = \gamma_G \cdot (\gamma_{1,k} \cdot h_1 \cdot h_4 + 0.5 \cdot \gamma_{2,k} \cdot h_4^2 - 2 \cdot c_{u,2,k} \cdot h_4) + \gamma_Q \cdot (q_k \cdot h_4).$$

Using the numbers from the example:

$$E_{ah4,d} = 1.00 \cdot (20 \cdot 4 \cdot 2 + 0.5 \cdot 15 \cdot 2^2 - 2 \cdot 17 \cdot 2) + 1.30 \cdot (25 \cdot 2)$$
$$= 187.0 \text{ kN/m}.$$

Soil strength parameter design values are required to analyse the resistances in the GEO limit state. In the example these are:

$$c_{u,2,d} = c_{u,2,k}/\gamma_{cu} = 20/1.25 = 16 \text{ kN/m}^2$$

$$\min c_{u,2,d} = \min c_{u,2,k}/\gamma_{cu} = 10/1.25 = 8 \text{ kN/m}^2.$$

In addition, the unit weight of the soft soil is adopted for determining the horizontal resisting force.

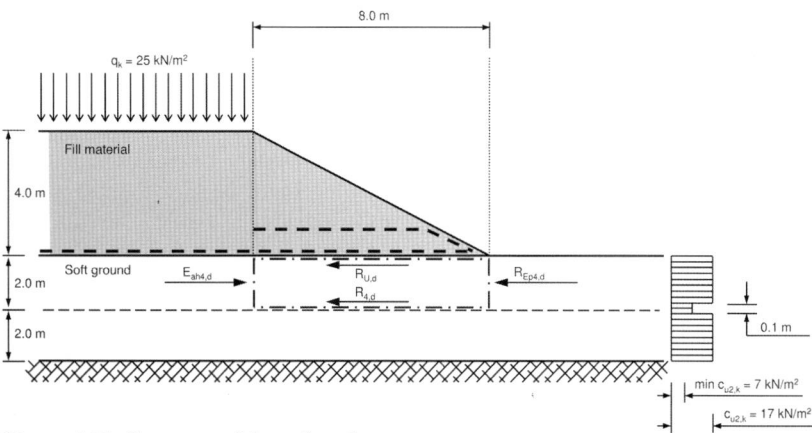

Figure 4.11 Geometry of the soil wedge

63

Where:

$\gamma_{2,d} = \gamma_{2,k}$.

The passive earth pressure design value in front of the soil wedge $R_{3,d}$ is:

$$R_{Ep4,d} = 0.5 \cdot \gamma_{2,d} \cdot h_4^2 + 2 \cdot c_{u,2,d} \cdot h_4$$
$$= 0.5 \cdot 15 \cdot 2^2 + 2 \cdot 16 \cdot 2 = 94.0 \text{ kN/m}.$$

The design value of the friction resistance on top of the soil wedge $R_{U,d}$ is:

$$R_{U,d} = c_{u,2,d} \cdot h_1 / \tan \beta$$
$$= 16 \cdot 4 / 0.5 = 128 \text{ kN/m}.$$

The design value of the friction resistance on the bottom of the soil wedge $R_{U,d}$ is:

$$R_{4,d} = \min c_{u,2,d} \cdot h_1 / \tan \beta$$
$$= 8 \cdot 4 / 0.5 = 64.0 \text{ kN/m}.$$

The conditions of Equation (4.13):

$$E_{a4,d} \leq R_{Ep4,d} + R_{U,d} + R_{4,d}$$
$$187 \leq 94 + 128 + 64 = 286.0 \text{ kN/m}$$

are thus adhered to.

The design strength or the pull-out resistance of the geosynthetics shall also be analysed in analogy to equation (4.14)

$$R_{U,d} \leq \min (R_{B,d}; R_{A,d})$$
$$R_{B,d} \text{ or } R_{A,d} \geq 128 \text{ kN/m}$$

A governing anchorage length of L_{Ai} of approx. 7.5 m shall be demonstrated, without considering the action of the reinforcement wrap-around.

4.7.5 Analysing Bearing Capacity

The deep-seated stability of fill slopes on soft soil is always investigated to DIN **1054**. Either the bearing capacity is analysed as for shallow foundations to **DIN 4017** or the global stability to **DIN 4084**, depending on soil strata and other boundary conditions (e.g. geosynthetics in the base plane of embankments). In this example bearing capacity is not the governing factor, because the soft ground is only 4 m thick and the ground below this is regarded as stable. The soft layer is therefore several times less thick than the embankment contact area.

4.7.6 Selecting the Geosynthetics

4.7.6.1 Analysing Reinforcement Failure

The following minimum geosynthetics design resistance values result from the analyses as described in Sections 4.7.1 to 4.7.5:

Table 4.1 Required design resistances $R_{B,d}$

Design resistance from Section	Initial state	Final state
4.7.1	42.0 kN/m	5.0 kN/m
4.7.2	132.0 kN/m	–
4.7.3	–	30.3 kN/m
4.7.4	128.0 kN/m	–

This gives the required short-term strength of geosynthetics using the following equation:

$$R_{B,k0} = \text{max. } R_{B,d} \cdot (A_1 \cdot A_2 \cdot A_3 \cdot A_4 \cdot A_5 \cdot \gamma_M)$$

a) Initial state

The maximum design resistance value of the geosynthetics based on the resistance deficit in the overall stability analysis (here: general wedge method) for the initial state is $R_{B,d} = 132.0$ **kN/m** (see Section 4.7.2). The initial state is temporary inasmuch as the geotechnical boundary conditions may be enhanced by consolidation of the soft strata.

Using geosynthetics produced from a polyester raw material as an example, situated on soft soil and with soil group SE embankment fill material, and assuming $A_1 = 1.4$ (low value due to approx. 1.5 years consolidation duration), $A_2 = 1.2$ and $A_3 = A_4 = 1.0$ (manufacturer's data, $\gamma_M = 1.3$ due to assumed LC2), the following required characteristic short-term strength is acquired:

$$R_{B,k0} = 132 \cdot (1.4 \cdot 1.2 \cdot 1 \cdot 1 \cdot 1.3) = \underline{288.3 \text{ kN/m.}}$$

b) Final state

The maximum design strength (reinforcement tensile force design value) resulting from the analysis of sliding below the geosynthetics in the final state is $R_{B,d} = 30.3$ **kN/m** in the GEO limit state as described in Section 4.7.3.3.

Using geosynthetics produced from a polyester raw material as an example, situated on soft soil and with soil group SE embankment fill material, and assuming extensive investigation results provide lower coefficients than the minimum values, $A_1 = 2.5$, $A_2 = 1.2$, $A_3 = 1.0$ and $A_4 = 1.4$ provide the following required characteristic short-term strength for permanent structures, for example:

$$R_{B,k0} = 30.3 \cdot (2.5 \cdot 1.2 \cdot 1 \cdot 1.4 \cdot 1.4) = \underline{178.2 \text{ kN/m.}}$$

In this case, then, the required, governing short-term strength in the temporary state is 288.3 kN/m.

4.7.6.2 Analysing Reinforcement Pull-out

The pull-out resistance of the reinforcement layer shall be analysed on the active (sliding) and the passive side (facing the embankment) of the embankment. In this example the governing anchorage length L_{Ai} is always located on the active, sliding side.

The following minimum values for the pull-out resistance $R_{A,d}$ and anchorage lengths L_{Ai} of the geosynthetics, without consideration of reinforcement wrap-around, are given by the analyses described in 4.7.1 to 4.7.5:

Table 4.2 Required pull-out resistances req. $R_{A,d}$ and governing anchorage lengths L_{Ai}

Analysis in Section	Initial state		Final state	
	Pull-out resistance working $R_{A,d}$	Anchorage length L_{Ai}	Pull-out resistance working $R_{A,d}$	Anchorage length L_{Ai}
4.7.1	42.0 kN/m	12.5 m	5.0 kN/m	6.5 m
4.7.2	132.0 kN/m	9.5 m	–	–
4.7.3	–	–	30.3 kN/m	7.5 m
4.7.4	128.0 kN/m	9.5 m	–	–

The pull-out resistance is a function of the limit state shear stresses mobilised between the reinforcement element and the fill soil. Shear stresses on both the top $R_{A,1g,d}$ and the bottom $R_{A,2g,d}$ can be activated as resistances. In addition, a pull-out resistance $R_{A,Um,d}$ provided by the reinforcement wrap-around acts. The following condition shall be met at all times:

$$R_{A,1g,d} + R_{A,2g,d} + R_{A,Um,d} \geq R_{A,d}.$$

Reinforcement pull-out shall be analysed for both the initial and the final state. The pull-out resistance provided by the reinforcement wrap-around is always the same and is calculated using:

$$R_{A,Um,d} = (2 \cdot G_{Um,k} \cdot f_{1g,k})/\gamma_B,$$

where:

$G_{Um,k}$ the soil dead weight acting on the reinforcement wrap-around,
$f_{1g,k}$ the characteristic value of the friction coefficient between the embankment fill material and the geosynthetics,
γ_B the partial safety factor for the pull-out resistances of flexible reinforcement elements in the GEO limit state.

In this example $R_{A,Um,d}$ is:

$$R_{A,Um,d} = (2 \cdot 204.8 \cdot 0.35)/1.4 = 102.4 \text{ kN/m}$$

where:

$G_{Um,k} = 0.5 \cdot h_3 \cdot l_3 \cdot \gamma_{1,k} = 0.5 \cdot 3.2 \cdot 6.4 \cdot 20 = 204.8$ kN/m,

$f_{lg,k} = 0.5 \cdot \tan \varphi'_{1,k} = 0.5 \cdot \tan (35°) = 0.35$.

In the GEO limit state the following applies for the upper side of the geosynthetics facing the embankment fill material (granular soil) in the **initial state:**

$R_{A,1g,d} = (G_{LAi,k} \cdot f_{lg,k})/\gamma_B$

and for the lower side of the geosynthetics, facing the ground (cohesive soil):

$R_{A,2g,d} = (a_k \cdot L_{Ai})/\gamma_B$,

where:

$G_{LAi,k}$ the soil dead weight acting on the anchorage length L_{Ai},

$f_{lg,k}$ the characteristic value of the friction coefficient between the embankment fill material and the geosynthetics,

a_k the characteristic value of the adhesion between the soft, cohesive ground and the geosynthetics (here: $a_k = c_{u,k}$),

γ_B the partial safety factor for the pull-out resistances of flexible reinforcement elements in the GEO limit state.

The effective pull-out resistances for the individual analyses are given below:

Analysis for Section	$L_{Ai} =$	For $L_{Ai} > \tan (\beta) \cdot h_1$: $G_{LAi,k} = 0.5 \cdot (L_{Ai} + L_{Ai} - 8) \cdot h_1 \cdot \gamma_{1,k}$
4.7.1	12.5 m	$0.5 \cdot (12.5 + 12.5 - 8) \cdot 4 \cdot 20 = 680$ kN/m
4.7.2	9.5 m	$0.5 \cdot (9.5 + 9.5 - 8) \cdot 4 \cdot 20 = 440$ kN/m
4.7.3	–	–
4.7.4	9.5 m	$0.5 \cdot (9.5 + 9.5 - 8) \cdot 4 \cdot 20 = 440$ kN/m

Analysis for Section	$R_{A,1g,d} =$	$R_{A,2g,d} =$	$R_{A,Um,d} =$
4.7.1	$(680 \cdot 0.35)/1.4$ $= 170.0$ kN/m	$(20 \cdot 12.5)/1.4$ $= 178.6$ kN/m	102.4 kN/m
4.7.2	$(440 \cdot 0.35)/1.4$ $= 110.0$ kN/m	$(20 \cdot 9.5)/1.4$ $= 135.7$ kN/m	102.4 kN/m
4.7.3	–	–	–
4.7.4	$(440 \cdot 0.35)/1.4$ $= 110.0$ kN/m	$(20 \cdot 9.5)/1.4$ $= 135.7$ kN/m	102.4 kN/m

Analysis for Section	$R_{A,1g,d} + R_{A,2g,d} + R_{A,Um,d}$	req. $R_{A,d}$	
4.7.1	451.0 kN/m	42.0 kN/m	Analysis verified
4.7.2	348.1 kN/m	132.0 kN/m	Analysis verified
4.7.3	–	–	–
4.7.4	348.1 kN/m	128 kN/m	Analysis verified

For the **final state** the expressions for the pull-out resistance above the geosynthetics $R_{A,1g,d}$ and the resistances from reinforcement wrap-around $R_{A,Um,d}$ remain unaltered compared to the initial state. For the resistance $R_{A,2g,d}$:

$$R_{A,2g,d} = (G_{LAi,k} \cdot f_{2g,k}) / \gamma_B$$

applies with the characteristic value of the friction coefficient between the embankment fill material and the geosynthetics $f_{2g,k}$. This coefficient is given by:

$$f_{1g,k} = 0.5 \cdot \tan \varphi'_{2,k} = 0.5 \cdot \tan (20°) = 0.18.$$

In analogy to the calculations for the initial state this gives:

Analysis for Section	$L_{Ai} =$	For $L_{Ai} \leq \tan (\beta) \cdot h_1$: $G_{LAi,k} = 0.5 \cdot \tan (\beta) \cdot L_{Ai} \cdot L_{Ai} \cdot \gamma_{1,k}$
4.7.1	6.5 m	$0.5 \cdot 0.5 \cdot 6.5 \cdot 6.5 \cdot 20 = 211.2$ kN/m
4.7.2	–	–
4.7.3	7.5 m	$0.5 \cdot 0.5 \cdot 7.5 \cdot 7.5 \cdot 20 = 281.2$ kN/m
4.7.4	–	–

Analysis for Section	$R_{A,1g,d} =$	$R_{A,2g,d} =$	$R_{A,Um,d} =$
4.7.1	$(211.2 \cdot 0.35)/1.4$ = 52.8 kN/m	$(211.2 \cdot 0.18)/1.4$ = 27.1 kN/m	102.4 kN/m
4.7.2	–	–	–
4.7.3	$(281.2 \cdot 0.35)/1.4$ = 70.3 kN/m	$(281.2 \cdot 0.18)/1.4$ = 36.1 kN/m	102.4 kN/m
4.7.4	–	–	–

Analysis for Section	$R_{A,1g,d} + R_{A,2g,d} + R_{A,Um,d}$	req. $R_{A,d}$	
4.7.1	182.3 kN/m	5.0 kN/m	Analysis verified
4.7.2	–	–	–
4.7.3	208.8 kN/m	30.3 kN/m	Analysis verified
4.7.4	–	–	–

5 Reinforced Foundation Pads

5.1 Definitions

A reinforced foundation pad refers to a reinforced earth structure (see Figure 5.1), replacing a soft soil to a given depth and with an upper surface forming the subgrade surface for a rigid foundation with a flat, horizontal base.

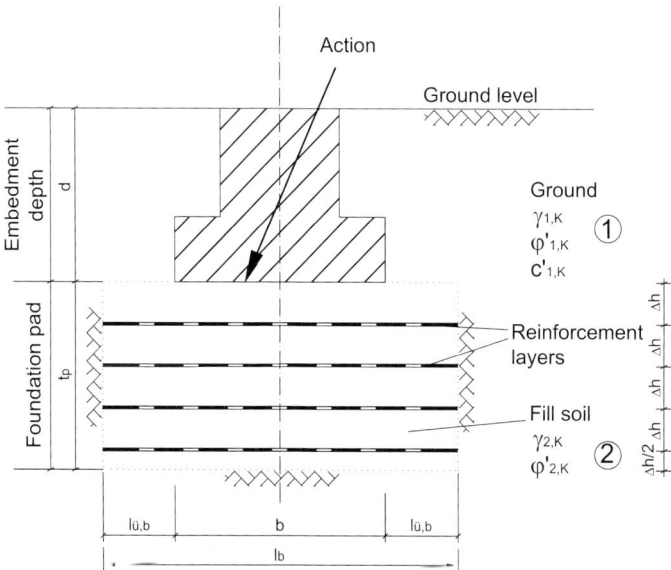

Figure 5.1 Reinforced foundation pad with foundation and cover fill

5.2 Application and Modus Operandi

A foundation pad in the terms of these Recommendations is created by inserting geosynthetics into the fill soil below footing and strip foundations. It displays enhanced bearing capacity and lower deformations compared to unreinforced soil replacement.

5.3 Design and Engineering Notes

5.3.1 Construction Principle

Reinforced foundation pads are generally installed in areas where soil replacement is planned. These areas are filled in layers with the fill soil and the reinforcement installed as described in Section 5.3.2.

Recommendations for Design and Analysis of Earth Structures using
Geosynthetic Reinforcements (EBGEO). German Geotechnical Society.
© 2011 Ernst & Sohn GmbH & Co. KG.
Published by Ernst & Sohn GmbH & Co. KG.

5.3.2 Reinforcement Configuration

The number n_B of reinforcement layers arranged within the foundation pad depends on structural requirements. However, at least 2 reinforcement layers shall be used ($n_B \geq 2$). For foundations with $b/a \leq 0.2$ the analyses in Section 5.5 shall be carried out for the smaller foundation side b. For foundations with $b/a > 0.2$ they shall be carried out separately for each side of the foundation. When using geosynthetics with different design strengths in the longitudinal and transverse directions the reinforcement design strengths are aligned as dictated by structural requirements. The vertical distance Δh between individual reinforcement layers should be equal ($\Delta h = const$). The following limits should also be observed. The governing value is the smaller of the two:

$$0.15 \text{ m} \leq \Delta h < 0.40 \text{ m} \qquad\qquad \text{Eq. (5.1)}$$

$$\Delta h \leq 0.50 \text{ b} \qquad\qquad \text{Eq. (5.2)}$$

5.3.3 Reinforcement Lengths

The length of all reinforcement layers installed in each direction should be the same. The following minimum dimensions apply:

– parallel to the foundation width b:

$$(b + 4 \cdot \Delta h) < l_b \leq 2 \text{ b} \qquad\qquad \text{Eq. (5.3)}$$

– parallel to the foundation length a:

$$(a + 4 \cdot \Delta h) < l_a \leq a + b \qquad (b/a > 0.2) \qquad \text{Eq. (5.4)}$$

$$l_a = a \qquad\qquad\qquad (b/a \leq 0.2) \qquad \text{Eq. (5.5)}$$

Where l_a and l_b are the dimensions of the foundation pad parallel to the longer side a and shorter side b of the foundation (also see Figure 5.3).

5.3.4 Foundation Pad Dimensions

The dimensions in plan are selected as described in Section 5.3.3. The pad depth t_p in accordance with the Recommendations is given by:

$$t_p = (n_B + 0.5) \cdot \Delta h \qquad\qquad \text{Eq. (5.6)}$$

where:

n_B number of reinforcement layers.

Here, too, minimum and maximum dimensions shall be observed:

$$\min t_p = 2.5 \cdot \Delta h \qquad\qquad \text{Eq. (5.7)}$$

$$\max t_p = (b/2) \cdot \tan (45° + \varphi'_{2.k} / 2) \qquad\qquad \text{Eq. (5.8)}$$

72

Based on Equation (5.8) the maximum recommended depth of the foundation pad max. t_p is dependent on the characteristic friction angle of the fill material $\varphi'_{2,k}$.

5.3.5 Building Materials

Refer to Section 2 for details of the materials used (fill soil/geosynthetic reinforcement). The fill soils of foundations pads are installed with $D_{pr} \geq 100\%$ to reduce intrinsic deformations.

5.4 Actions and Resistances

Actions include permanent and variable loads as defined in DIN 1054 and Section 1.2. Resistances include the shear strength and stiffness of the foundation pad and the ground, the tensile strength and axial stiffness of the geosynthetics and the shear behaviour between the geosynthetics and the ground, taking the described application into particular consideration.

5.5 Analysing the Reinforced Foundation Pad

5.5.1 General Recommendations

The bearing capacity of reinforced foundation pads is calculated to DIN 1054 and DIN 4017 and the settlement to DIN 4019 in conjunction with these Recommendations. Design generally follows the descriptions in Section 5.3.

Note: The analysis is based on the idea that the stability of reinforced foundation pads can be analysed using the same failure model as conventional foundations. This theory is supported by observations on small-scale tests ([1], [2]). The effect of the stronger fill soil compared to the existing ground is taken into consideration using corrected bearing capacity coefficients in the bearing capacity analysis [3]. The analysis described here applies to reinforcement installed from sheets only. If strip reinforcements are employed the design shall be modified accordingly [2].

5.5.2 Effects

The characteristic effects are determined for all governing construction conditions and the system's final state. The final state of the system includes consideration of the completed structure with the planned actions for the design working life. The time-dependence of the behaviour of the reinforcement material and the ground shall be taken into consideration where appropriate.

In systems predominantly subjected to dynamic effects, e.g. foundation pads below machine footings, the dynamic effects are adopted for design as described in Section 12 of these Recommendations.

5.6 Analysis and Design

5.6.1 Analysing Bearing Capacity

Bearing capacity analyses are performed to DIN 1054 and DIN 4017.

In an initial step the analysis is performed for a situation without foundation pad. The bearing capacity analyses are first performed for the unreinforced foundation pad and then for the reinforced pad. In the latter case it is assumed that any possible failure mechanisms intersect both the reinforced earth structure and the reinforcement layers.

The following bearing capacity analyses are performed:

- analysis of sliding safety to DIN 1054 (STR),
- analysis of the location of the bearing pressure resultant to DIN 1054 (EQU and SLS),
- analysis of bearing capacity to DIN 4017 (STR).

Note: Inclined, eccentric loads, layered ground, etc., are taken into consideration compliant with this standard.

- Analysis of global stability for foundations on or in slopes or on terraces to DIN 4084 (GEO).

5.6.1.1 Analysing Sliding Safety (STR)

Analysis of sliding safety on top of the foundation pad is performed to DIN 1054.

5.6.1.2 Analysing Bearing Capacity (STR)

The bearing capacity of the reinforced foundation pad is analysed in analogy to DIN 4017, where the dimensionless bearing capacity coefficients N_b, N_d and N_c of the ground are multiplied by the correction factors k_b, k_d and k_c. The correction factors k_b, k_d and k_c are given by:

$$k_b = C \cdot k_{b,\delta} + 1 \qquad\qquad \text{Eq. (5.9)}$$

$$k_d = C \cdot k_{d,\delta} + 1 \qquad\qquad \text{Eq. (5.10)}$$

$$k_c = C \cdot k_{c,\delta} + 1 \qquad\qquad \text{Eq. (5.11)}$$

The coefficient C, which is a function of the friction angle of the ground $\varphi'_{1,k}$ and the fill soil $\varphi'_{2,k}$, is given by:

$$C = \left[\frac{2}{\varphi'_{1,k}} \cdot \sqrt{40° - \varphi'_{2,k}} \cdot \left(\frac{\varphi'_{2,k}}{\varphi'_{1,k}} \right)^{0,7} + 1 \right]^{-1} \qquad\qquad \text{Eq. (5.12)}$$

C = 1.0 is adopted for fill soils with friction angles $\varphi'_{F,k} \geq 40°$. The correction factors $k_{b,\delta}$, $k_{d,\delta}$ and $k_{c,\delta}$ are taken from Figures 5.2 and 5.3.

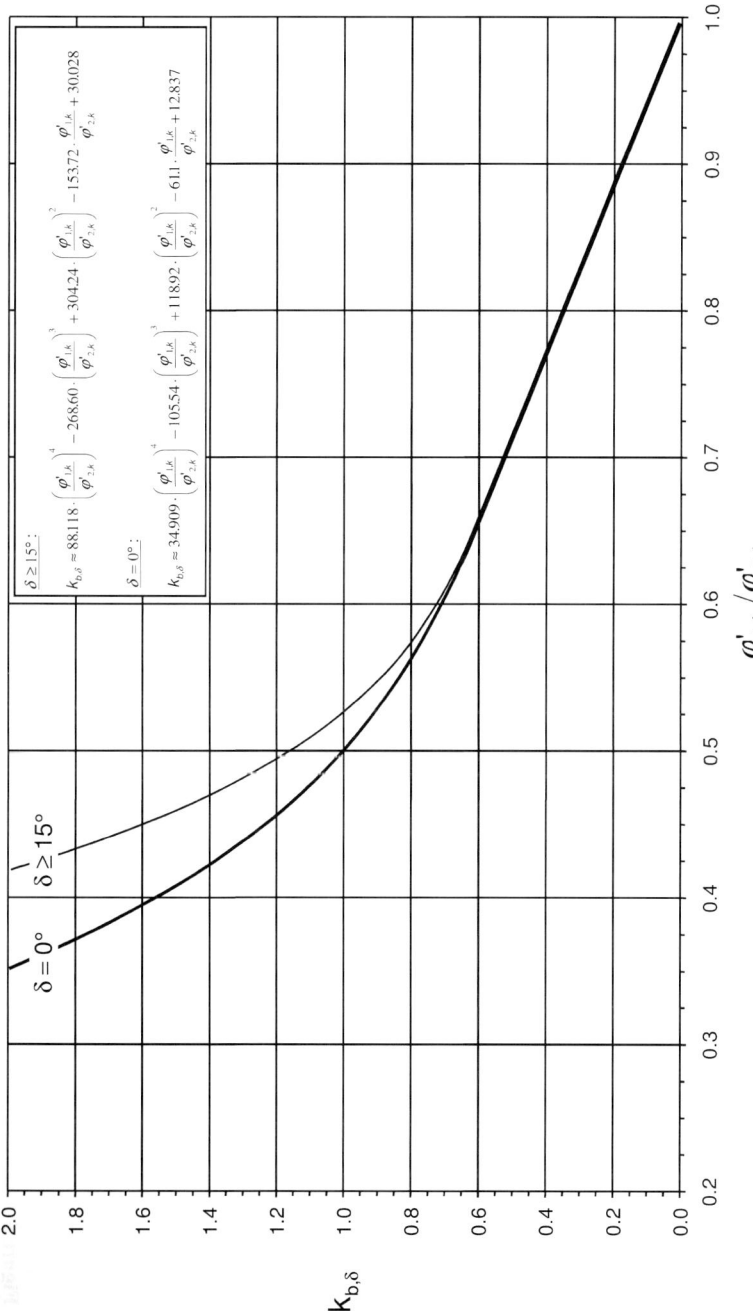

Figure 5.2 Correction factor $k_{b,\delta}$ as a function of $\varphi'_{1,k}/\varphi'_{2,k}$ and load inclination angle δ in [°]

The following appears within the figure:

$\delta \geq 15°$:

$$k_{b,\delta} = 88.118 \cdot \left(\frac{\varphi'_{1,k}}{\varphi'_{2,k}}\right)^4 - 268.60 \cdot \left(\frac{\varphi'_{1,k}}{\varphi'_{2,k}}\right)^3 + 304.24 \cdot \left(\frac{\varphi'_{1,k}}{\varphi'_{2,k}}\right)^2 - 153.72 \cdot \frac{\varphi'_{1,k}}{\varphi'_{2,k}} + 30.028$$

$\delta = 0°$:

$$k_{b,\delta} = 34.909 \cdot \left(\frac{\varphi'_{1,k}}{\varphi'_{2,k}}\right)^4 - 105.54 \cdot \left(\frac{\varphi'_{1,k}}{\varphi'_{2,k}}\right)^3 + 118.92 \cdot \left(\frac{\varphi'_{1,k}}{\varphi'_{2,k}}\right)^2 - 61.1 \cdot \frac{\varphi'_{1,k}}{\varphi'_{2,k}} + 12.837$$

Axis labels: $\varphi'_{1,k}/\varphi'_{2,k}$ (horizontal), $k_{b,\delta}$ (vertical)

Curve labels: $\delta = 0°$, $\delta \geq 15°$

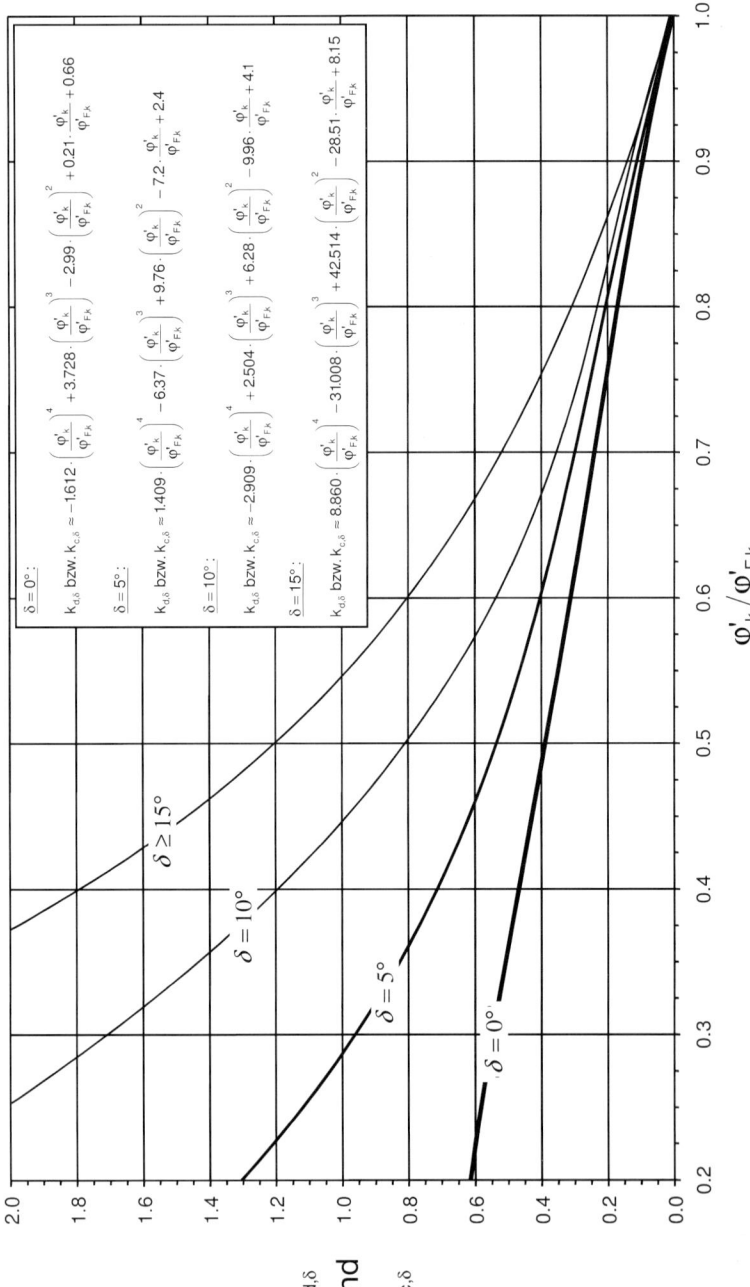

Figure 5.3 Correction factors $k_{d,\delta}$ and $k_{c,\delta}$ as functions of $\varphi'_k / \varphi'_{F,k}$ and load inclination angle δ in [°]

Note: *The correction factors k take into consideration that the governing slip plane to DIN 4017 does not pass through homogeneous ground, but instead through the foundation pad with high shear strength values on the one hand and through the underlying and adjacent ground with lower shear strengths on the other. The data was identified in comparative analyses using slip planes consisting of linear and logarithmic spirals. In contrast to DIN 4017 the strength values for the ground and the foundation pad were differentiated and adopted for the slip plane sections within and outside of the foundation pad. The calculated bearing capacity data produce suitable correction factors in terms of the bearing capacity factors in DIN 4017.*

They apply for:

$$t_{p,\delta} = \frac{\sin \vartheta_{a,\delta} \cdot \cos (\vartheta_{a,\delta} - \varphi'_{2,k})}{\cos \varphi'_{2,k}} \cdot b \qquad\qquad \text{Eq. (5.13)}$$

where:

$$\vartheta_{a,\delta} = \text{arc cot} \left[\sqrt{(1 + \tan^2 \varphi'_{2,k}) \cdot \frac{\tan \varphi'_{2,k} - \tan \delta}{\tan \varphi'_{2,k} + \tan \delta}} - \tan \varphi'_{2,k} \right] \qquad \text{Eq. (5.14)}$$

where:

$\vartheta_{a,\delta}$ slip plane angle of failure wedge (Figure 5.4)),
$t_{p,\delta}$ theoretical pad thickness for a load inclination $\delta \neq 0$,
t_p theoretical pad thickness for a load inclination $\delta = 0$,
$\varphi'_{2,k}$ characteristic value of the friction angle of the foundation pad fill soil,
δ load inclination in [°].

For working $t_p < t_{p,\delta}$ the correction factors k'_b, k'_d and k'_c are used instead of k_b, k_d and k_c. The k and k' relationships are:

$$k'_b = 1 + (k_b - 1) \cdot (\text{work. } t_p / t_{p,\delta}) \qquad\qquad \text{Eq. (5.15)}$$

$$k'_d = 1 + (k_d - 1) \cdot (\text{work. } t_p / t_{p,\delta}) \qquad\qquad \text{Eq. (5.16)}$$

$$k'_c = 1 + (k_c - 1) \cdot (\text{work. } t_p / t_{p,\delta}). \qquad\qquad \text{Eq. (5.17)}$$

The bearing resistance $R'_{n,k}$ of the foundation can now be calculated using the correction factors k'_b, k'_d and k'_c or k_b, k_d and k_c taking the fill soil into consideration:

$$R'_{n,k} = a' \cdot b' \cdot (\gamma_{2,k} \cdot b' \cdot N_b \cdot k_b + \gamma_{1,k} \cdot d \cdot N_d \cdot k_d + c'_{2,k} \cdot N_c \cdot k_c) \qquad \text{Eq. (5.18)}$$

or:

$$R'_{n,k} = a' \cdot b' \cdot (\gamma_{2,k} \cdot b' \cdot N_b \cdot k'_b + \gamma_{1,k} \cdot d \cdot N_d \cdot k'_d + c'_{2,k} \cdot N_c \cdot k'_c) \qquad \text{Eq. (5.19)}$$

where:

a' and b'   reduced foundation dimensions,
d   embedment depth.

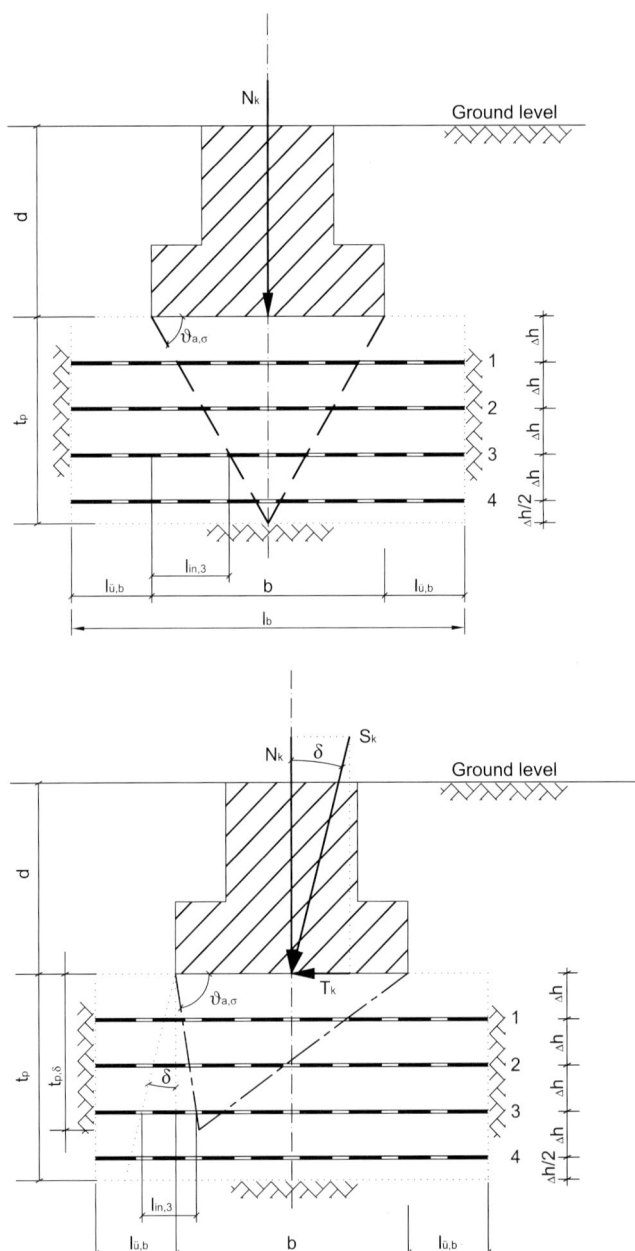

Figure 5.4 Reinforced foundation pad – failure wedges for vertical (top) and inclined loads (bottom)

The bearing capacity factors N_b, N_d and N_c are given by the bearing capacity analysis to DIN 4017 without consideration of the fill soil.

In addition to increasing the bearing capacity of the compacted fill soil, which displays better shear parameters than the natural ground, additional resistances occur as a result of the reinforcing action of the geosynthetic layers. The increased bearing capacity component $\Delta R_{n,k}$ resulting from the reinforcement forces is given by:

$$\Delta R_{n,k} = \frac{\cos \varphi'_{2,k} \cdot \cos \delta}{\cos(\vartheta_{a,\delta} - \delta)} \cdot \sum_{i=1}^{n} R_{i,k} \,, \qquad \text{Eq. (5.20)}$$

where $R_{i,k}$ is described either by the characteristic resistance $R_{Bi,k}$ or the characteristic pull-out resistance $R_{Ai,k}$.

The bearing capacity of the reinforced foundation pad working $R_{n,k}$ is then:

$$\text{work. } R_{n,k} = R'_{n,k} + \Delta R_{n,k} \,. \qquad \text{Eq. (5.21)}$$

The following are analysed to DIN 4017:

$$\text{work. } R_{n,d} = \text{work. } R_{n,k} / \gamma_{Gr} \qquad \text{Eq. (5.22)}$$

$$\text{work. } R_{n,d} \geq E_d \qquad \text{Eq. (5.23)}$$

5.6.1.3 Analysing Global Stability (GEO)

The global stability of foundations on or in slopes or on terraces is analysed in the GEO limit state to DIN 4084. Observe the notes in Section 3 for slip planes intersecting the reinforcement. The reinforcement's resisting forces can be taken into consideration, as described in the respective sections.

5.6.1.4 Analysing Reinforcement Failure (STR)

The design strength of a reinforcement layer is given by the specifications in Section 3.3.1 and is determined in the STR limit state as described there.

5.6.1.5 Analysing Reinforcement Pull-out Resistance (STR)

The possible reinforcement pull-out resistance design value $R_{Ai,k}$ in any reinforcement layer i resulting from friction against the fill soil for a planar reinforcement parallel to the foundation width b is:

$$R_{Ai,k} = 2 \cdot f_{sg,k} \cdot (N_k / b \cdot l_{in,i} + \sigma_{v,i} \cdot l_{ü,b}) \qquad \text{Eq. (5.24)}$$

where:

N_k characteristic vertical force,
$f_{sg,k}$ see Section 3.3.3.1,
$l_{in,i}$ length between failure wedge and foundation footprint edge in the nth layer (see Figure 5.5):

$$l_{in,n} = (\cot \vartheta_{a,\delta} + \tan \delta) \, \Delta h \cdot i \qquad\qquad \text{Eq. (5.25)}$$

where:

$\vartheta_{a,\delta}$ slip plane angle of failure wedge,
δ inclination of the foundation load,
Δh vertical distance between reinforcement layers,
n   number of reinforcement layer under consideration and
$\sigma_{v,n}$ stress from ground surcharge in the nth layer.

$$\sigma_{v,i} = \gamma_{2,k} \cdot \Delta h \cdot i + \gamma_{1,k} \cdot d \qquad\qquad \text{Eq. (5.26)}$$

where:

$\gamma_{2,k}$ unit weight of fill soil,
$\gamma_{1,k}$ unit weight of ground,
d   embedment depth,
$l_{\ddot{u},b}$ length of reinforcement for foundation width b protruding over the foundation footprint.

$$l_{\ddot{u},b} = 1/2 \, (l_b - b) \qquad\qquad \text{Eq. (5.27)}$$

where:

l_b overall reinforcement length parallel to foundation side b.

For analysis parallel to foundation side a, a and b are swapped and $l_{\ddot{u},a}$ adopted in place of $l_{\ddot{u},b}$.

5.6.2 Serviceability Limit State Analysis

The serviceability analyses are performed to DIN 1054 for the serviceability limit state:

- deformations to DIN 4019 (SLS),
- analysis of bearing pressure resultant (SLS),
- if necessary, horizontal displacement (SLS).

Settlements are analysed to DIN 4019 for the foundation bottom-foundation top-foundation pad interface. Serviceability analyses are generally carried out to DIN 1054 in the serviceability limit state. Settlement and any tilting of the foundation are generally determined using suitable settlement modelling applications. Under the condition that no further interfaces exist in the foundation's stress influence zone, an approximate average stiffness for the foundation pad and the ground can be identified using the following approach. Settlement modelling for a monolayer system can then be carried out to DIN 4019. With reference to [1]:

$$E'_{s,k} = E_{s,k} \cdot \left(1 - \frac{N_k}{\text{work. } R_{n,k}} \right) \cdot \frac{\text{work. } R_{n,k}}{R'_{n,k}} \qquad\qquad \text{Eq. (5.28)}$$

Note: Settlement modelling within the reinforced foundation pad can generally be dispensed with. During further analysis the foundation pad is assumed to be a solid body and the reinforcing effect is not adopted in conventional analyses.

5.7 Notes on Execution

When installing and compacting the fill soil in layers and installing the reinforcement the water table shall be at least 0.50 m below the excavation level. Relative compactions of $D_{pr} \geq 100\%$ are demanded for the fill soil compliant with *ZTV E-StB*, similar to made ground.

5.8 Bibliography

[1] Vogler, R. (1981): Beitrag zur Ermittlung des Tragverhaltens bewehrter Gründungen bei Variation von Erdstoff, Lastart, Lastgeometrie und Geometrie der Bewehrung, Dissertation Universität Rostock.

[2] Strate, R. (1986): Untersuchungen zur Verwendung von schwachbindigen Erdstoffen im Rahmen der Bauweise der bewehrten Erde, Dissertation at the University of Rostock.

[3] Wendt, D. (1990): Berechnung bewehrter und unbewehrter Gründungspolster nach TGL 11 464/01 und/02, Bauplanung-Bautechnik, 444. Jg. pp. 274–277, Heft 6.

5.9 Example of a Reinforced Foundation Pad below a Strip Foundation

5.9.1 Geometry, Loads and Soil Mechanics Parameters

Geometry:

- strip foundation (see Figure 5.5),
- width: $b = 1.50$ m,
- embedment depth: $d = 1.00$ m.

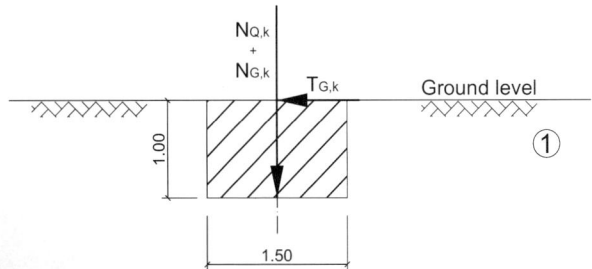

Figure 5.5 Geometry/load

81

Load:

- normal
  - permanent load $F_{gv,k} = 250$ kN/m
  - variable load $F_{pv,k} = 75$ kN/m
- horizontal
  - permanent load $F_{gh,k} = 40$ kN/m

Ground:

Flood plain loam (TM/UM) from 0.00 to 10.0 m below ground level

$E_{s1,k} = 8.0$ MN/m^2

$\gamma_{1,k} = 18.0$ kN/m^3

$\varphi'_{1,k} = 25.0°$

$c'_{1,k} = 5.0$ kN/m^2

No groundwater or perched water.

Fill Soil:

Broken gravel wit the following properties is planned for the foundation pad:

Soil group: GW with U > 15

$\gamma_{2,k} = 20.0$ kN/m^3

$\varphi'_{2,k} = 40.0°$

$c'_{2,k} = 0.0$ kN/m^2

Analyses:

The following analyses shall be performed as part of the dimensioning process:

- bearing capacity failure, unreinforced case   STR,
- bearing capacity failure, reinforced case   STR,
- sliding failure, reinforced case   STR,
- position of resultant (eccentricity)   SLS/EQU,
- settlements   SLS,
- tilting   SLS,
- failure of the geosynthetic reinforcement   STR.

Only the analyses discussed above are introduced in this example.

5.9.2 Analysing Bearing Capacity

5.9.2.1 Design without Foundation Pad

a) Load/effect:

$$N_{G,k} = F_{gv,k} = 250 \text{ kN/m}$$

$$N_{Q,k} = F_{pv,k} = 75 \text{ kN/m}$$

$$T_{G,k} = F_{gh,k} = 40 \text{ kN/m}$$

$$E_d = N_{G,k} \cdot \gamma_G + N_{Q,k} \cdot \gamma_Q \qquad \text{Eq. (5.29)}$$

Note: γ_G or γ_Q from DIN 1054, Table 2

$$E_d = 250 \cdot 1.35 + 75 \cdot 1.50 = \underline{450 \text{ kN/m}} \qquad \text{Eq. (5.30)}$$

b) Determining the equivalent width of the foundation:

$$e = \Sigma M / (N_{G,k} + N_{Q,k}) \qquad \text{Eq. (5.31)}$$

$$\Sigma M = T_{G,k} \cdot 1.0 + E_{Q,k} \cdot h / 3$$

$$= 40 \cdot 1.0 + 1 / 2 \cdot 18 \cdot 1^2 \cdot \tan^2(45° - 25° / 2) \cdot 1 / 3 = 41.22 \text{ kNm/m}$$

$$e = 41.22 / 325 = 0.13 \text{ m}$$

$$b' = b - 2 \cdot e = 1.5 - 2 \cdot 0.13 = \underline{1.25 \text{ m}}$$

c) Load inclination:

$$\tan \delta = T_{G,k} / (N_{G,k} + N_{Q,k}) \qquad \text{Eq. (5.32)}$$

$$N_k = N_{G+Q,k} + b' \cdot c'_{1,k} \cdot \cot \varphi'_{1,k} \qquad \text{Eq. (5.33)}$$

$$N_k = 325 + 1.24 \cdot 5 \cdot \cot 25° = 338.3 \text{ kN/m} \qquad \text{Eq. (5.34)}$$

$$\tan \delta = 40.0 / (250 + 75) = \underline{0.123} \qquad \text{Eq. (5.35)}$$

$$\delta = 7.02° \qquad \text{Eq. (5.36)}$$

d) Bearing capacity factors N_d, N_c, N_b (also see DIN 4017):

$$N_d = N_{d0} \cdot v_d \cdot i_d \cdot \lambda_d \cdot \xi_d \qquad \text{Eq. (5.37)}$$

$$N_c = N_{c0} \cdot v_c \cdot i_c \cdot \lambda_c \cdot \xi_c \qquad \text{Eq. (5.38)}$$

$$N_b = N_{b0} \cdot v_b \cdot i_b \cdot \lambda_b \cdot \xi_b \qquad \text{Eq. (5.39)}$$

Influence of foundation depth:

$$N_{d0} = e^{\pi \cdot \tan \varphi'_{1,k}} \cdot \tan^2(45° + \varphi'_{1,k} / 2) \qquad \text{Eq. (5.40)}$$

Influence of cohesion c:

$$N_{c0} = (N_{d0} - 1) / \tan \varphi'_{1,k} \qquad \text{Eq. (5.41)}$$

Influence of foundation width:

$$N_{b0} = (N_{d0} - 1) \cdot \tan \varphi'_{1,k}$$ Eq. (5.42)

$$N_{d0} = e^{\pi \cdot \tan 25} \cdot \tan^2 (45° + 25° / 2) = \underline{10.66}$$ Eq. (5.43)

$$N_{c0} = (10.66 - 1) / \tan 25° = \underline{20.72}$$ Eq. (5.44)

$$N_{b0} = (10.66 - 1) \cdot \tan 25° = \underline{4.50}$$ Eq. (5.45)

Shape factors ν:

$$\nu_d = \nu_b = \nu_c = 1.00 \quad \text{(strip)}$$ Eq. (5.46)

Load inclination factors:

$$m = 1.56 \quad \text{(cf. DIN 4017, Section 7.2.4)}$$ Eq. (5.47)

$$i_d = (1 - \tan \delta_E)^m$$ Eq. (5.48)

$$i_c = (i_d \cdot N_{d0} - 1) / (N_{d0} - 1)$$ Eq. (5.49)

$$i_b = (1 - \tan \delta_E)^{m+1}$$ Eq. (5.50)

$$i_d = (1 - 0.123)^{1.56} = \underline{0.815}$$ Eq. (5.51)

$$i_c = (0.870 \cdot 10.66 - 1) / (10.66 - 1) = \underline{0.796}$$ Eq. (5.52)

$$i_b = (1 - 0.123)^{1.56+1} = \underline{0.715}$$ Eq. (5.53)

Ground and base inclination factors λ and ξ all equal 1.0, therefore:

$$N_d = 10.66 \cdot 1 \cdot 0.815 \cdot 1 \cdot 1 = 8.691$$ Eq. (5.54)

$$N_c = 20.72 \cdot 1 \cdot 0.796 \cdot 1 \cdot 1 = 16.494$$ Eq. (5.55)

$$N_b = 4.50 \cdot 1 \cdot 0.715 \cdot 1 \cdot 1 = 3.221$$ Eq. (5.56)

e) Analysing bearing capacity:

Analysis of the STR limit state is performed with the characteristic values of the shear parameters (to DIN 1054, Table 3):

$$R_{n,k} = b' \cdot (c'_{1,k} \cdot N_c + \gamma_{1,k} \cdot d \cdot N_d + \gamma_{1,k} \cdot b' \cdot N_b)$$ Eq. (5.57)

$$R_{n,k} = 1.25 \, (5 \cdot 16.494 + 18 \cdot 1.0 \cdot 8.691 + 18 \cdot 1.24 \cdot 3.221)$$

$$= 390.69 \text{ kN/m}$$ Eq. (5.58)

$$R_{n,d} = R_{n,k} / \gamma_{Gr}$$ Eq. (5.59)

$$R_{n,d} = \underline{279.06} < 450 \text{ kN/m} = E_d$$ Eq. (5.60)

$$R_{n,d} / E_d = 0.62 < 1 \quad \text{Analysis not verified!}$$ Eq. (5.61)

5.9.2.2 Design with Foundation Pad – Foundation Pad Geometry

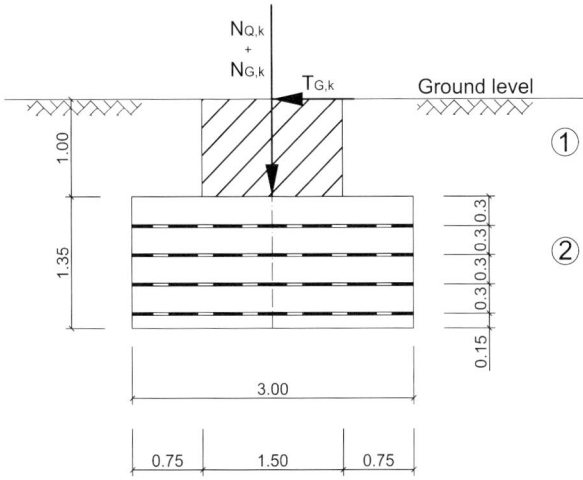

Figure 5.6 Foundation pad geometry

Depth of foundation pad:

$$\min t_p = 2.5 \cdot \Delta h = 2.5 \cdot 0.3 = 0.7 \text{ m} \qquad \text{Eq. (5.62)}$$

$$\max t_p = (b/2) \cdot \tan (45° + \varphi'_{2,k}/2)$$
$$= (1.50/2) \cdot \tan (45° + 40°/2) = 1.61 \text{ m} \qquad \text{Eq. (5.63)}$$

$$t_p = (n_B + 0.5) \cdot \Delta h = (4 + 0.5) \cdot 0.30 = 1.35 \text{ m} \qquad \text{Eq. (5.64)}$$

$$\min t_p = 0.75 < t_p = 1.35 \le \max t_p = 1.61 \text{ m} \qquad \text{Eq. (5.65)}$$

Determining $\vartheta_{a,\delta}$ and $t_{p,\delta}$:

$$\vartheta_{a,\delta} = \text{arc cot} \left[\sqrt{(1 + \tan^2 \varphi'_{2,k}) \cdot \frac{\tan \varphi'_{2,k} - \tan \delta}{\tan \varphi'_{2,k} + \tan \delta}} - \tan \varphi'_{2,k} \right]$$

$$= \text{arc cot} \left[\sqrt{(1 + \tan^2 40°) \cdot \frac{\tan 40° - \tan 7°}{\tan 40° + \tan 7°}} - \tan 40° \right] = 74° \qquad \text{Eq. (5.66)}$$

$$t_{p,\delta} = \frac{\sin \vartheta_{a,\delta} \cdot \cos (\vartheta_{a,\delta} - \varphi'_{2,k})}{\cos \varphi'_{2,k}} \cdot b$$

$$= \frac{\sin 74° \cdot \cos (74° - 40°)}{\cos 40°} \cdot 1.5 = 1.56 \text{ m} \qquad \text{Eq. (5.67)}$$

85

Possible reinforcement configuration:

$0.15 \text{ m} \le \Delta h = \underline{0.30 \text{ m}}$ (planned: 4 layers) $< 0.4 \text{ m}$

$\Delta h \le 0.5 \cdot b = 0.5 \cdot 1.50 = 0.75 \text{ m}$

Possible reinforcement lengths:

$1.5 + 4 \cdot 0.30 < l_b = \underline{3.00 \text{ m}} \le 2 \cdot b = 2 \cdot 1.50$

5.9.2.3 Analysing Bearing Capacity of the Unreinforced Foundation Pad

In an initial step, analysis (STR for Load Case 1) is performed at the top edge of an unreinforced foundation pad below the foundation, similar to Section 5.9.2.1.

a) Load:

as for Section 5.9.2.1 a),

b) Determining eccentricity:

as for Section 5.9.2.1 b),

c) Load inclination:

$\tan \delta = T_{g,k} / N_k = 40.0 / 325 = \underline{0.123}$

$\delta = \underline{7.09°}$ Eq. (5.68)

d) Bearing capacity factors N_d, N_c, N_b:

See Section 5.9.2.1 d) for

Calculating the correction factors for the foundation pad:

The bearing capacity factors N_c, N_d and N_b are multiplied by correction factors, which reflect the impact of the foundation pad on the bearing capacity.

The correction factors $k_{c,\delta}$, $k_{d,\delta}$ and $k_{b,\delta}$ are taken from Figure 5.2 and Figure 5.3 for $\delta = 7°$ and for:

$\varphi'_{1,k} / \varphi'_{2,k} = 25° / 40° = 0.625$ Eq. (5.69)

where:

$k_{d,\delta} = 0.42 = k_{c,\delta}$ Eq. (5.70)

$k_{b,\delta} = 0.66$.   Eq. (5.71)

The correction factors then give:

$k_d = C \cdot k_{d,\delta} + 1 = 1.0 \cdot 0.42 + 1 = 1.42 = k_c$ Eq. (5.72)

$k_b = C \cdot k_{d,\delta} + 1 = 1.0 \cdot 0.66 + 1 = 1.66$,   Eq. (5.73) where the factor

$$C = 1.00 \qquad\qquad\text{Eq. (5.74)}$$

is adopted.

Because $t_{p,\delta} = 1.56 > t_p = 1.35$, the k′ correction factors are adopted in the calculation in place of k.

The correction factors k'_c, k'_b and k'_d are:

$$k'_c = 1 + (k_c - 1)\,(t_p / t_{p,\delta}) = 1 + 0.42 \cdot 1.35 / 1.56 = \underline{1.36} \qquad\qquad\text{Eq. (5.75)}$$

$$k'_d = 1 + (k_d - 1)\,(t_p / t_{p,\delta}) = 1 + 0.42 \cdot 1.35 / 1.56 = \underline{1.36} \qquad\qquad\text{Eq. (5.76)}$$

$$k'_b = 1 + (k_b - 1)\,(t_p / t_{p,\delta}) = 1 + 0.66 \cdot 1.35 / 1.56 = \underline{1.57}\,. \qquad\qquad\text{Eq. (5.77)}$$

e) Analysing bearing capacity:

Analysis of the STR limit state is performed with the characteristic values of the shear parameters (to DIN 1054, Table 3). Taking the k′ correction factors into consideration the modified bearing capacity equation is then:

$$R'_{n,k} = b' \cdot (c'_{1,k} \cdot N_c\, k'_c + \gamma_{1,k} \cdot d \cdot N_d \cdot k'_d + \gamma_{1,k} \cdot b' \cdot N_b \cdot k'_b) \qquad\qquad\text{Eq. (5.78)}$$

$$R'_{n,k} = 1.24\,(5 \cdot 16.497 \cdot 1.36 + 20 \cdot 1.0 \cdot 8.693 \cdot 1.36$$

$$+ 18 \cdot 1.24 \cdot 3.221 \cdot 1.57) = 544.07 \text{ kN/m} \qquad\qquad\text{Eq. (5.79)}$$

$$R'_{n,d} = R'_{n,k} / \gamma_{Gr} = 544.07 / 1.4 = 388.62 \text{ kN/m} \qquad\qquad\text{Eq. (5.80)}$$

$$R'_{n,d} = \underline{388.62 \text{ kN/m}} < 450 \text{ kN/m} = E_d \qquad\qquad\text{Eq. (5.81)}$$

$$R'_{n,d} / E_d = 0.86 < 1 \qquad \text{Analysis not verified!} \qquad\qquad\text{Eq. (5.82)}$$

Taking only the soil replacement into consideration, the bearing capacity increases stability, but does not yet satisfy the limiting conditions. The bearing capacity-increasing effect of the geosynthetics is now considered.

5.9.2.4 Design with Reinforced Foundation Pad

Analysis for LC 1 in the STR limit state

a) Geometry:

The geometry of the foundation pad is taken from Section 5.9.2.2 and Figure 5.6

b) to d):

See Section 5.9.2.3

Selection of Geosynthetics:

Note: To achieve optimum reinforcement utilisation even during preliminary design we recommend operating with product-specific parameters.

An imaginary geosynthetic with design strength $F_{B,k}$ and subsequent imaginary coefficients A_i are selected for foundation pad reinforcement:

$F_{B,k} = 200 \text{ kN/m}$

$A_1 = 2.5; A_2 = 1.5; A_3 = A_4 = A_5 = 1.0$

The elongation at failure is $\varepsilon = 8\%$ and the characteristic friction coefficient between reinforcement and fill soil is adopted as $f_{sg,k} = 0.50 \cdot \tan \varphi'_{2,k} = 0.42$.

Design Strength of a Reinforcement Layer:

The design strength of the geosynthetics is then calculated using the partial safety factor for geosynthetics $\gamma_B = 1.40$ in Load Case 1:

$$R_{B,d} = F_{B,k} / (A_1 \cdot A_2 \cdot A_3 \cdot A_4 \cdot A_5 \cdot \gamma_B) \qquad \text{Eq. (5.83)}$$

$$R_{B,d} = 200 / (2.5 \cdot 1.5 \cdot 1.0 \cdot 1.0 \cdot 1.0 \cdot 1.4) = 38.1 \text{ kN/m} \qquad \text{Eq. (5.84)}$$

Design Value of the Pull-out Resistance of a Reinforcement Layer:

Calculation of the design value of the pull-out resistance of a reinforcement layer is carried out to Section 3.3.3, where the characteristic pull-out resistance is first determined for the STR limit state:

$$R_{Ai,k} = 2 \cdot f_{sg,k} \cdot (N_k / b \cdot l_{in,i} + \sigma_{v,i} \cdot l_{ü,b}) \qquad \text{Eq. (5.85)}$$

The overburden stress $\sigma_{v,i}$ is determined from the permanent characteristic normal force N_k (here: 250 kN/m) and the soil unit weights $\gamma_{1,k}$ and $\gamma_{2,k}$:

$$N_k / b = 250 / 1.5 = 166.7 \text{ kN/m} \qquad \text{Eq. (5.86)}$$

$$l_{in,i} = (\cot \vartheta_a + \tan \delta) \cdot \Delta h \cdot i = (0.292 + 0.120) \cdot 0.30 \cdot i = \underline{0.1236 \cdot i} \qquad \text{Eq. (5.87)}$$

$$\sigma_{v,i} = \gamma_{2,k} \cdot \Delta h \cdot i + \gamma_{1,k} \cdot d = 20 \cdot 0.30 \cdot i + 18 \cdot 1.0 = \underline{6 \cdot i + 18} \qquad \text{Eq. (5.88)}$$

$$l_{ü,b} = 0.5 \cdot (l_b - b) = 0.5 \cdot (3.0 \text{ m} - 1.50 \text{ m}) = \underline{0.75 \text{ m}} \qquad \text{Eq. (5.89)}$$

This then gives:

$$R_{Ai,k} = 0.84 \cdot (166.7 \cdot l_{in,i} + 0.75 \cdot \sigma_{v,i}) \qquad \text{Eq. (5.90)}$$

$$R_{Bi,k} = R_{Bi,d} \cdot \gamma_B = 38.1 \cdot 1.40 = 53.34 \text{ kN/m (for each layer i)}. \qquad \text{Eq. (5.91)}$$

The analysis results are shown in the table below for the individual reinforcement layers.

Layer i	$l_{in,i}$	$\sigma_{v,n}$	$R_{Ai,k}$	$R_{Bi,k}$
[–]	[m]	[kN/m^2]	[kN/m]	[kN/m]
1	0.123	24.0	**32.31**	53.34
2	0.246	30.0	**53.29**	53.34
3	0.369	36.0	74.27	**53.34**
4	0.492	42.0	95.25	**53.34**

88

In this case the pull-out resistance of layers 1 and 2 and the failure strength of the reinforcement for layers 3 and 4 are the governing factors. The total force $\Sigma\,R_{Bi,k}$ is therefore adopted for the following analyses:

$$\Sigma\,R_{Bi,k} = 32.31 + 53.29 + 53.34 + 53.34 = \underline{192.27\ kN/m} \qquad \text{Eq. (5.92)}$$

Note: If failure wedges are activated by other $\vartheta_{a,\delta}$, which no longer intersect certain reinforcement layers, this shall be taken into consideration in the calculation of $\Sigma R_{Bi,k}$.

e) See Section 5.9.2.3 for analysis of the bearing capacity of the foundation pad and also:

Analysis of the increase in bearing capacity due to the reinforcement:

$$\Delta R_{n,k} = \frac{\cos\varphi'_{2,k}\cdot\cos\delta}{\cos(\vartheta_{a,\delta}-\delta)}\cdot\sum_{i=1}^{n} R_{Bi,k} \qquad \text{Eq. (5.93)}$$

$$\Delta R_{n,k} = 1.94\cdot 192.27 = 373.65\ kN/m \qquad \text{Eq. (5.94)}$$

f) Analysing bearing capacity:

$$\text{work.}\,R_{n,k} = R'_{n,k} + \Delta R_{n,k} \qquad \text{Eq. (5.95)}$$

$$\text{work.}\,R_{n,k} = 544.07 + 373.65 = 917.72\ kN/m \qquad \text{Eq. (5.96)}$$

$$R_{n,d} = \text{work.}\,R_{n,k}\,/\,\gamma_{Gr} = 917.72\,/\,1.4 = 655.51\ kN/m \qquad \text{Eq. (5.97)}$$

$$R_{n,d} = 655.51\ kN/m > 450.0\ kN/m = E_d \qquad \text{Eq. (5.98)}$$

$$\text{and}\ \ E_d\,/\,R_{n,d} = 0.68 < 1 \quad \text{Analysis satisfied!} \qquad \text{Eq. (5.99)}$$

5.9.3 Analysing Sliding Stability

Analysis of sliding stability is performed to DIN 1054. The analysis requires that both the plane below the foundation and the plane at the base of the foundation pad or intermediate slip planes are investigated at the level of the geosynthetic layers.

5.9.4 Serviceability Limit State Analysis

Serviceability analyses are generally carried out to DIN 1054 in the serviceability limit state. Settlement analyses are carried out to DIN 4019. Settlement/tilting estimates are generally carried out using the methods described above. Under the condition that no further stratum exists in the foundation's stress influence zone, an approximate average stiffness for the foundation pad and the ground can be determined using the following approach. Settlement modelling can then be carried out for a monolayer system to DIN 4019.

The stiffness of the foundation pad is adopted for consideration in the analysis as described in Section 5.6.1. The following applies:

$$E'_{s,k} = E_{s,k} \cdot \left(1 - \frac{N_k}{\text{work. } R_{n,k}}\right) \cdot \frac{\text{work. } R_{n,k}}{R'_{n,k}} \qquad \text{Eq. (5.100)}$$

$$E'_s = 8000.0 \cdot [1 - (325 / 917.72)] \cdot (917.72 / 544.07) \qquad \text{Eq. (5.101)}$$

$$E'_s = 8524 \text{ kN/m}^2 \qquad \text{Eq. (5.102)}$$

Note: The characteristic values for ground stiffness may be adopted to DIN 1054. Alternatively, analysis can be performed conventionally for a multi-layer system.

6   Transport Routes

6.1   General Recommendations

Geosynthetics contribute to improved load-bearing behaviour in a variety of functions in non-stabilised base and protective courses, both on transport routes and in their subgrades. The complexity of the various overlapping functions and mechanisms of geosynthetics plays a governing role when assessing the modus operandi of the geosynthetics in the non-stabilised layers of transport routes.

These functions or modes can include:

- the reinforcement function, by means of which load-bearing behaviour is improved and – depending on the specific application – the depth of ruts limited, driveability improved or the required thickness of a soil replacement layer lessened; local weak zones can be bridged and differential settlements or variable bearing capacities compensated,
- the stabilising function, which lessens or prevents grain redistribution or displacement caused by the dynamic actions of vehicles in the base and protective courses by means of friction or interlocking between the geosynthetics and the ground, thus stabilising the base and protective courses,
- the separating function, which prevents different soils mixing in the long-term and thus retains their individual properties,
- the filter function, which retains the structure of the load-bearing ground and simultaneously allows the passage of water when the interface is subjected to hydraulic pressure, and grain mobility in one or both contact soils is given (including in both directions),
- the drainage function, which preserves bearing capacity by accepting and discharging groundwater or precipitation and which may ease accelerated consolidation without causing erosion.

No general regulations yet exist for product independent design of superstructure on base courses for roads and trafficked areas with low allowable deformations such as working levels, storage and assembly areas and soil replacement layers below the subgrade in accordance with *ZTV E-StB* [1], either for reinforced or unreinforced systems.

The manufacturers possess empirical data for various geosynthetics products. They are reflected in specific design procedures and in publications. Their applicability shall be examined on a case-by-case basis.

The stability of working levels and storage areas with non-stabilised pavements only occasionally trafficked and subjected to high static distributed loads can be analysed approximately with the aid of global stability analyses as described in Section 4 (Embankments on Soft Soils) (DIN 4084).

The recommendations in this section deal solely with the use of geosynthetics in combination with non-stabilised fill materials for trafficked areas with large

Recommendations for Design and Analysis of Earth Structures using
Geosynthetic Reinforcements (EBGEO). German Geotechnical Society.
© 2011 Ernst & Sohn GmbH & Co. KG.
Published by Ernst & Sohn GmbH & Co. KG.

allowable deformations (e.g. construction roads, access routes, forestry routes, etc.), which are directly trafficked.

Geosynthetics used as separating layers **without** reinforcement function are selected in accordance with the *'Merkblatt über die Anwendung von Geokunststoffen im Erdbau des Straßenbaus'* [5].

6.2 Trafficked Areas with Non-stabilised Pavement and Large Allowable Deformations

6.2.1 Applications

The serviceability of trafficked areas is improved with the aid of geosynthetics, where the primary aims are as follows:

– reducing the thickness of the trafficked fill [3], [6],
– reducing deformations (e.g. rut depth) for the same load transfer index,
– extending design working life (number of passages) [3],
– increasing possible axle loads.

In certain cases driveability is only made possible at all by the use of geosynthetic reinforcements [4].

6.2.2 Design Concept

The following design example is based on the Giroud/Noiray method [3] (geosynthetic membrane function). The required unreinforced layer thickness is first determined as a function of the anticipated bearing capacity of the ground under a given load (typical application range for this design method 30 kN/m^2 < c_u < 90 kN/m^2) and the number of traverses with an axle load of 100 kN.

The allowable reduction in fill thickness resulting from the geosynthetic reinforcement is then determined and from this the required reinforced layer thickness. Diagrams are given below as a function of the mineral aggregate used for rut depths of 7.5 cm to 10.0 cm. A maximum strain of 2% for a geosynthetic resistance $R_{B,d,\varepsilon = 2\%}$ = 8 kN/m was assumed to determine the fill layer thickness for reinforced structures.

The geosynthetic resistance $R_{B,d,\varepsilon = 2\%}$ given above is given by the force transmitted in the short tensile tests at 2% strain, taking the reduction factors and the partial safety factor discussed in Section 3.3 into consideration.

Note: A reduced reduction factor A$_1$, which relates to the true duration of actions from traffic, may be adopted in contrast to the recommendations in Section 2.2.4.5 for temporary works roads subjected to live loads.

A safety factor for unfavourable, variable actions was taken into consideration to compile the diagrams, which was adopted as γ_Q = 1.2 for Load Case 2, construction stage based on DIN 1054.

The following diagrams are based on publications by Giroud/Noiray [3] and Holtz/Sivakugan [7] and apply to **crushed stone base course material** compliant with ZTV SoB-StB [2] (Figure 6.1) or to **gravel base course material** (Figure 6.2).

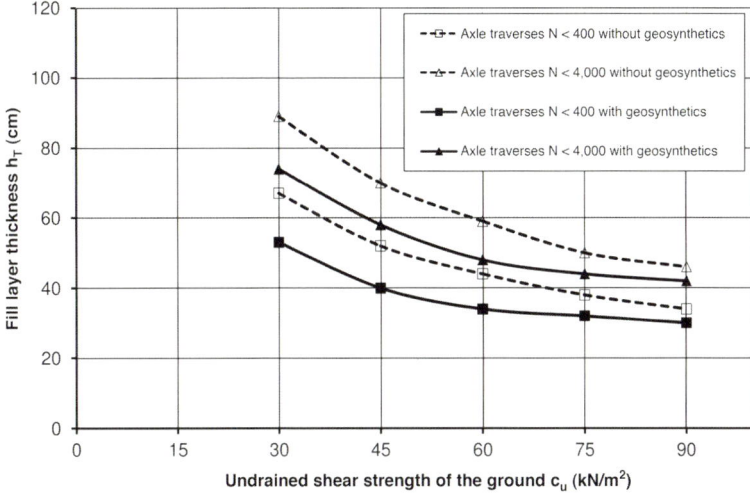

Figure 6.1 Required fill layer thicknesses as a function of the number of traverses (100 kN axle load) and the undrained shear strength of the ground for a rut depth of 7.5 cm to 10 cm and using crushed stone base course material

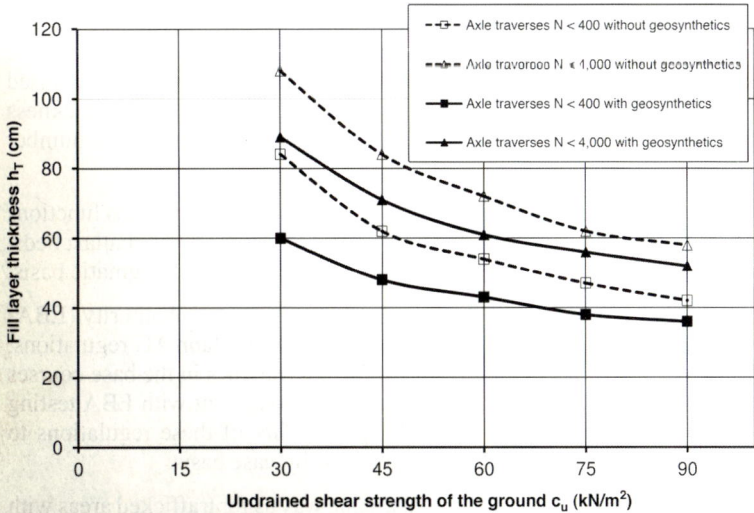

Figure 6.2 Required fill layer thicknesses as a function of the number of traverses (100 kN axle load) and the undrained shear strength of the ground for a rut depth of 7.5 cm to 10 cm and using gravel stone base course material

It may be expedient to install a second geosynthetic layer or to form a pad with complete wrap-around of the reinforcement layers where the required fill thickness is greater than 50 cm, depending on the fill material used and the strength of the ground.

This design approach conservatively ignores product-specific favourable effects leading to a reduction in the layer thicknesses given in the diagrams. If appropriate empirical data for similar boundary conditions are available they can be evaluated on a project-specific basis and taken into consideration for design.

6.3 Trafficked Areas in Railways Engineering

Geosynthetics are used in a variety of functions in railway engineering on **routes with ballast beds**. Nonwovens or composites, and occasionally wovens, may be considered for improving separating and filter stability below protective layers of granular soils above soft, cohesive soils. In many cases this alone considerably improves the service behaviour of railways lines on soft ground. Additionally, the use of geosynthetics with reinforcement functions may prove practicable as bridging structures above local weak points [8].

Geogrids are also used for higher dynamic actions and on weak, soft ground to stabilise the ground by their interlocking properties in or below unstabilised base courses or in soil replacement applications. A similar effect can also be provided by wovens, but to a lesser extent and only utilising friction with the ground, due to the closed structure. The stabilising effect leads to stiffer load-bearing behaviour in the protective layers and a reduction in the transfer of shear stresses to the ground, with the result that deformations and vibrations in the track are reduced. The improvement is generally taken into consideration by an increased static modulus of deformation or a surcharge to the thickness of the thickness of the protective layers. This has been confirmed by field testing in a number of cases [8].

The relevant rail operator's regulations control the design of the various functions of geosynthetics used to reinforce the base course of railways with ballast beds, generally from an engineering perspective and on an empirical-pragmatic basis.

ELTB [11] and EBRL [12] published by the Federal Railway Authority (EBA) apply in the case of federal railways, as well as Deutsche Bahn AG regulations, Guideline Ril 836 for the conditions for using geosynthetics in the base courses below ballast beds [9], with product requirements compliant with EBA testing requirements for geosynthetics [10]. The applicability of these regulations to non-federal railways shall be examined on a case-by-case basis.

The above notes apply to **slab tracks** in the same way as for trafficked areas with low allowable deformations. No well-founded, general application conditions or design methods using geosynthetics to improve the load-bearing or service behaviour of slab tracks are available.

6.4 Installation and Emplacement Notes

In terms of highways engineering the requirements of *ZTV E-StB* [1] and the notes in the bulletin [5] regarding fill material, compaction, emplacement and installation of the geosynthetics shall be observed.

To ensure that the reinforcement effect of the geosynthetics is given the products shall be wrapped around at the edges where necessary or an embedded width of at least 0.5 m to 1.0 m provided adjacent to the trafficked area.

The appropriate technical regulations shall be observed for other transport routes.

6.5 Bibliography

[1] Zusätzliche Technische Vertragsbedingungen und Richtlinien für Erdbau im Straßenbau *ZTV E-StB*, Forschungsgesellschaft für Straßen- und Verkehrswesen e. V., Cologne.

[2] Zusätzliche Technische Vertragsbedingungen und Richtlinien für den Bau von Schichten ohne Bindemittel im Straßenbau ZTV SoB-StB, Forschungsgesellschaft für Straßen- und Verkehrswesen e. V., Cologne.

[3] Giroud, J. P., Noiray, L. (1981): Geotextile-Reinforced Unpaved Road Design. Proceedings ASCE, Vol. 107, GT 9.

[4] Müller-Rochholz, J. (2008): Geokunststoffe im Erd- und Verkehrswegebau. Werner Verlag Cologne, 2nd edition.

[5] Forschungsgesellschaft für Straßen- und Verkehrswesen e. V. (2005): Merkblatt über die Anwendung von Geokunststoffen im Erdbau des Straßenbaus, Cologne.

[6] Rüegger, R., Hufenus, R. (2003): Bauen mit Geokunststoffen – Ein Handbuch für den Geokunststoff-Anwender. Schweizerischer Verband für Geokunststoffe SVG, St. Gallen 2003.

[7] Holtz, R. D., Sivakugan, N. (1987): Design Charts for Roads with Geotextiles. Geotextiles and Geomembranes 5.

[8] Göbel, C., Lieberenz, K. (2004): Handbuch Erdbauwerke der Bahnen. Eurailpress Hamburg.

[9] Richtlinie 836 "Erdbauwerke und sonstige geotechnische Bauwerke – planen, bauen und instand halten" der Deutschen Bahn AG.

[10] Prüfungsbedingungen für Geokunststoffe, Federal Railways Authority, EBA.

[11] Eisenbahnspezifische Liste Technischer Baubestimmungen (ELTB), Federal Railways Authority, EBA.

[12] Eisenbahnspezifische Bauregellisten (EBRL) und Eisenbahnspezifische Ergänzungen und Anlagen zu den Bauregellisten A, B und der Liste C des DIBt, Federal Railways Authority, EBA.

7   Retaining Structures

A retaining structure in terms of these Recommendations is an earth structure reinforced by geosynthetics for temporary or permanent stabilisation of a terrace, slope or hillside. The reinforced earth structure is required if the ground alone cannot guarantee the required stability.

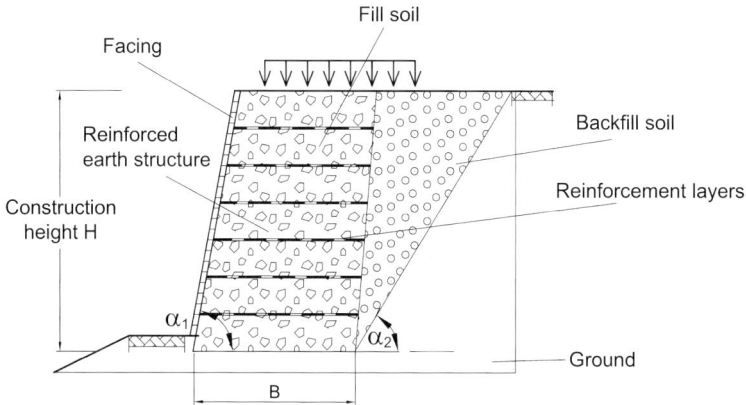

Figure 7.1 Retaining structure: designations and geometry

7.1   Definitions

Global failure in terms of these Recommendations is partial or complete slipping of a terrace stabilised by a retaining structure. Failure occurs because the shear strength of the ground, the interlock bond between the reinforcement and the ground or the resistance of the reinforcement layers to tensile stress are exceeded. The shear plane passes through the backfill zone, the ground and/or the structure.

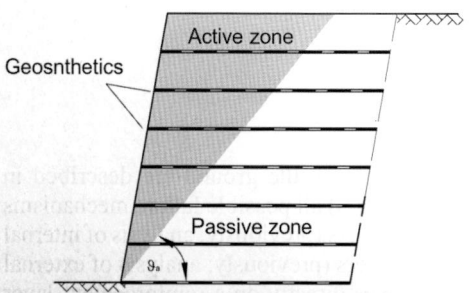

Figure 7.2 Active and passive zones of a reinforced earth structure (example)

Recommendations for Design and Analysis of Earth Structures using
Geosynthetic Reinforcements (EBGEO). German Geotechnical Society.
© 2011 Ernst & Sohn GmbH & Co. KG.
Published by Ernst & Sohn GmbH & Co. KG.

The **active zone** of a reinforced earth structure is the sliding component.

The **passive zone** of a reinforced earth structure is the undeformed or only very slightly deformed resisting component.

Reinforcement layers in terms of these Recommendations are non-prestressed tension elements according to the definition given in DIN 4084 and DIN 1054.

7.2 Design Notes

7.2.1 Demands and Boundary Conditions

The following boundary conditions shall be observed during draft design and preliminary investigations for a reinforced retaining structure. They impact the geometry, dimensioning and engineering design:

- ground conditions below and behind the retaining structure,
- location of the groundwater table,
- impacts from perched water,
- any excavation battering angle or existing slopes,
- height and inclination of the reinforced retaining structure,
- design and requirements of facing,
- planned design working life,
- actions on the structure (e.g. live loads),
- allowable deformations,
- properties of the intended materials.

7.2.2 Geometry

A reinforcement length of 70% of the structure height H may be adopted for preliminary drafts. The vertical distance between reinforcement layers is usually between 0.3 m and 0.6 m.

These rules of thumb have proved reliable for normal ground conditions and approximately horizontal terrain. Substantial deviations in the design may occur under different boundary conditions.

7.3 Analysis Principles

7.3.1 General Principles

Load transfer mechanisms for geosynthetics in the ground are described in Section 3.1. In the **Ultimate Limit State (ULS)** all possible failure mechanisms and slip planes intersecting reinforcement layers (previously: analysis of internal stability), not intersecting reinforcement layers (previously: analysis of external stability) and where the sliding body moves directly on a reinforcement layer are investigated.

The resistance in intersected reinforcement layers is the smaller of the two following resistances:

- the design resistance of each reinforcement layer (reinforcement failure: STR),
- the design value of the pull-out resistance of each reinforcement layer from the surrounding fill soil on both sides of the respective slip plane (pull-out: GEO).

The resistances of reinforcement connections, junctions, seams and any connections to structural elements compared to effects shall be considered (STR).

The analyses for the **serviceability limit state (SLS)** are performed to DIN 1054, 12.5, e.g.:

- analysis of compatible structural deformations: the deformation of the structure as a consequence of characteristic, permanent and variable actions is estimated using characteristic values of the soil parameters. The compatibility of these deformations with installed elements/facing structures, etc. shall be analysed,
- settlement modelling based on **DIN 4019**,
- position of the bearing pressure resultant to DIN 1054, Section 7.6.1 (position of the bearing pressure resultant in the first kernel width).

7.3.2   Slip Planes and Failure Mechanisms

All possible slip planes shall be considered and the most unfavourable failure mechanism identified.

Slip planes completely enclosing the reinforced retaining structure, intersecting the reinforcement layers or passing through the geosynthetics/ground contact planes are investigated. Additionally, slip planes passing through the reinforced earth structure without intersecting a reinforcement layer are considered (*cf.* Figure 7.3 and Figure 7.4, for example).

For geosynthetic-reinforced retaining structures it is common to investigate the following failure mechanisms (also see DIN 4084):

- for geosynthetic-reinforced retaining structures it is common to investigate the following failure mechanisms (also see DIN 4084):

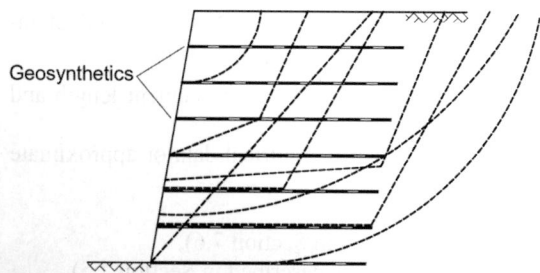

Figure 7.3 Possible slip planes through a retaining structure

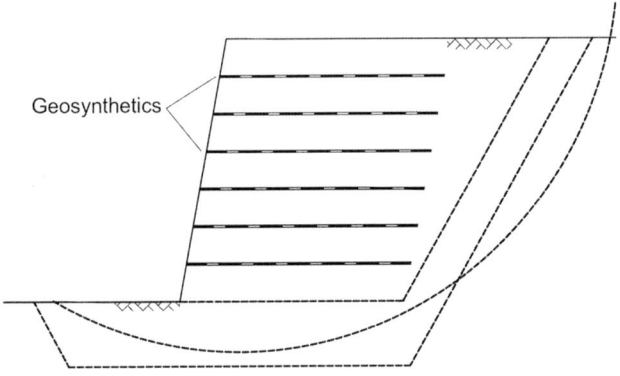

Figure 7.4 Possible slip planes around a retaining structure

- failure masses with circular slip planes,
- failure bodies with logarithmic spirals as slip planes,
- composite failure mechanisms with at least two failure masses and planar slip planes.

7.3.3 Analysis Overview

The analyses are performed for the limit states defined in DIN 1054, also see Section 3.1. Analyses of the limit equilibrium (Section 7.4) and the connection to the facing (Section 7.6) are performed for the ultimate limit state (ULS).

Analyses of the prevalent deformations and settlements are performed (Section 7.5) for the serviceability limit state (SLS).

The analyses described in this section are only relevant to retaining structures allowing a planar boundary at the end of the reinforcement element, such that a geometrically defined rear wall occurs. They can be regarded as quasi-monoliths for the analyses described in Section 7.4. Refer to the notes in Section 7.4.8 for structures deviating from this principle.

For illustration one possible procedure for iterative design of a reinforced retaining structure with uniform reinforcement length is presented here:

- define the geometry as described in Section 7.2 (reinforcement length and layer spacing),
- select the reinforcement elements (based on empirical data or approximate calculations),
- analyse in the ultimate limit state (ULS),
- design the connections/facing (as described in Section 7.6),
- analyse the serviceability limit state (SLS as described in Section 7.5),
- examine the geometry and the reinforcement elements.

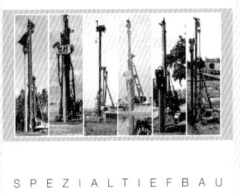

S P E Z I A L T I E F B A U
KOMPENDIUM
VERFAHRENSTECHNIK UND GERÄTEAUSWAHL

Ernst & Sohn   LIEBHERR
BVV

Band I:
Ramm- und Bohrgeräte (LRB)
2008. 380 S., 300 Abb., Gb.
€ 129,– / sFr 204,–
ISBN: 978-3-433-02904-6

Liebherr-Werk Nenzing GmbH (Hrsg.)
Spezialtiefbau
Kompendium Verfahrenstechnik
und Geräteauswahl

Die Verfahren und die Gerätetechnik des Spezialtiefbaus haben sich in den letzten Jahren rasant fortentwickelt. Die Anwendung der komplexen Techniken erfordert spezielle Kenntnisse und praktischeErfahrung. So ist es heute sowohl für Anwender als auch für Hersteller von Spezialtiefbaugeräten schwierig geworden,den Überblick über den Stand der Technik auf diesem Gebiet zu behalten. Das vorliegende Kompendium gibt eine umfassende Übersicht über die Verfahren und ihre Anwendungsgebiete. Im Einzelnen werden die Herstelltechniken von Gründungskonstruktionen und ihre Anwendungsbereiche mit denentsprechenden Gerätekomponentenaufgezeigt. Dabei wird im Detail auf die Besonderheiten der Verfahren und die Wahl der Gerätetechnik eingegangen. Aus der intensiven Zusammenarbeit von Ingenieuren, Technikern, Geräteherstellern und Anwendern entstand somit ein Hilfsmittel für die Planung und die Ausführung von Grundbaumaßnahmen.

S P E Z I A L T I E F B A U
KOMPENDIUM BAND II
VERFAHRENSTECHNIK UND GERÄTEAUSWAHL
BOHRGERÄTE UND HYDROSEILBAGGER

Ernst & Sohn   LIEBHERR
BVV

Band II: Bohrgeräte und
Hydroseilbagger (LB und HS)
2009. 336 S., 247 Abb.,
43 Tab., Gb.
€ 129,–* / sFr 204,–
ISBN: 978-3-433-02933-6

Erscheint auch in Englisch im September 2009,
ISBN 978-3433-02932-6

Set-Preis fuer Band I und Band II
zum Sonderpreis
ISBN 978-3-433-02934-3
€ 189,- / sFr 299,-

* Der €-Preis gilt ausschließlich für Deutschland. Irrtum und Änderung vorbehalten. 0108509016_my

Ernst & Sohn
A Wiley Company
www.ernst-und-sohn.de

Ernst & Sohn Verlag für Architektur und technische Wissenschaften GmbH & Co. KG
Für Bestellungen und Kundenservice: Verlag Wiley-VCH, Boschstraße 12, D-69469 Weinheim
Tel.: +49(0)6201 606-400, Fax: +49(0)6201 606-184, E-Mail: service@wiley-vch.de

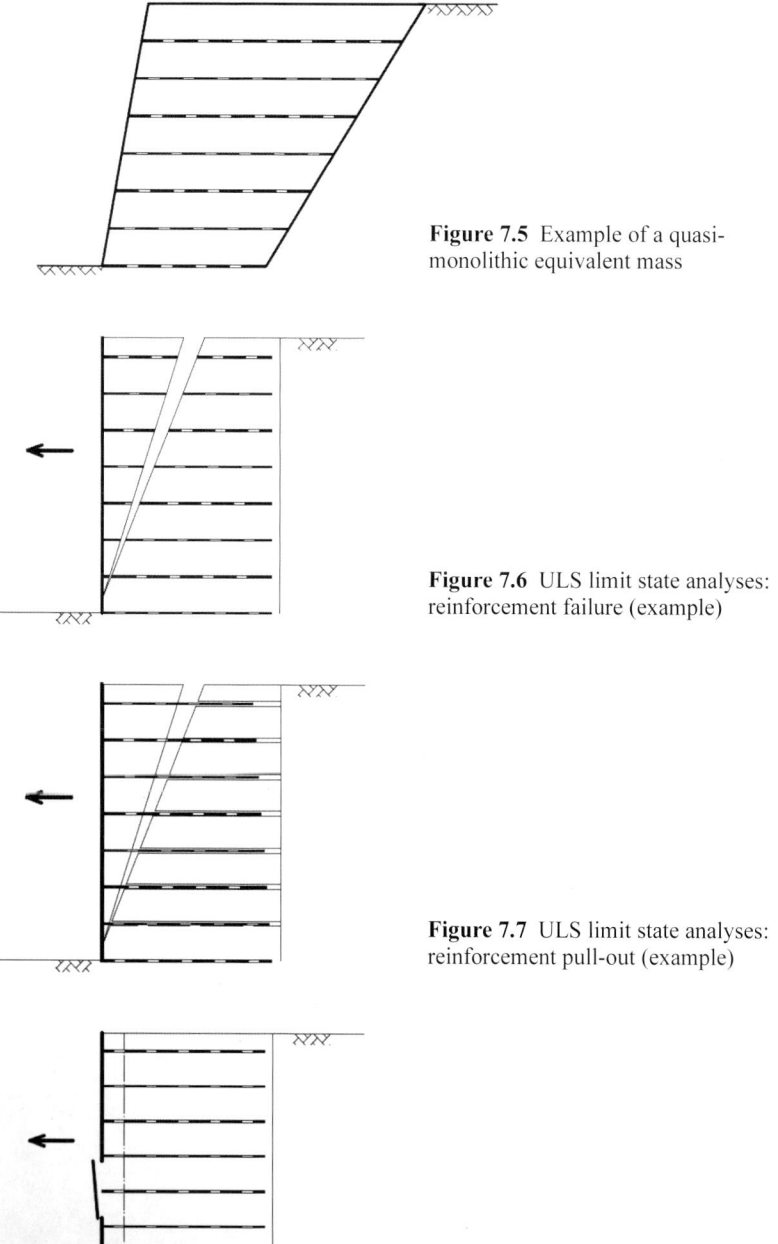

Figure 7.5 Example of a quasi-monolithic equivalent mass

Figure 7.6 ULS limit state analyses: reinforcement failure (example)

Figure 7.7 ULS limit state analyses: reinforcement pull-out (example)

Figure 7.8 ULS limit state analyses: analysis of connections/facing (example)

101

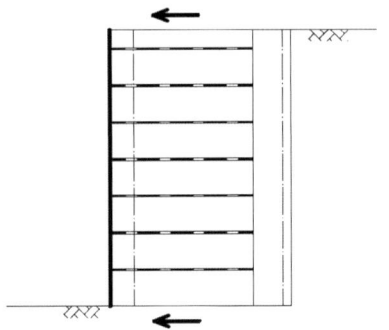

Figure 7.9 ULS limit state analyses: sliding (example)

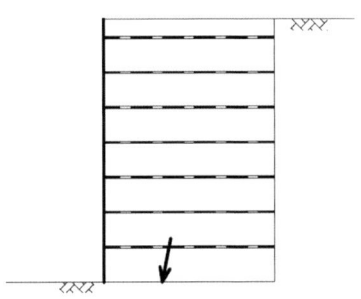

Figure 7.10 ULS limit state analyses: allowable eccentricity (example)

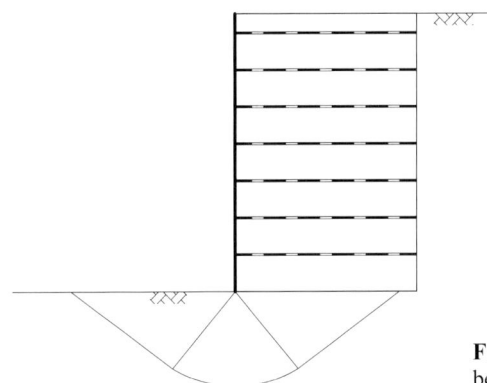

Figure 7.11 ULS limit state analyses: bearing capacity (example)

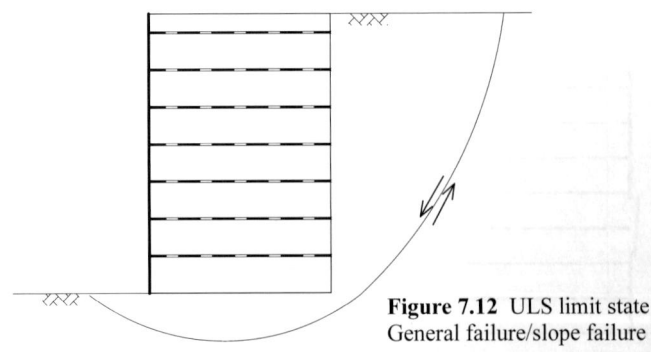

Figure 7.12 ULS limit state analyses: General failure/slope failure (example)

Table 7.1 Analysis overview

Analysis	LS	Section
Ultimate limit state		
General failure/slope failure	GEO	7.4.4
Bearing capacity failure	STR	7.4.5
Sliding	STR	7.4.6
Position of bearing pressure resultant	EQU	7.4.7
Failure on slip planes penetrating the retaining structure	GEO	7.4.4
Design strength of reinforcement	STR	7.4.3
Pull-out resistance of reinforcement	GEO	7.4.3
Analysis of connections	STR	7.6
Analysis of reinforcement overlapping/joining (reinforcement junctions)	STR	7.6
Serviceability limit state		
Position of bearing pressure resultant	SLS	7.5.2
Deformation of the structure	SLS	7.5
Settlement in the contact zone	SLS	7.5

7.4   Analyses in the Ultimate Limit State (ULS)

7.4.1   General Recommendations

If the reinforced retaining structure fails on a slip plane passing through the reinforced soil mass and intersects or at least touches the reinforcement, the sliding earth structure (Figure 7.13, prism 1) shall be held in equilibrium. This is achieved by shear forces in the slip plane or forces from an anchorage in the passive zone. A failure mass forms behind the sliding earth structure (prism 2), which moves relative to prism 1. In this simplified model approach it is assumed that the relative movements between prism 1 and prism 2 are adequate to justify adopting a maximum earth pressure angle $\delta_a = 2/3 \cdot \varphi'$ for determining the active earth pressure E_a.

The following limit state condition shall be met for all failure mechanisms for analysis of the ultimate limit state (ULS):

$$E_d \leq \Sigma R_d$$

Eq. (7.1)

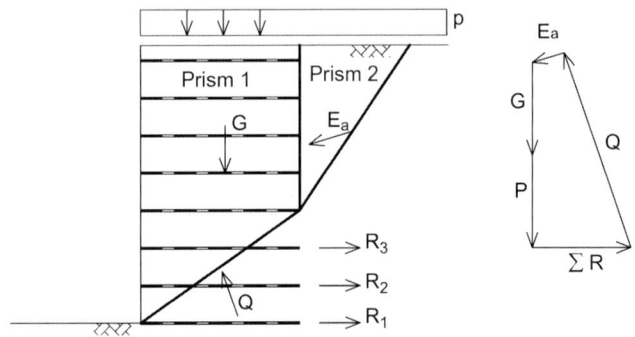

Figure 7.13 Distribution of forces in a reinforced retaining structure (example)

where:

ΣR_d sum of the resistances in all intersected reinforcement layers, governing per layer min R_d either
 – pull-out resistance $R_d = R_{Ai,d}$ (GEO)
 or
 – material strength of the reinforcement element $R_d = R_{Bi,d}$ (STR).

7.4.2   Actions and Effects

Geosynthetic-reinforced retaining structures can be impacted by actions such as dead weight, and vertical and horizontal loads. The design effects on the tension elements are determined to DIN 1054, 12.4.2:

– from the force equilibrium deficit of sliding masses bounded by failure mechanisms with planar or curved slip planes,
– to DIN 4084 in the GEO limit state, where the slip planes intersect some of the tension elements when varied (*cf.* Section 7.4.4).

7.4.3   Resistances

On the resistance side, reinforcement failure or pull-out in the reinforcement zone shall be analysed.

The **design strength** is the design tensile strength value of a reinforcement layer $R_{Bi,d}$ as described in Section 3.3

The **design pull-out resistance value** $R_{Ai,d}$ results from the interaction between the reinforcement elements and the fill soil, see Section 3.3.

7.4.4 Analysing General Failure/Slope Failure (GEO)

Adequate safety against general failure/slope failure shall be demonstrated. This is done by demonstrating that the DIN 4084 limit state conditions are adhered to adopting the GEO limit state partial safety factors (see section 7.3.2) for the failure mechanisms involved (DIN 1054, 12.3 and DIN 4084) in the construction and final states.

$$E_d \leq R_d \qquad \text{Eq. (7.2)}$$

where:

E_d design value of the resultant effect parallel to the slip surface or the design value of the moment of actions around the centre of rotation,

R_d design value of the resistance parallel to the slip surface or the design value of the moment of the resistance around the centre of rotation.

7.4.5 Analysing Bearing Capacity (STR)

Adequate bearing capacity in the STR limit state shall be demonstrated based on DIN 4017 for a quasi-monolith to DIN 1054, 12.4.4 and 7.5.2. All governing combinations of permanent and variable actions are investigated (Figure 7.11).

The following DIN 1054 limit state condition shall be met:

$$N_d \leq R_{n,d} \qquad \text{Eq. (7.3)}$$

where:

N_d design value of the effect normal to the foundation base,

$R_{n,d}$ design value of the bearing resistance.

7.4.6 Analysing Sliding Safety (STR)

Adequate sliding safety in the STR limit state shall be demonstrated for a quasi-monolith to DIN 1054, 7.5.3. All governing combinations of permanent and variable actions are investigated.

The following DIN 1054 limit state condition shall be met:

$$T_d \leq R_{t,d} + E_{p,d} \qquad \text{Eq. (7.4)}$$

where:

T_d design value of the effect normal to the foundation base,

$R_{t,d}$ design value of the sliding resistance,

$E_{p,d}$ design value of the pull-out resistance.

Note: Passive earth pressure in front of the reinforced earth structure may only be taken into consideration in the analysis if excavation in front of the structure can be ruled out.

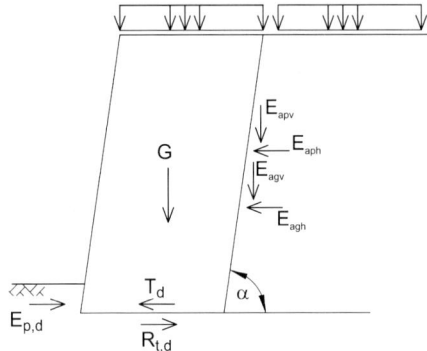

Figure 7.14 Forces adopted for sliding analysis

The characteristic value of the base friction angle $\delta_{S,k}$ is determined to DIN 1054, Section 5.2.3.5 by:

$$\tan \delta_{S,k} = f_{sg,k} = \lambda \cdot \tan \varphi'_k \qquad \text{Eq. (7.5)}$$

7.4.7 Position of Bearing Pressure Resultant

The position of the bearing pressure resultant shall be analysed analogous to DIN 1054, 7.5/7.6. The second kernel width (maximum foundation gap to the centroid) is analysed in the EQU limit state (DIN 1054, 7.5.1). This analysis is performed for a quasi-monolithic structure. All governing combinations of permanent and variable actions are investigated.

Analysis of the position of the bearing pressure resultant is performed for both the air and the ground sides.

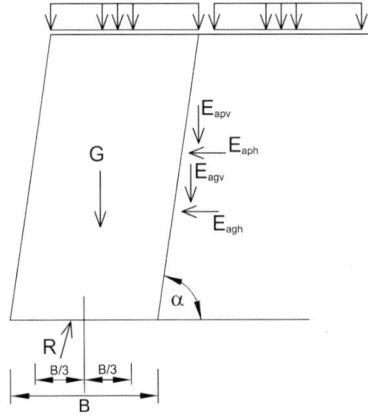

Figure 7.15 Forces adopted for analysis of the position of the resultant

Note: *If the resultant on the earth side lies outside the allowable range this means the* **retaining structure** *leans into the backfill and exerts an additional load on it. This is allowable if it is shown that the reaction forces can be transferred with an appropriate factor of safety. Where appropriate the adopted earth pressure angle* δ_a *or the angle* α *of the calculated rear face of the structure shall be examined.*

Analysis of the first kernel width (no foundation gap) is performed to DIN 1054, 7.6.1 in the serviceability limit state in the course of serviceability analyses, see Section 7.5.

7.4.8   Special Regulations

The analyses described in Sections 7.4.4 to 7.4.7 cannot be applied directly to a geosynthetic-reinforced retaining structure if it cannot be regarded as a quasi-monolithic mass in terms of Section 7.3.3. Such a retaining structure can be modelled as a combination of several quasi-monolithic masses. The stability of the respective individual masses and the composite mass shall be investigated (*cf.* Figure 7.16, for example).

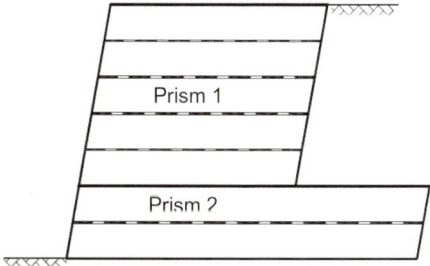

Figure 7.16 Example of a combination of several quasi-monolithic masses

7.5   Serviceability Limit State (SLS) Analyses

7.5.1   General Recommendations

Serviceability analyses comprise analysis of the position of the bearing pressure resultant analogous to DIN 1054, Section 7.6.1 and analysis of the compatibility of deformations and displacements of the retaining structure. Ground settlement, intrinsic settlement of the fill material, frontal displacement of the retaining structure and the resulting surface displacement (shear deformations) are all taken into consideration.

The magnitude of the allowable deformations is determined by the structure's use and the engineering design, for example at the front of the structure (rigid or flexible facing). Reinforced retaining structures themselves are regarded as structures insensitive to settlement.

The results of extensive laboratory testing (e.g. [1], [2], [3], [7], [8], [9], [16], [17]), and instrument-monitored temporary (e.g. [3], [8, [9]) and permanent retaining structures (e.g. [5], [6], [10], [14], [17], [18]) are now available to help estimate the deformation behaviour of reinforced retaining structures. In [4] an empirical value for the horizontal movements of the front of a reinforced retaining structure is given as a maximum of approx. 1% to 2% of the wall height H.

Note: *The interaction of fill soil and geosynthetics can affect the deformation behaviour of geosynthetic-reinforced retaining earth structures. The ensuing composite material displays substantially smaller deformations than would be anticipated based solely on the load-extension behaviour of the geosynthetics.*

Allocation to Geotechnical Categories (GC 1–3) follows Table 3.2.

If, in straightforward cases (e.g. GC 1), its use places no particular demands on deformation behaviour and empirical data ([4]) on deformations in the serviceability limit state are acceptable, precise numerical analysis of the deformations of the reinforced retaining structure may be dispensed with.

If particular demands are made on deformation behaviour as a result of the use of reinforced retaining structures, or if no specific empirical data is available for the planned materials, the deformations of the retaining structure and the overall system, including the ground, shall be determined. In cases allocated to GC 3 the deformation behaviour of the retaining wall should be monitored by instruments (observational method analogous to DIN 1054) in addition to using deformation forecasts.

The following deformation components on a retaining structure shall be considered:

v_U ground settlement,
v_E intrinsic settlement of the fill material,
v_{hi} horizontal displacement of the slope front at the level of reinforcement layer i,
v_S shear deformation.

If the demands placed on the structure are great or large surcharges are applied more detailed analysis of the deformation components during and after manufacture is required.

The individual deformation components can be estimated using the simple approaches described below of the deformation behaviour of the overall system be numerically determined using the finite element method (FEM), for example. It is extremely important that the results of a numerical analysis are checked for plausibility.

108

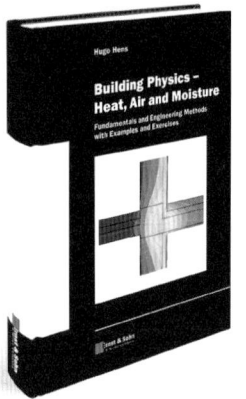

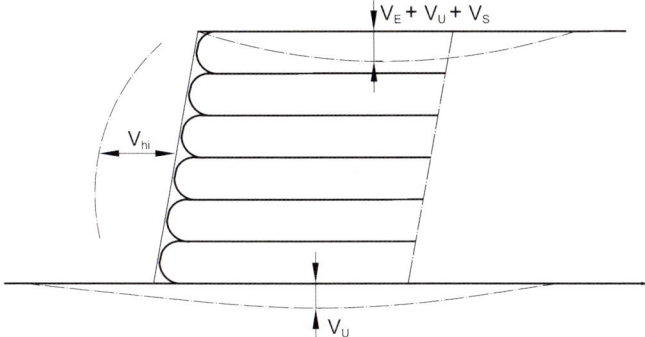

Figure 7.17 Deformation components on a retaining structure

7.5.2 Analysing the Position of the Bearing Pressure Resultant

In analogy to DIN 1054, 7.6.1 it shall be demonstrated that no foundation gap occurs in the foundation base plane as a result of permanent actions (resultant remains within the first kernel width).

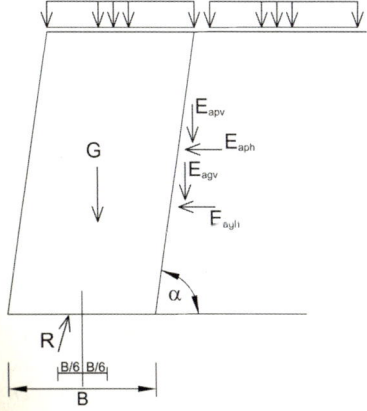

Figure 7.18 Forces adopted for analysis of the position of the resultant

7.5.3 Displacements in the Base Plane

The displacement in the base plane is determined to DIN 1054, 7.6.2.

7.5.4 Ground Settlement v_U

Ground settlement is determined to DIN 1054, 7.6.3 and DIN 4019.

The regulations of DIN 4019 and the DGGT 'Recommendations; Deformation of the subsoil below structures' – EVB, apply when determining ground settlements v_U (Figure 7.17) resulting from the dead weight of the reinforced earth structure and any surcharges. The reinforced retaining structure may be adopted as a flexible load area.

These settlements may be a governing factor on soft ground. Special attention shall be paid to the evolution of the settlements with time (consolidation).

Note: *A reinforced retaining structure offers the advantage that it is itself insensitive to settlement. Differential settlements can generally be compensated by the structure. This construction method is therefore particularly favoured on soft ground, given an appropriate facing.*

7.5.5   Intrinsic Settlement of Fill Soil v_E

It is necessary to adequately compact the fill soil when dealing with reinforced retaining structures. Empirical data indicates that intrinsic settlements of 0.2% to 1.0% of the structure height are normal [4]. Intrinsic settlement v_E of the fill soil predominantly occurs during the construction period.

Fill soil settlement as a result of surcharges on the reinforced earth structure can be approximately determined using elastic analysis methods.

7.5.6   Horizontal Displacements of the Slope Front v_{Hi}

The deformation behaviour of the composite structure consisting of soil and reinforcement is complex and can only be described approximately.

The fill soil itself is deformed as a result of the manufacturing process and by actions. The reinforcement is installed flexibly, such that it undergoes strain until an equilibrium is achieved between the effect and the tensile force.

Note: *Load testing to failure (e.g. [7], [8], [9]) and numerical post analysis of these tests (e.g. [1]) have shown that reinforced retaining structures display approximately linear load-deformation behaviour in the service load range. Pronounced plastic deformation of the soil and sliding between the ground and the reinforcement were not identified in the service load range, at least for well compacted, granular fill soil. Strain in the reinforcement increases approximately linearly with load. From this it was derived that that the facing displacements of the retaining structure can be determined in the service load range from the changes in length of the individual reinforcement layers [1]. In addition to the tensile force on the reinforcement (determined as described in Section 7.5) knowledge of the load-extension behaviour of the reinforcement materials in the ground is also required to determine the changes in length of the reinforcement layers.*

The following analysis steps are necessary to estimate the horizontal facing displacements v_{hi}:

- analysis of the tensile forces and their distribution in all reinforcement layers for the serviceability limit state,
- determine the associated axial stiffness of the reinforcement layers,
- determine the strains/strain distributions for all reinforcement layers,
- integrate the strains in all reinforcement layers to determine the change in length of each layer.

The failure mechanism governing the equilibrium of each reinforcement layer described in Section 7.4.4 is determined iteratively for analysis of the tensile forces in all reinforcement layers, taking the SLS partial safety factors into consideration. This gives a tensile force necessary for equilibrium for each failure mechanism and intersected reinforcement layer. The strains involved are then identified on this basis.

Note: This maximum value can be conservatively adopted as a constant for the entire length of the reinforcement. See [3], [8] for example, for other tensile force distributions.

For reinforcement materials with low structural strains, such as wovens or geogrids, the load-extension behaviour (axial stiffness) for these tensile forces can be approximately adopted from tensile testing to DIN EN ISO 10310. Linear behaviour is generally adequate for wovens/geogrids.

Note: The load-extension behaviour of nonwovens is altered by the soil. The results of tensile testing to DIN EN ISO 10319 generally provide conservative axial strengths, thereby overestimating the strains on the reinforcement and thus the deformations in the reinforced retaining structure. Investigations have shown that the load-extension behaviour of nonwovens depends on the soil and the surcharge ([1], [3]) and that the axial stiffness increases with the surcharge. More recent investigations report that similar behaviour is also observed in wovens and geogrids [9].

The changes in length are acquired by integrating the strains along the reinforcement layers. These correspond approximately to the frontal displacement v_{hi} at the level of layer i.

Note: Highly variable changes in length in the reinforcement layers indicate an unfavourable reinforcement configuration.

If product-typical isochrones are adopted for the design service life of the structure (see Section 2.2.4.5) to determine deformations as described above, instead of the load-extension curves in DIN EN ISO 10319 (short-term test), long-term deformations can also be determined.

7.5.7 Shear Deformation in the Retaining Structure v_S

Shear deformations v_S result predominantly from the strains in the reinforcement layers necessary to achieve an equilibrium condition in the reinforced retaining structure. These horizontal displacements cause vertical displacements on the surface.

Comparative analyses have shown that in the case of a uniform surcharge and uniform reinforcement distribution shear deformation causes an additional settlement component in the reinforced earth structure at a magnitude of approx. 30% to 50% of the frontal displacement $v_{hi,max}$.

7.5.8 Vertical Displacements at the Surface v_O

The vertical, surface displacements v_O of a retaining structure result from the settlement of the ground v_U, the intrinsic settlement of the fill material v_E and the shear deformation v_S of the reinforced earth structure.

7.5.9 Numerical Methods

The deformation behaviour of the overall system can be determined using numerical analysis methods (e.g. the finite element method (FEM)). Because inadequate data is available on the composite material it is usual to model the ground and the reinforcement separately. Direct shear tests are suitable for identifying the behaviour of the composite. Modelling can be performed with spring elements parallel to the reinforcement similar to the methods developed for use in rock mechanics. These models should simulate the elastic-plastic shear force transfer, as determined in direct shear tests.

The reinforcement itself can be modelled using elastic spring or rod elements. Linear-elastic material laws may be adequate, or others may be necessary, depending on the load-extension behaviour of the reinforcement materials.

When checking analysis results care shall be taken that no unrealistic tensile stresses occur in the ground as a result of differential stiffnesses in the ground-reinforcement contact zone. Analysis results should always be examined for plausibility using simple comparative analyses (e.g. equilibrium of vertical and horizontal forces).

7.6 Facing Analyses

The following facing options are currently used (DIN EN 14475, Appendix C):

- non-deformable (rigid) facing elements:
  panels or blocks, usually of precast concrete, with low vertical compressibility and high flexural stiffness,

– partially deformable facing elements:
preformed steel wire mesh section, preformed steel element or gabions filled with rock material with higher vertical compressibility and low flexural stiffness,

– deformable (flexible) facing elements:
facing with no flexural stiffness, where the fill material is enclosed by a geogrid or geotextile, such as padded walls of wrap-around geosynthetics (temporary formwork may be required); lightweight, abandoned formwork, for example using sandbags; an external frame with no structural function for the retaining structure may be used as protection and additional design element.

Type of facing element	Description
Non-deformable facing elements	
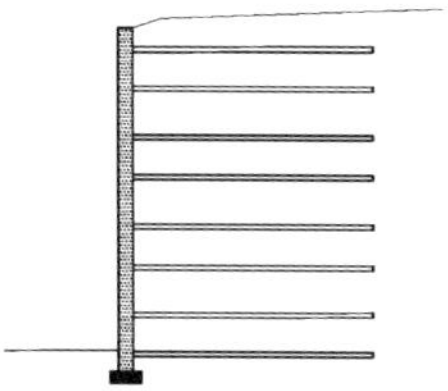 Full height panels	The elements are pre-manufactured such that they cover the entire height of the reinforced retaining structure in one piece.
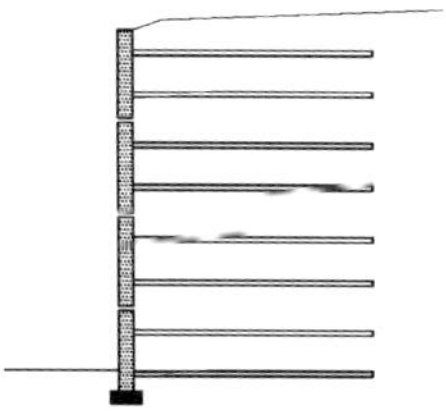 Partial height panels	Partial height panels are usually 1 m to 2 m high and 100 mm to 200 mm thick.
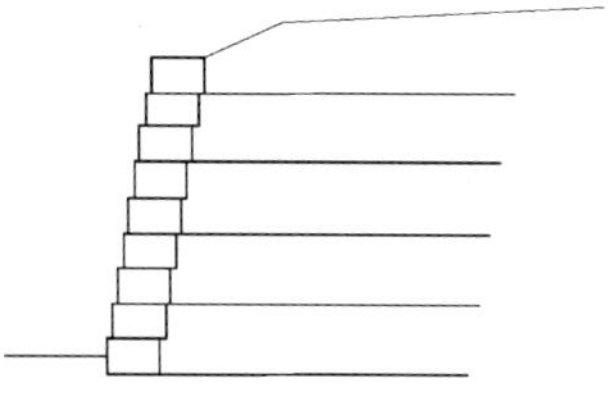 Block elements, moulded bricks	Facing elements of precast concrete blocks (e.g. modular blocks, segmented blocks) or connected natural stones.

Figure 7.19 Facing elements

113

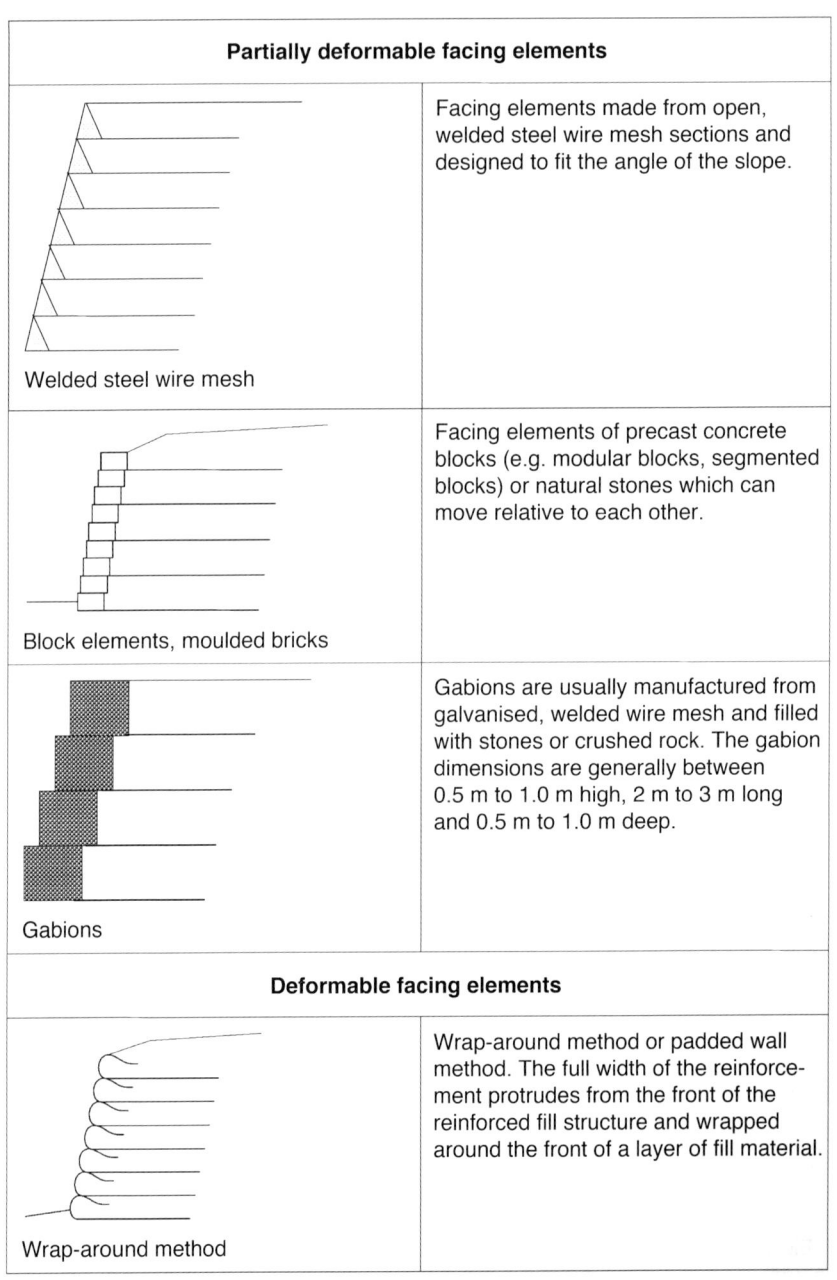

Partially deformable facing elements	
Welded steel wire mesh	Facing elements made from open, welded steel wire mesh sections and designed to fit the angle of the slope.
Block elements, moulded bricks	Facing elements of precast concrete blocks (e.g. modular blocks, segmented blocks) or natural stones which can move relative to each other.
Gabions	Gabions are usually manufactured from galvanised, welded wire mesh and filled with stones or crushed rock. The gabion dimensions are generally between 0.5 m to 1.0 m high, 2 m to 3 m long and 0.5 m to 1.0 m deep.
Deformable facing elements	
Wrap-around method	Wrap-around method or padded wall method. The full width of the reinforcement protrudes from the front of the reinforced fill structure and wrapped around the front of a layer of fill material.

Figure 7.19 (continued)

114

The various facing designs are connected to the reinforcement layers using a variety of connection methods and systems, which are product- or system-specific. The demands placed on the connections depend on the facing design (non-deformable, partially deformable, deformable).

It is not generally possible to precisely determine the horizontal stress on the facing. The active earth pressure to DIN 4085 is used as the reference variable for effects in the following notes.

It is not always necessary to adopt the active earth pressure for the full height of a geosynthetic-reinforced structure when analysing the geosynthetic connection to the facing elements. The magnitude of the horizontal stresses is critically influenced by the properties of the composite reinforced earth structure and the deformability of the facing.

An examination of whether the deformations are acceptable for both the structure and the surrounding ground shall be performed. Vertical differential deformations between the facing elements and the reinforced earth structure, in particular, shall be avoided (e.g. differences in the facing and the reinforced earth structure supports, inadequate compaction behind the facing). The horizontal movement of the facing which the design is based on shall be guaranteed for the entire height (including the toe).

Calibration factors for reducing the connected forces are given for various systems in Table 7.2.

The geosynthetic force to be transmitted at the facing is calculated as the active earth pressure with a layer thickness l_v at the depth H_i.

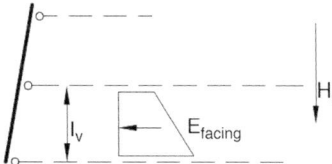

Figure 7.20 Earth pressure

Analysis:

E_{facing} characteristic earth pressure to DIN 4085

The earth pressure on the external skin is:

$$e_{facing} = \eta_g \cdot K_{agh,k} \cdot \gamma_k \cdot H_i \cdot \gamma_G + \eta_q \cdot K_{aqh,k} \cdot q \cdot \gamma_Q \qquad \text{Eq. (7.6)}$$

$$E_{facing} = e_{facing} \cdot l_v \qquad \text{Eq. (7.7)}$$

Analysis of STR for facing connection:

$$R_{Bi,d} \quad \text{or} \quad R_{Ai,d} \geq E_{facing} \qquad \text{Eq. (7.8)}$$

where:

$R_{Bi,d}$ design value of the long-term tensile strength of the geosynthetics in the nth reinforcement layer,

$R_{Ai,d}$ design value of the entire pull-out resistance provided by friction or as a connection force (design value determined using γ_B),

η_g, η_q calibration factor as given in Table 7.2

Note: If adequate pull-out resistance of the wrap-around can be demonstrated for a facing using the wrap-around method, the earth pressure E_{facing} can be distributed equally on the reinforcement layer and the wrap-around.

Table 7.2 Calibration factor

	Calibration factor			Earth pressure angle
	η_g		η_q	δ
	$0 < h \leq 0.4\,H$	$0.4\,H < h \leq H$		
Non-deformable facing elements	1.0	1.0	1.0	Analogous to DIN 4085
Partially deformable facing elements	1.0	0.7	1.0	$1/3\,\varphi'$ to $1.0\,\varphi'$ (see [11])
Deformable facing elements	1.0	0.5	1.0	0

Notes: 1) In the contact zone h = H.
2) The calibration factors are derived from literature evaluations and large-scale tests [18].

Earth pressures from bounded surcharges are also considered without redistribution. Compaction earth pressures to DIN 4085 are adopted additionally for design (also see [17]).

7.7 Bibliography

[1] Lehrstuhl und Prüfamt für Grundbau, Bodenmechanik und Felsmechanik at the Technische Universität München, Heft 26, 1997.

[2] Bauer, A., Bräu, G., Floss, R. (2000): In-soil-testing of geogrids with low construction deformations. Proceedings of the Second European Geosynthetics Conference EUROGEO 2000.

[3] Bauer, A., Bräu, G. (2002): Entwicklung eines Bemessungsverfahrens für die Bodenbewehrung mit Vliesstoffen basierend auf Zugversuchen im Bodenkontakt. Federal Ministry of Transport, Building and Urban Affairs, Forschung Straßenbau und Straßenverkehrstechnik, Heft 831, 2002.

[4]   Floss, R. (1997): Zusätzliche Technische Vertragsbedingungen und Richtlinien für Erdarbeiten im Straßenbau, Kommentar mit Kompendium Erd- und Felsbau, Kirschbaum-Verlag Bonn, 1997.

[5]   Floss, R., Stiegeler, R. (2000): Design and measurements of a reinforced steep slope under motorway Nürnberg–Berlin. Proceedings of the Second European Geosynthetics Conference EUROGEO 2000.

[6]   Herold, A. (2001): Das erste Straßenbrückenwiderlager in Deutschland als Permanentkonstruktion in der Bauweise kunststoffbewehrte Erde. Tagungsband der 7. Informations- und Vortragstagung über 'Kunststoffe in der Geotechnik', March 2001, Munich, Special Edition of the DGGT's Geotechnik Journal.

[7]   Nimmesgern, M. (1998): Untersuchungen über das Spannungs-Verformungs-Verhalten von mehrlagigen Kunststoffbewehrungen in Sand. Lehrstuhl und Prüfamt für Grundbau, Bodenmechanik und Felsmechanik at the Technische Universität München, Heft 27, 1997.

[8]   Matichard, Y., Thamm, B. R. (1992): Performance of geotextile reinforced earth under surface loading; French-German Cooperation in Highway Research, Bast-LCPC Seminar 1992.

[9]   Bussert, F. (2006): Verformungsverhalten geokunststoffbewehrter Erdstützkörper – Einflussgrößen zur Ermittlung der Gebrauchstauglichkeit. Institut für Geotechnik und Markscheidewesen, Technical University of Clausthal, Heft 13.

[10]   Herold, A. (2004): Geokunststoffbewehrte Großbauwerke – Verformungsmessungen und Rückrechnung – Wie kann das Verformungsverhalten von KBE-Konstruktionen optimal prognostiziert werden? 4th Österreichische Geotechniktagung, Vienna.

[11]   Merkblatt über Stützkonstruktionen aus Betonelementen, Blockschichtungen und Gabionen, Ausgabe 2003, Forschungsgesellschaft für Straßen- und Verkehrswesen, Arbeitsgruppe Erd- und Grundbau, Cologne, FGSV Heft No. 555.

[12]   Merkblatt über die Anwendung von Geokunststoffen im Erdbau des Straßenbaus, Ausgabe 2005, Forschungsgesellschaft für Straßen- und Verkehrswesen, Arbeitsgruppe Erd- und Grundbau, Cologne, FGSV Heft No. 535.

[13]   DIN EN 14475: 'Execution of Special Geotechnical Work – Reinforced Fill', German Edition EN 14475:2006.

[14]   Köhler, U. (2003): Der 'Erddruckfänger' aus kunststoffbewehrter Erde für Instandsetzung und Neubau von Trockenmauern. Tagungsband der 8. Informations- und Vortragstagung über 'Kunststoffe in der Geotechnik', March 2003, Munich.

[15]   Alexiew, D.: Belastungsversuche eines geogitterbewehrten Brückenwiderlagers. Huesker Synthetic GmbH, Gescher.

[16]   Lieberenz, K., Großmann, S., Göbel, C. (2007): Stützbauwerke aus geokunststoffbewehrter Erde – Weiterentwicklung des Systems für vorrangig dynamische Einwirkungen. Tagungsband der 10. Informations- und Vortragstagung über 'Kunststoffe in der Geotechnik', March 2007, Munich.

[17]   Naciri, O., Bussert, F. (2007): Erfahrungen aus Verformungsmessungen an geokunststoffbewehrten Stützkonstruktionen. Tagungsband der 10. Informations- und Vortragstagung über 'Kunststoffe in der Geotechnik', March 2007, Munich.

117

[18] Pachomow, D., Vollmert, L., Herold, A. (2007): Der Ansatz des horizontalen Erd-druckes auf die Front von KBE-Systemen. Tagungsband der 10. Informations- und Vortragstagung über 'Kunststoffe in der Geotechnik', March 2007, Munich.

[19] Franke, D. (2008): Verdichtungserddruck bei leichter Verdichtung. Bautechnik 85/2008, Heft 3.

7.8 Retaining Structure Design Example

A design example will be used for illustration. It will demonstrate possible design procedure for a reinforced retaining structure with equal reinforcement lengths.

7.8.1 Geometry, Soil Properties and Load Assumptions

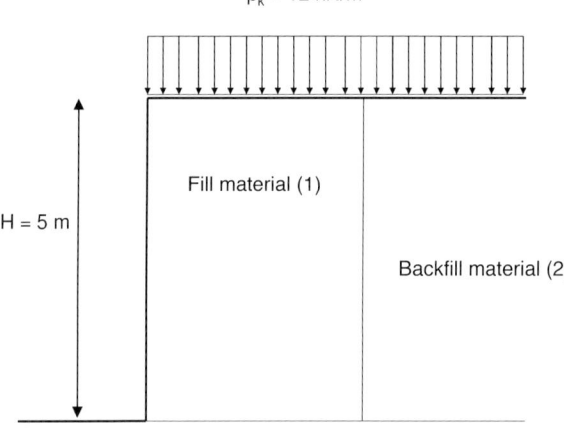

Figure 7.21 Geometry and soil properties of the retaining structure

Table 7.3 Soil properties

	Layer x	Unit weight $\gamma_{x,k}$ $[kN/m^3]$	Friction angle $\varphi_{x,k}$ $[°]$	Cohesion $c_{x,k}$ $[kN/m^2]$
Fill material	1	20	35	0
Ground	2	18	30	0
Backfill material	3	19	30	0

Design of reinforced earth structures to EBGEO, Section 7.2.2 is based on a reinforcement length of 70% of the structure height:

B = 0.7 · 5 m = 3.5 m.

118

The following analyses are performed using this dimension.

Note: Initial investigation of the bearing capacity and general stability using the preliminary draft geometry is recommended.

The analysis example is based on Load Case 1.

7.8.2 Determining the Characteristic Actions

Earth pressure is determined to DIN 4085:

$\varphi_{3,k} = 30°$; $\delta_{3,k} = 2/3 \cdot \varphi_{3,k} = 20°$; $\alpha = 0°$; $\beta = 0°$

$$K_{ah,k} = \frac{\cos^2(30°)}{\left[1 + \sqrt{\dfrac{\sin(30° + 20°) \cdot \sin 30°}{\cos 20°}}\right]^2} = 0.279$$

Characteristic earth pressure from soil dead weight:

$E_{agh,k} = 0.5 \cdot K_{ah,k} \cdot \gamma_{3,k} \cdot H^2 = 0.5 \cdot 0.279 \cdot 19 \cdot 5^2 = 66.26$ kN/m

$E_{agv,k} = E_{agh,k} \cdot \tan(\delta - \alpha) = 66.26 \cdot \tan 20° = 24.12$ kN/m

Characteristic earth pressure from variable load:

$E_{aph,k} = K_{ah,k} \cdot p_k \cdot H = 0.279 \cdot 12 \cdot 5 = 16.74$ kN/m

$E_{apv,k} = E_{aph,k} \cdot \tan(\delta - \alpha) = 16.74 \cdot \tan 20° = 6.09$ kN/m

Characteristic action from dead weight:

$G_k = H \cdot B \cdot \gamma_{1,k} = 5 \cdot 3.5 \cdot 20 = 350$ kN/m

Characteristic variable action:

$P_k = B \cdot p_k = 3.5 \cdot 12 = 42$ kN/m

7.8.3 Analysis in the Ultimate Limit State (ULS)

7.8.3.1 Analysing Sliding Safety

It is assumed for analysis of sliding safety that the lowest reinforcement layer is placed at the base of the earth structure. That is, the lowest value for the friction angle above ($\varphi_{1,k\ above}$) and below ($\varphi_{2,k\ below}$) the reinforcement is adopted as $\varphi_{k,\ governing}$.

The characteristic frictional coefficient between the reinforcement and the ground is determined by $f_{sg,k} = \tan \delta_{s,k} = \lambda \cdot \tan \varphi_{k,\ governing}$. $\lambda = 0.8$ is assumed for this example. The value is confirmed by testing.

The variable action is adopted conservatively behind the retaining structure for analysis of sliding safety only.

To ensure adequate sliding safety it shall be demonstrated to DIN 1054, Section 7.5.3 for the STR limit state that the

$$T_d \leq R_{t,d} + E_{p,d}$$

condition is met. Where:

T_d design value of the effect parallel to the foundation base,
$R_{t,d}$ design value of the sliding resistance,
$E_{p,d}$ design value of the passive earth pressure parallel to the base, is ignored here.

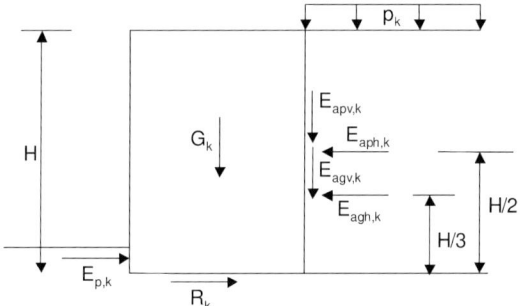

Figure 7.22 Distribution of forces for analysis of sliding safety

Determining the design values of effects:

$N_{G,k} = E_{agv,k} + G_k = 24.12 + 350 = 374.12$ kN/m

$N_{Q,k} = E_{apv,k} = 6.09$ kN/m

$T_{G,k} = E_{agh,,k} = 66.26$ kN/m

$T_{Q,k} = E_{aph,k} = 16.74$ kN/m

Note: $\gamma_G, \gamma_Q, \gamma_{GL}$ from DIN 1054, Tables 2 and 3

$T_d = 66.26 \cdot 1.35 + 16.74 \cdot 1.5 = 114.56$ kN/m

$R_{t,d} = (N_{G,k} + N_{Q,k}) \cdot \tan \delta_{s,k} / \gamma_{GL} = 380.21 \cdot (0.8 \cdot \tan 30°) / 1.1 = 159.65$ kN/m

$T_d = 114.56 \leq R_{t,d} = 159.65$ kN/m

Analysis verified.

Utilisation factor $\mu = \dfrac{T_d}{R_{t,d}} = \dfrac{114.56}{159.65} = 0.72$

7.8.3.2 Position of Bearing Pressure Resultant

(to DIN 1054, 7.5.1)

The forces relevant to determining the position of the bearing pressure resultant are shown in the figure below:

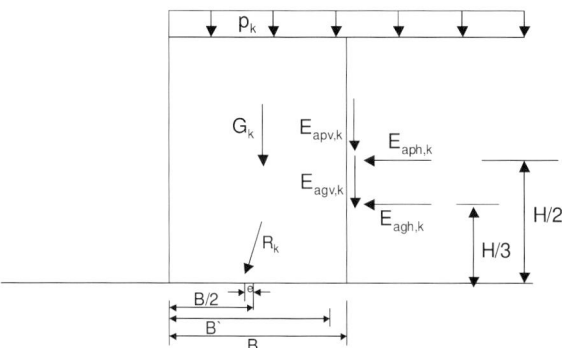

Figure 7.23 Distribution of forces for the position of the bearing pressure resultant

Determining the characteristic effect in the base plane

– with variable action:

$$M_k = \left(E_{agh,k} \cdot \frac{H}{3} \right) + \left(E_{aph,k} \cdot \frac{H}{2} \right) - \left\{ E_{agv,k} \cdot \frac{B}{2} \right\} - \left\{ E_{apv,k} \cdot \frac{B}{2} \right\}$$

$$= \left(66,26 \cdot \frac{5}{3} \right) + \left(16,74 \cdot \frac{5}{2} \right) - \left\{ 24,12 \cdot \frac{3,5}{2} \right\} - \left\{ 6,09 \cdot \frac{3,5}{2} \right\} = 99.42 \text{ kNm/m}$$

Characteristic effect normal to the base:

$$N_{G,k} + N_{Q,k} = 350 + 42 + 24.12 + 6.09 = 422.21 \text{ kN/m}$$

– without variable action:

$$M_k = \left(E_{agh,k} \cdot \frac{H}{3} \right) - \left\{ E_{agv,k} \cdot \frac{B}{2} \right\} = \left(66,26 \cdot \frac{5}{3} \right) - \left\{ 24,12 \cdot \frac{3,5}{2} \right\}$$

$$= 68.22 \text{ kNm/m}$$

$$N_{G,k} = 350 + 24.12 = 374.12 \text{ kN/m}$$

Analysis of the position of the bearing pressure resultant (DIN 1054, Section 7.5.1: all actions) EQU, allowable: foundation gap as far as centre of foundation (second kernel width).

$$e = \frac{M_k}{N_k} = \frac{99.42}{422.21} = 0.235 < \frac{b}{3} = \frac{3.5}{3} = 1.16\ m \quad \text{Analysis verified!}$$

Note: See analysis of the serviceability limit state (SLS) for analysis of the first kernel width.

7.8.3.3 Analysing Bearing Capacity

Adequate bearing capacity failure safety is given if the following condition is met:

$$N_d \le R_{n,d} = R_{n,k} / \gamma_{Gr}$$

Analysis of the STR limit state is performed to DIN 1054 and DIN 4017.

N_d design value of the effect normal to the foundation base,
$R_{n,d}$ bearing resistance design value,
$R_{n,k}$ characteristic bearing resistance,
γ_{Gr} bearing resistance, see DIN 1054, Table 3.

Characteristic bearing resistance $R_{n,k}$:

$$
\begin{aligned}
R_{n,k} = a' \cdot b' \cdot (&c_{2,k} \cdot N_{c0} \cdot v_c \cdot i_c \cdot \lambda_c \cdot \xi_c \\
&+ \gamma_{2,k} \cdot d \cdot N_{d0} \cdot v_d \cdot i_d \cdot \lambda_d \cdot \xi_d \\
&+ \gamma_{2,k} \cdot b' \cdot N_{b0} \cdot v_b \cdot i_b \cdot \lambda_b \cdot \xi_b)
\end{aligned}
$$

Because no cohesion, depth, ground or base inclination need be adopted for the example and the structure is regarded as strips, all shape coefficients are adopted as equal to 1; this simplifies the equation to:

$$R_{n,k} = b' \cdot (\gamma_{2,k} \cdot b' \cdot N_{b0} \cdot i_b)$$

Determining bearing capacity coefficients:

$$N_{d0} = e^{\Pi \cdot \tan \varphi_{2,k}} \cdot \tan^2 \left(45 + \frac{\varphi_{2,k}}{2} \right) = e^{\Pi \cdot \tan 30} \cdot \tan^2 \left(45 + \frac{30}{2} \right) = 18.4$$

$$N_{b0} = (N_{d0} - 1) \cdot \tan \varphi_{2,k} = (18.4 - 1) \tan 30° = 10.05$$

Determining load inclination coefficients:

$m = 2.0$ (*cf.* DIN 4017, Section 7.2.4)

$$i_b = (1 - \tan \delta)^{m+1}$$

$$\tan \delta = \frac{T_k}{N_k} = \frac{66.26 + 16.74}{422.21} = 0.196$$

$$i_b = (1 - 0.196)^{2.0+1} = 0.519$$

$$N_b = N_{b0} \cdot i_b = 10.05 \cdot 0.519 = 5.22$$

Equivalent width:

$b' = B - 2 \cdot e = 3.5 - 2 \cdot 0.235 = 3.03$ m

$R_{n,k} = 3.03 \cdot 18 \cdot 3.03 \cdot 10.05 \cdot 0.519 = 861.97$ kN/m

$R_{n,d} = R_{n,k} / \gamma_{Gr} = 861.97 / 1.4 = 615.70$ kN/m

$N_d = N_{G,k} \cdot \gamma_G + N_{Q,k} \cdot \gamma_Q = 374.12 \cdot 1.35 + 48.09 \cdot 1.5 = 577.20$ kN/m

Analysis:

$N_d = 577,20 < R_{n,d} = 615.70$ Analysis verified.

Utilisation factor $\mu = \dfrac{N_d}{R_{n,d}} = \dfrac{577.20}{615.70} = 0.94$

7.8.3.4   Analysing General Failure

Partial safety factors for the GEO limit state, LC 1:

Actions:
Permanent actions: $\qquad\qquad \gamma_G = 1.0$
Unfavourable variable actions: $\quad \gamma_Q = 1.3$

Resistances:
Friction coefficient tan φ' $\qquad \gamma_\varphi = 1.25$
Cohesion c' $\qquad\qquad\qquad\quad \gamma_c = 1.25$
Flexible reinforcement elements
(pull-out resistance) $\qquad\qquad \gamma_B = 1.4$

Design values:

Design value of the variable action:

$p_d = p_k \cdot \gamma_Q = 12 \cdot 1.3 = 15.6$ kN/m^2

Design values of the shear strength:

Angle of friction

$\tan \varphi_{i,d} = (\tan \varphi_{i,k}) / \gamma_\varphi$

$\varphi_{i,d} = \text{arc tan} [(\tan \varphi_{i,k}) / \gamma_\varphi]$

Cohesion

$c_{i,d} = c_{i,k} / \gamma_c$

Design values of unit weights

$\gamma_{i,d} = \gamma_{i,k} \cdot \gamma_G$

The design values of the soil properties can be taken from the following table:

Table 7.4 Design values of soil properties (GEO)

		$\varphi_{i,d}\ [°]$	$c_{i,d}\ [kN/m^2]$	$\gamma_{i,d}\ [kN/m^3]$
1	Fill Soil	29.25	0	20
2	Ground	24.79	0	18
3	Backfill material	24.79	0	19

All possible slip planes shall be considered in an analysis of the bearing capacity and the most unfavourable failure mechanism investigated. Slip planes completely enveloping reinforcement layers and those intersecting or at least touching the reinforcement layers are all investigated. Additionally, slip planes passing through the reinforced earth structure without intersecting a reinforcement layer are considered.

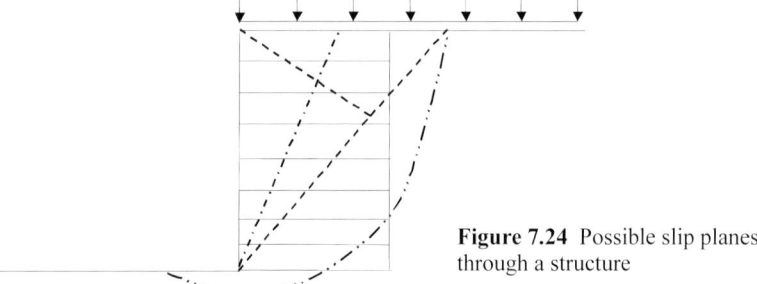

Figure 7.24 Possible slip planes through a structure

Various slip planes are investigated for the example. The slip planes are analysed using several angles ϑ, starting at the toe of the retaining structure. The slip planes penetrate the entire reinforced structure as far as the rear face of the reinforcement and then move to the surface. All (acting) mobilising forces are compared to the (resisting) restraining force of the reinforcement.

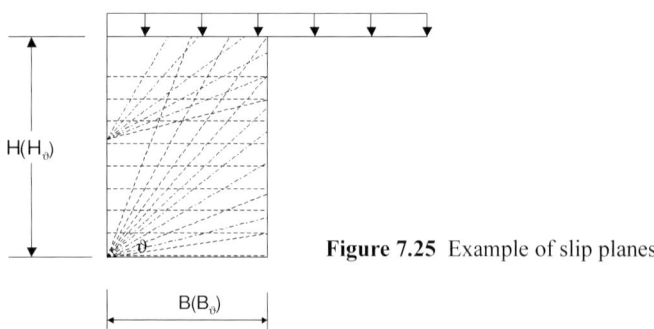

Figure 7.25 Example of slip planes

The following failure systems, which penetrate the structure at its foot, are considered as examples for $\vartheta = 40°$ and $\vartheta = 45° + \varphi/2$, in both cases with permanent and variable actions.

Case 1: $\vartheta = 40°$

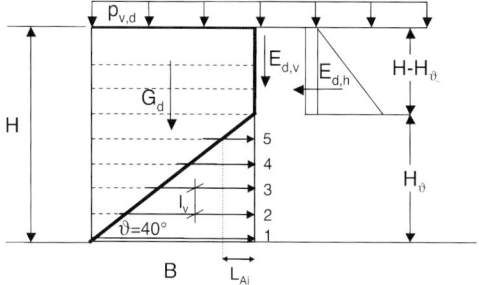

Case 2: $\vartheta = 40 + \varphi/2$

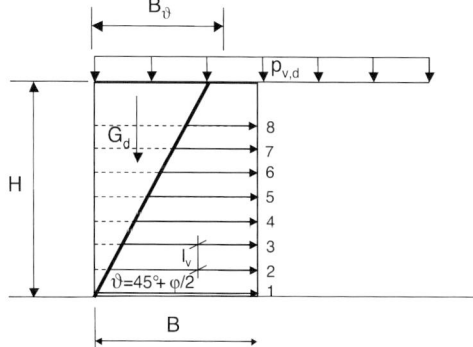

Figure 7.26 Distribution of forces for slip planes

The effect is:

$$\Sigma F_{d(\vartheta)} = (G_{d(\vartheta)} + P_{v,d(\vartheta)} + E_{v,d(\vartheta)}) \cdot \tan(\vartheta - \varphi_{1,d}) + E_{h,d(\vartheta)},$$

where:

$$G_{d(\vartheta)} = \frac{1}{2} \cdot \gamma_{1,d} \cdot B_{(\vartheta)} \cdot H_{(\vartheta)} + \gamma_{1,d} \cdot B_{(\vartheta)} \cdot (H - H_\vartheta)$$

$$P_{v,d(\vartheta)} = p_{v,d} \cdot B_{(\vartheta)}$$

$$E_{gh,d(\vartheta)} = \frac{1}{2} \cdot \gamma_{v,d} \cdot k_{ah,g,d} \cdot (H - H_\vartheta)^2$$

$$E_{gv,d(\vartheta)} = E_{gh,d(\vartheta)} \cdot \tan(\delta - \alpha)$$

$$E_{ph,d(\vartheta)} = k_{ah,p,d} \cdot (H - H_\vartheta) \cdot p_{v,d}$$

125

$E_{pvd(\vartheta)} = E_{phd(\vartheta)} \cdot \tan(\delta - \alpha)$

$E_{h,d(\vartheta)} = E_{gh,d(\vartheta)} + E_{ph,d(\vartheta)}$

$E_{v,d(\vartheta)} = E_{gv,d(\vartheta)} + E_{pv,d(\vartheta)}.$

Design assumptions for geosynthetics:

An imaginary geosynthetic with the required, characteristic, short-term strength and the following imaginary coefficients is selected per reinforcement layer i for the reinforcement of the retaining structure's reinforced earth structure:

$A_1 = 2.5; A_2 = 1.2$ and $A_3 = A_4 = A_5 = 1.0$.

Note: The reduction factors are product-specific and shall be analysed.

Determining the design strength of a reinforcement layer i:

(with partial safety factor $\gamma_M = 1.4$)

$$R_{B,d} = \frac{R_{B,k0}}{2.5 \cdot 1.2 \cdot 1.0 \cdot 1.0 \cdot 1.0 \cdot 1.4} = \frac{R_{B,k0}}{4.2}$$

Eight layers of geosynthetic reinforcement with constant vertical spacing as described in Section 7.2.2 where $0.3 \leq l_v = 0.6 \text{ m} \leq 0.6 \text{ m}$ are selected for the planned earth structure.

The following types of geosynthetics are used:

– 4 layers of geosynthetics at 80 kN/m (bottom)
  and design strength $R_{B,d} = 19.04$ kN/m,
– 4 layers of geosynthetics at 50 kN/m (top)
  and design strength $R_{B,d} = 11.9$ kN/m.

The configuration and the types of reinforcement are shown below:

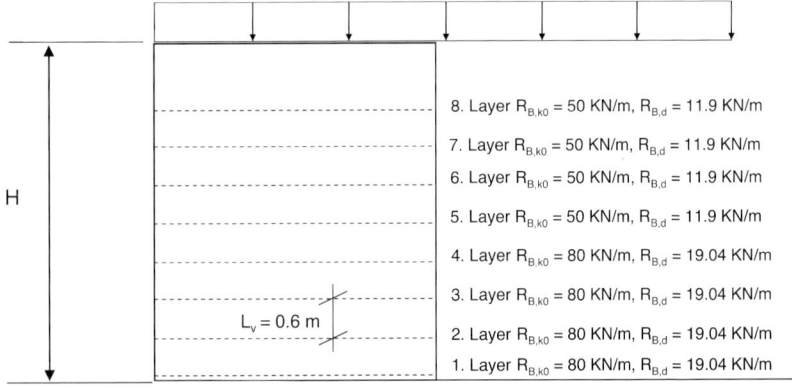

Figure 7.27 Reinforcement configuration

126

Case 1: $\vartheta = 40°$

The earth pressure coefficient is first determined.

For the GEO limit state:

$\varphi_{1,d} = 29.25°$; $\delta_{1,d} = 2/3 \cdot \varphi_{1,d} = 19.5°$; $\alpha = 0°$; $\beta = 0°$

$k_{ah,d} = 0.29$

$G_{d(40°)} = 0.5 \cdot 20 \cdot 3.5 \cdot 2.94 + 20 \cdot 3.5 \cdot 2.06 = 247.1$ kN/m

$P_{v,d(40°)} = 15.6 \cdot 3.5 = 54.6$ kN/m

$E_{gh,d(40°)} = 0.5 \cdot 20 \cdot 0.29 \cdot (2.06)^2 = 12.31$ kN/m

$E_{gv,d(40°)} = 12.31 \cdot \tan 19.5° = 4.36$ kN/m

$E_{ph,d(40°)} = 0.29 \cdot 2.06 \cdot 15.6 = 9.32$ kN/m

$E_{pv,d(40°)} = 9.32 \cdot \tan 19.5° = 3.3$ kN/m

$E_{h,d(40°)} = 12.31 + 9.32 = 21.63$ kN/m

$E_{v,d(40°)} = 4.36 + 3.3 = 7.66$ kN/m

$\Sigma F_{(40°)} = (247.1 + 54.6 + 7.66) \cdot \tan(40° - 29.25°) + 21.63 = 80.36$ kN/m

The magnitude of the resisting reinforcement forces is determined by the sum of the design values of the intersected reinforcement layers (resistance), taking the pull-out resistances into consideration.

Failure by rupture and pull-out of the reinforcement elements shall be examined on the resistance side.

Equilibrium is generally given if the following limit state condition is met:

$\Sigma E_{i,d(\vartheta)} \leq \min(\Sigma R_{Bi,d}; \Sigma R_{Ai,d})$.

The governing value for each layer is the smaller one,

where:

$R_{Ai,d} = 2 \cdot \sigma_{v,di} \cdot L_{Ai} \cdot (f_{sg,k}/\gamma_B)$ with weighted $f_{sg,k} = 0.8 \cdot \tan \varphi_{v,k}$
$\quad = 2 \cdot \sigma_{v,di} \cdot L_{Ai} \cdot (0.8 \cdot \tan 35°/1.4)$
$\quad = 2 \cdot 20 \cdot h_i \cdot L_{Ai} \cdot (0.8 \cdot \tan 35°/1.4)$
$\quad = 16 \cdot h_i \cdot L_{Ai}$

or

$R_{B,d} = \dfrac{R_{B,k0}}{4.2}$.

Table 7.5 Analysis for $\vartheta = 40°$

$\vartheta = 40°$	Analysis of rupture $R_{B,d}$ [kN/m]	Analysis of pull-out $R_{Ai,d}$ [kN/m]	Governing [kN/m]
1st layer	19.04	280	19.04
2nd layer	19.04	195.7	19.04
3rd layer	19.04	125.85	19.04
4th layer	19.04	69.376	19.04
5th layer	11.90	26.62	11.90
		$\Sigma R_{d,l} =$	88.06 kN/m

Analysis:

$\Sigma F_i = 80.36$ kN/m $< \Sigma R_{d,i} = 88.06$ kN/m.

Case 2: $\vartheta = 45° + \varphi/2$

$\vartheta = 45° + 29.25°/2 = 59.62°$

$G_{d(59.62°)} = 0.5 \cdot 20 \cdot 5 \cdot 2.931 = 146.55$ kN/m

$P_{(9)} = 15.6 \cdot 2.931 = 45.72$ kN/m

$F_{d(9)} = (G_{(9)} + P_{(9)}) \cdot \tan \cdot (\vartheta - \varphi) = (146.55 + 45.72) \cdot \tan (59.62° - 29.25°)$
$= 112.67$ kN/m

Table 7.6 Analysis for $\vartheta = 59.62°$

$\vartheta = 59.62°$	Analysis of rupture R_B [kN/m]	Analysis of pull-out R_{Ai} [kN/m]	Governing [kN/m]
1st layer	19.04	280.0	19.04
2nd layer	19.04	221.76	19.04
3rd layer	19.04	170.24	19.04
4th layer	19.04	125	19.04
5th layer	11.90	87	11.90
6th layer	11.90	55.68	11.90
7th layer	11.90	31.14	11.90
8th layer	11.9	13.31	11.9
		$\Sigma R_{di} =$	123.76 kN/m

Analysis:

$\Sigma F_i = 112.67$ kN/m $< \Sigma R_{d,i} = 123.76$ kN/m.

7.8.2.5 Analysing Facing for Partially Deformable Facing Elements

The final state (LC 1) is analysed.

The earth pressure coefficient is first determined for the STR limit state:

where:

$\phi_{1,k} = 35°$; $\delta_{1,k} = 2/3 \cdot \phi_{1,k} = 23.33°$, assuming that: $\alpha = 0°$; $\beta = 0°$,

we get:

$K_{agh,k} = K_{aqh,k} = 0.224$.

The earth pressure on the facing is:

$$e_{facing} = \eta_g \cdot K_{agh,k} \cdot \gamma_k \cdot H_i \cdot \gamma_G + \eta_q \cdot K_{aqh,k} \cdot q \cdot \gamma_Q$$

$$E_{facing} = e_{facing} \cdot l_v$$

Table 7.7 Facing analysis

	Z_i [m]	H_i [m]	η_g	η_q	l_v [m]	E_{facing} [kN/m]	$R_{Bi,d}$ or $R_{Ai,d}$ $\geq E_{facing}$
Layer 1	5.0	4.7	0.7	1.0	0.6	14.38	19.04
Layer 2	4.4	4.1	0.7	1.0	0.6	12.86	19.04
Layer 3	3.8	3.5	0.7	1.0	0.6	11.33	19.04
Layer 4	3.2	2.9	0.7	1.0	0.6	9.80	19.04
Layer 5	2.6	2.3	0.7	1.0	0.6	8.28	11.90
Layer 6	2.0	1.7	1.0	1.0	0.6	8.60	11.90
Layer 7	1.4	1.1	1.0	1.0	0.6	6.42	11.90
Layer 8	0.8	0.4	1.0	1.0	0.8	5.20	11.90

Note: Only the design strength ($R_{Bi,d}$) has been adopted for this example. $R_{Ai,d}$ is dependent on the system (various facing design options) and cannot therefore be specified as a generic value.

7.8.4 Serviceability Limit State (SLS) Analysis

(allowable position of the bearing pressure resultant to DIN 1054, 7.6.1)

7.8.4.1 Analysing the Position of the Bearing Pressure Resultant

(to DIN 1054, 7.6.1: permanent actions)

$$e = \frac{\sum M_k}{N_k} = \frac{68.22}{374.12} = 0.182 < \frac{b}{6} = \frac{3.5}{6} = 0.583 \text{ m}$$

7.8.4.2 Displacements in the Base Plane

(to DIN 1054, 7.6.2)

The passive earth pressure in front of the reinforced retaining structure was not taken into consideration for the analysis of sliding safety. However, the analysis is verified even without taking the passive earth pressure into consideration.

7.8.4.3 Settlements

(to DIN 1054, 7.6.3)

Settlement modelling is necessary, but is not carried out for this example.

8 Landfill Engineering – Reinforcement of Surface-parallel Stratified Systems

8.1 General Recommendations

This section was originally developed for landfill engineering purposes. Because these Recommendations are also used in other fields and are transferable, they also apply to liner systems used to protect groundwater and to installations for collecting, storing and discharging water and other fluids and substances where such surface-parallel, stratified systems require reinforcement to improve their stability.

Accordingly, they can be applied to multi-layered systems with interfaces parallel to the slope and without a liner function.

In terms of landfill engineering only regulations relevant to the drafting and design of geosynthetic reinforcement layers in reinforced earth and waste structures are provided in these Recommendations. Otherwise, the relevant regulations apply (among others: Technical Instructions on Waste (*TA-Abfall*) [1], Technical Instructions on Domestic Waste (*TA-Siedlungsabfall*) [2], DGGT recommendations: 'Geotechnical Aspects of Landfill and Brownfield Sites' (*Geotechnik der Deponien und Altlasten*) GDA E2 to E5 [3], [8]), Landfill Regulations (*Deponieverordnung*) [4]).

Load conditions requiring the use of geosynthetic reinforcement layers may predominantly occur in liner systems during construction and operation of waste landfills. Reinforcements with temporary functions and those with permanent functions are differentiated. Under certain landfill engineering boundary conditions an environment harmful to the durability of geosynthetics is anticipated in cases where the reinforcement is located within the landfill area enclosed by the liner system. Reduction factors for chemical and/or biological actions are therefore adopted for the reinforcement within the zone enclosed by the liner system.

Considerably increased temperatures may occur within the body of the landfill as a result of chemical-biological processes. They affect the load-bearing behaviour of the reinforcement and reduce its durability. This shall be taken into consideration in reinforcement dimensioning. GDA E2-14 assumes long-term temperatures of 15 °C to 40 °C at the base of domestic landfills without waste pre-treatment. Temperatures of 60 °C and more have been recorded over several years in waste incinerator residue landfill.

In terms of permanent applications geosynthetic reinforcements are predominantly used in landfill capping systems (Application 1 in Figure 8.1). If the calculated stability of an unreinforced slope in the final state cannot be guaranteed, the reinforcement shall be designed for the operational life of the liner system. A more detailed description is given below, the necessary stability analyses introduced and general design and engineering notes provided.

Recommendations for Design and Analysis of Earth Structures using Geosynthetic Reinforcements (EBGEO). German Geotechnical Society.
© 2011 Ernst & Sohn GmbH & Co. KG.
Published by Ernst & Sohn GmbH & Co. KG.

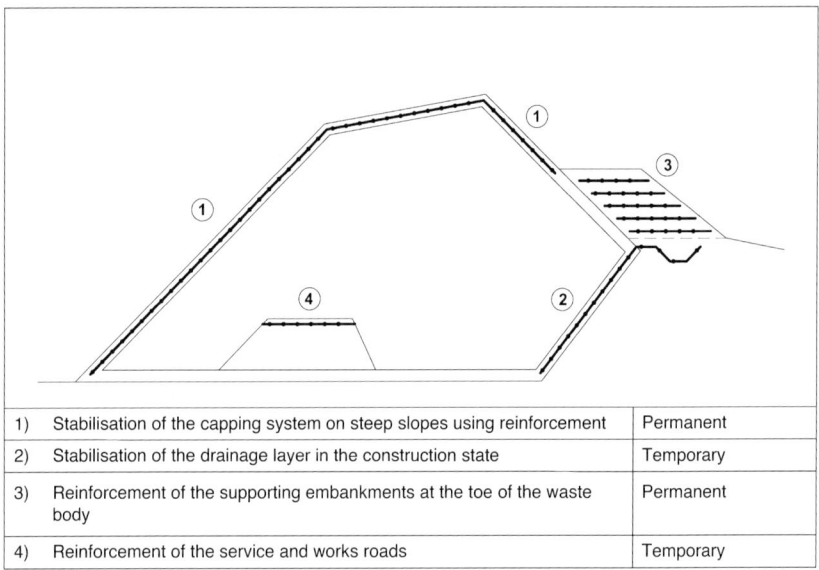

1)	Stabilisation of the capping system on steep slopes using reinforcement	Permanent
2)	Stabilisation of the drainage layer in the construction state	Temporary
3)	Reinforcement of the supporting embankments at the toe of the waste body	Permanent
4)	Reinforcement of the service and works roads	Temporary

Figure 8.1 Possible applications of geosynthetic reinforcement layers in landfill engineering

Typical applications are shown in Figure 8.1. Design is governed by the appropriate sections of EBGEO. These Recommendations can be applied accordingly for use with base or intermediate liner systems.

As described in GDA E2-3 and E2-4 liner systems used in landfill engineering are composed of several functional layers. In addition to regular liner systems the Landfill Regulations (*Deponieverordnung*) also allow alternative liner systems. Geosynthetic liners, which may not normally be subjected to tensile forces, represent a primary sealing element in regular liner systems used in landfill class II (domestic waste as classified by the waste disposal regulations (*Abfallablagerungsverordnung*) [5]) and landfill class III (hazardous wastes as classified by the waste disposal regulations [5]). The same applies to geosynthetic clay liners, which are often used as alternative liner components.

8.2 Design and Engineering Notes

The reinforcement is usually located at the base or in the lower zone of a recultivation layer, a topsoil, or the mineral aggregate drainage layer. If geosynthetic drainage elements are used, the reinforcement is located above this drainage layer.

For technical reasons only a single layer of reinforcement should be planned here. Special attention shall be paid to uninterrupted interlocking and adequate permeability of the soil and the geogrid where adjacent reinforcement webs overlap.

132

Where necessary, enhanced demands shall be placed on geometrical aspects of the geogrid mesh size.

Because of the primarily uniaxial load, reinforcements with anisotropic tensile strength properties (longitudinal axis equals principal load axis) are used. The reinforcements are unrolled according to a laying plan and overlap by approx. 20 cm for engineering reasons. Longitudinal joints shall be avoided. Where necessary, *hidden* anchoring planes are provided on the slope.

The reinforcement in the anchor trench is generally located as shown in Figure 8.3. Ponding in the anchor trench shall be avoided. A liner element, e.g. a geosynthetic liner including protective layer, generally ends before the trench base; a drainage mat ends before the anchor trench if the water cannot be discharged.

If the slope is heavily rounded in plan, the slope forces may be concentrated. In this case the trench geometry shall be designed for the higher effect.

Reinforcement on slopes interrupted by berms should be dimensioned separately for each slope section, including anchorage. The reinforcement is anchored in the berms. Installing reinforcement across berms to the next highest slope section shall be avoided (uplift forces).

8.3   Analyses

8.3.1   Principles

Multi-layer systems, predominantly parallel to the slope or ground surface, e.g. liner systems of soil and/or geosynthetics, may posses governing shear strength properties in any layer boundary or within elements (e.g. in bentonite and drainage mats). This means that stability analyses shall be carried out, in principle, for every layer boundary.

The overall stability in the GEO limit state is also analysed for potential failure planes within a layered system. If a layer boundary does not have the required stability or an element the required shear strength, they can be increased by using reinforcement layers. The overall stability is then analysed adopting the design value of the shear strength and the design resistance of the reinforcement layer as a load-bearing structural element. Geosynthetic reinforcement layers can be used to transfer slope pull-down forces, either wholly or partially.

The minimum required design resistance of a reinforcement layer is determined by rearranging the limit state equation. An application-specific correction factor η_M is introduced to modify the safety level for determining the characteristic value of the short-term strength of the reinforcement. This is:

- LC 1: $\eta_M = 1.10$,
- LC 2: $\eta_M = 1.05$,
- LC 3: $\eta_M = 1.00$.

$$R_{B,k0} = R_{B,k} \cdot \gamma_M \cdot \eta_M \qquad \text{Eq. (8.1)}$$

In addition, adequate overall stability of the slope or ground upon which the layered system rests shall be demonstrated in the GEO limit state to DIN 1054, or is assumed for the following deliberations.

The forces acting on the reinforcement layers are determined from the analysis of sliding in the construction state, taking construction equipment loads into consideration, and in the final state (see Section 8.3).

See Section 3.1 for details of estimating deformations in the serviceability limit state (SLS). The ability of the overall system to perform under the given deformations shall be considered when selecting the geosynthetic reinforcements.

Analysis of adequate anchorage of the geosynthetic reinforcement in the GEO limit state then follows.

Note: Generally, analyses of stability in the layer boundaries (system planes) of a liner system are performed using the shear parameters at failure obtained from direct shear tests compliant with GDA E3-8. However, the values of the shear parameters depend on the displacement. The displacement necessary to activate the maximum shear resistances shall be compatible with the system (cf. GDA E3-8). The geosynthetic/ground adhesion or the geosynthetic/geosynthetic adhesion may be adopted in certain cases (cf. GDA E3-8) when determining the force deficit to be transferred by a geogrid.

8.3.2   Analysing the Stability of the Inclined Liner System

GDA E2-7 requires numerical analysis to be carried out for all relevant load combinations in the construction and final states for liner systems on slopes. If the allowable utilisation factor is exceeded and there are no plans to adequately level off the slope angle, a reinforcement layer may be used. Its bearing capacity is adopted in the analysis. The stability analysis is then based on the limit equilibrium in the most unfavourable slip plane (Figure 8.2 a):

$$R_{t,d} + R_{B,d} - E_d \geq 0 \qquad \text{Eq. (8.2)}$$

where:

$R_{t,d}$ design value of the friction resistance,
E_d design value of the actions,
$R_{B,d}$ design resistance of the reinforcement.

Note: To determine the required tensile strength of the reinforcement all actions over the entire slope length are usually cumulated from the slope toe upwards, compared to the resistances, and the required reinforcement design strength calculated from this.

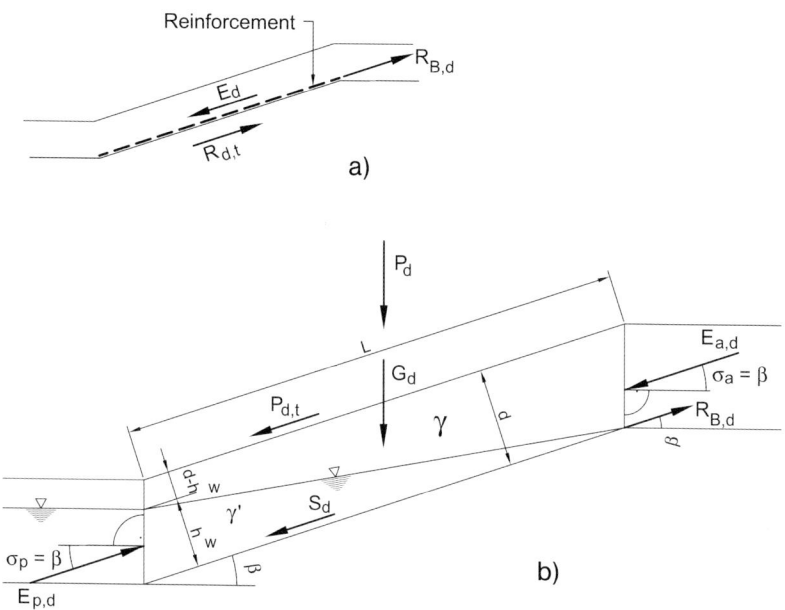

Figure 8.2 Section through the reinforced drainage layer in the slope

When using this equilibrium method live loads – e.g. from construction equipment traversing the landfill slope in the construction state load case, but also traversing berms during servicing and inspection work in the final state after the capping liner is complete – are regarded as local force actions. Snow loads are also taken into consideration in the final state. The most unfavourable load case combination governs reinforcement design.

The force deficit to be transferred can be determined on the basis of the equations described below. See Section 8.3.2.1 c) for details of adopting braking and accelerating forces from live loads.

The most unfavourable slip plane is determined using comparative analyses. For example, it may lie in the layer boundary between a geosynthetic liner and an aggregate sealing layer, in particular if the sealing layer is saturated and the undrained shear strength of the soil is low. It may lie between a protective geotextile layer or a geosynthetic drainage system and a geosynthetic liner, if only minor friction forces are activated between the two layers.

The actions are determined using the design values of the permanent and variable loads (using the partial safety factors in DIN 1054, Table 2).

The resistances are given by the design values of the shear parameters to DIN 1054, Table 3:

$$\gamma_\phi = \gamma_\delta \qquad \qquad \text{Eq. (8.3)}$$

$$\gamma_c = \gamma_a \qquad \qquad \text{Eq. (8.4)}$$

where:

γ_a partial safety factor for the adhesion of the ground,
γ_δ partial safety factor for the contact friction angle.

The actions E_d and the resistances $R_{t,d}$ are analysed as described below. All analyses are in terms of the entire slope length and a unit width of 1 m.

8.3.2.1   Actions and Effects

Actions (Figure 8.2 b) include the design values of the weight component G_d parallel to the slope of the layers above the governing slip plane, the component of any lateral load P_d parallel to the slope, braking and accelerating forces P_{dt}, seepage force S_d and the active earth pressure $E_{a,d}$ on the sliding body. The design value of the actions E_d is then calculated as follows:

$$E_d = G_d \cdot \sin\beta + P_d \cdot \sin\beta + P_{dt} + S_d + E_{a,d} \qquad \qquad \text{Eq. (8.5)}$$

where:

G_d design value of the weight,
S_d design value of the seepage,
P_{dt} design value of the braking and acceleration forces,
P_d design value of the vertical loads,
β slope angle,
$E_{a,d}$ design value of the active earth pressure.

a) Design value of the weight G_d

Where slopes are partially percolated by water, buoyancy shall be taken into consideration to calculate the characteristic value of the weight G_k of the soil layer above the liner. In addition, the weight of the water above the liner shall be taken into consideration for slip planes below the liner. For the phreatic line shown in Figure 8.2 b and a slip plane above the liner:

$$G_d = \gamma_G \cdot [\gamma_k \cdot (d - 1/2 \cdot h_w) + \gamma_k' \cdot 1/2 \cdot h_w] \cdot L \qquad \qquad \text{Eq. (8.6)}$$

where:

h_w thickness of layer below phreatic line,
L   slope length.

b) Design value of vertical force P_k from live loads

The vertical force P_k acting in the slip plane on a unit width of 1 m is:

$$P_d = \gamma_Q \cdot G_R / b_i \qquad \qquad \text{Eq. (8.7)}$$

where:

G_R dead weight of construction equipment [kN],
b_i imaginary width of the contact zone, relative to the slip plane and taking
a load distribution angle $\alpha = 30°$ to the vertical into consideration.

$$b_i = 2 \cdot (b_R + 2 \cdot d_i \cdot \tan \alpha) \qquad \text{Eq. (8.8)}$$

where:

b_R track width,
d_i thickness of the trafficked fill layer above the slip plane
(thickness of emplaced layer).

c) Design value of the braking and accelerating forces P_{dt} from construction equipment for the construction state

DIN 1054, Section 6.1.4 states that dynamic actions from site operations are generally covered by static equivalent loads. We recommend calculating the actions from braking and accelerating forces using a dynamic coefficient Φ as follows:

$$P_{dt} = \gamma_Q \cdot P_k \cdot \sin \beta \cdot (\Phi - 1) \qquad \text{Eq. (8.9)}$$

where:

P_{dt} design value of the braking and accelerating forces from construction
equipment for the construction state,
P_k vertical load,
Φ dynamic coefficient $\Phi = 1.4 - 0.1 \cdot d_i$.

Using this approach the effective live load is taken into consideration by the dynamic coefficient on one side and increased using the partial safety factor from DIN 1054 on the other.

Building procedures are specified in work instructions and monitored to ensure that adequate stability is given at all times. In terms of the selection of suitable construction equipment and procedures critical construction conditions are examined by establishing in-situ test sites.

Note: In terms of the analysis of horizontal actions from braking and accelerating forces, [6] requires that the prevalent shear forces be calculated taking the true service weight and the actual braking delay – determined as the quotient of travelling or shear velocity and braking duration – into consideration. Theoretically, this procedure promises greater accuracy than the simplified and approximate dynamic coefficient method introduced in EBGEO in 1997. However, its implementation is difficult, because the necessary equipment data is either unknown (in the planning and invitation to tender stages) or the corresponding specifications, e.g. on travel speeds or braking durations, either are not or cannot be monitored during construction (implementation phase).

Because no incidents which can be traced back to poor dimensioning of braking forces have been documented since publication of the EBGEO 1997, the Working Group has decided to retain this simplified analysis method.

Proposals for detailed analysis, in particular of local load concentration in the construction state, are given in [6], for example, or can be approximated on the basis of DIN 4084 using planar slip surfaces.

d) Design value of seepage S_d

For the phreatic line shown in Figure 8.2 b:

$$S_d = \gamma_G \cdot 1/2 \cdot \gamma_w \cdot i \cdot h_w \cdot L \qquad \qquad \text{Eq. (8.10)}$$

where:

i   hydraulic gradient ($i = \sin \beta$ on slopes with the angle β and phreatic line parallel to the slope).

Note: DIN 1054, Tab. 2 classifies the seepage force as a permanent action. The ponding height h_w in the drainage layer can be calculated using the GDA E2-20 specifications, for example. In practice, half of the thickness of the drainage layer is often adopted as a constant ponding height in aggregate drainage layers.

e) Design value of earth pressure $E_{a,d}$

Earth pressure at the top of the slope has little impact on long slopes with only thin layers and can therefore generally be ignored. If it is necessary to adopt the earth pressure the requirements of DIN 4085 apply.

8.3.2.2 Resistances

The friction resistance (Figure 8.2 b) is given by the ground reaction forces resulting from dead weight and live loads and the passive earth pressure on the sliding body. The weight and live loads are calculated as characteristic values in analogy to Section 8.3.2.1.

$$R_{t,d} = [(G_k + P_k) \cdot \cos\beta \cdot (\tan\delta_k)/\gamma_\delta + a_k/\gamma_a \cdot L] + E_{p,d} \qquad \text{Eq. (8.11)}$$

where:

a_k characteristic value of the adhesion between the ground and the geosynthetics, and between the geosynthetics and geosynthetics.

The passive earth pressure at the slope toe has only a minor impact on long slopes with only thin layers. It is therefore generally ignored. If it is adopted, the usual rules of DIN 4085 apply. Buoyancy shall be considered where necessary.

138

Different governing failure mechanisms for the support force at the toe of the slope or the berm must also be investigated, e.g. failure wedge shear at the lower end of the slope.

Note: *Adopting P_k assumes that the vertical load acts as an effective stress and is not transferred via excess porewater pressures. It can be assumed that this is predominantly correct for the material demands and installation notes for the mineral layers in capping systems discussed in the Landfill Regulations. Critical excess porewater pressures can only occur in liner layers comprising clay minerals if they are installed on the wet branch of the proctor curve and with low air void levels.*

The braking and accelerating forces are not adopted on the resistance side of the stability analysis.

The following values are adopted for the characteristic values of the shear parameters δ_k or φ_k and a_k or c_k, depending on the location of the investigated slip plane:

- for the contact zone between geosynthetic layers (see GDA E3-8):
  · friction angle δ_k (e.g. protective layer/geosynthetic liner $\delta_{gg,k}$ or ground/ geosynthetics $\delta_{sg,k}$),
  · adhesion a_k (only adopted in exceptional cases, *cf.* GDA E3-8).

- for soil layers in drained conditions:
  · friction angle of the drained soil φ'_k,
  · cohesion of the drained soil c'_k.

- for saturated, cohesive soil layers in undrained conditions:
  · friction angle of the undrained soil $\varphi_{u,k} = 0$,
  · cohesion of the undrained soil $c_{u,k}$.

8.3.3 Structural Resistance of Reinforcement

The following minimum load combinations shall be investigated for designing the tensile strength of the reinforcement layers in capping systems:

- for the construction state:
  liner system complete, up to and including the drainage layer,
  construction equipment loads,
  seepage force due to heavy rain event.

- for the final state:
  liner system complete, additional live load (e.g. snow load) and seepage force. When determining the phreatic line the storage and retention capacities of the soil strata above the drainage layers can be taken into consideration (also see GDA E2-20).

8.3.4 Anchorage

The standard cases with the reinforcement anchored on the top of the slope envisage the use of a trench or the reinforcement rolled out with appropriate cover fill. The analyses are performed for the GEO limit state to DIN 1054.

a) Safety against failure of the anchor trench

Failure of the anchor trench via the friction resistance along the nth section of the trench is analysed for adequate load transfer through the reinforcement (Figure 8.3). The limit state equation is:

$$R_{t,d} - R_{B,d} - E_{a,d} \geq 0 \qquad \text{Eq. (8.12)}$$

where:

$$R_{t,d} = \Sigma\, R_{ti,d} = \Sigma\, [(G_{i,d} \cdot \cos\beta_i \cdot \tan\delta_{i,d} + a_{i,d}) \cdot L_i]$$
$$+ [(G_{i,d} \cdot \cos\beta_i \cdot \lambda \cdot \tan\varphi'_{i,d} + c'_{i,d}) \cdot L_i]. \qquad \text{Eq. (8.13)}$$

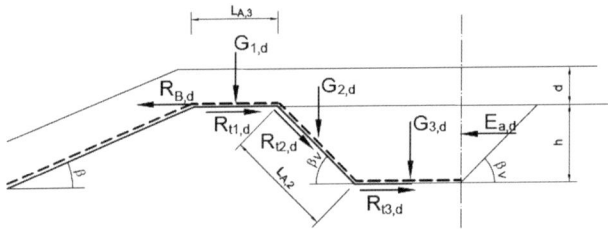

Figure 8.3 Analysis of failure of the anchor trench

Note: Thrust forces are ignored for this analysis. The thrust forces may lead to the soil in the anchor trench being lifted.

b) Analysing failure of the top of the slope

In addition, analysis of failure of the top of the slope along a potential shear plane as shown in Figure 8.4 shall be carried out. This is governed by the *failure on internal slip planes* GEO limit state. It is described by:

$$R_{t,d} - R_{B,d} \cdot \cos\beta - E_{a,d} \geq 0 \qquad \text{Eq. (8.14)}$$

where:

$$R_{t,d} = R_{t1,d} + R_{t2,d} = [(G_{1,d} \cdot (\tan\delta_{1,k}) / \gamma_\delta + a_{1,k} / \gamma_a) \cdot L_1]$$
$$+ [(G_{2,d} \cdot (\tan\varphi'_k) / \gamma_\varphi + c'_k / \gamma_c \cdot L_2]. \qquad \text{Eq. (8.15)}$$

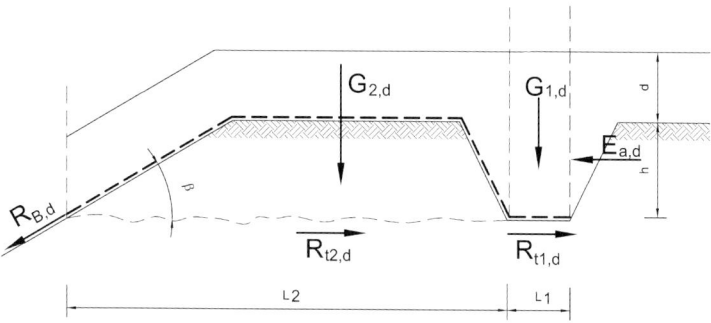

Figure 8.4 Analysis of failure of the slope top

8.4   Bibliography

[1]   TA Abfall, Zweite Allgemeine Verwaltungsvorschrift zum Abfallgesetz, Heymanns-Verlag, 1991.

[2]   TA Siedlungsabfall, Dritte Allgemeine Verwaltungsvorschrift zum Abfallgesetz, Bundesanzeiger-Verlag, 1993.

[3]   Recommendations of the Working Group 'Geotechnical Aspects of Landfill and Brownfield Sites', GDA, 2nd edition, Ernst & Sohn, 1993.

[4]   Verordnung zur Vereinfachung des Deponierechts, Verordnung über Deponien und Langzeitlager (Deponieverordnung DepV), 27.04.2009, Federal Law Gazette 2009, Part I No. 22.

[5]   Abfallablagerungsverordnung, Verordnung über die umweltverträgliche Ablagerung von Siedlungsabfällen, Federal Law Gazette I 2001, 305.

[6]   Saathoff, F., Werth, K. (2005): Standsicherheitsnachweise für Oberflächenabdichtungssysteme – Anmerkungen zum Lastfall Einbau geschichteter Systeme mit Geokunststoffen. 21. SKZ-Tagung 'Die sichere Deponie', Süddeutsches Kunststoffzentrum, Würzburg, Eigenverlag.

[7]   Syllwasschy, O., Sobolewski, J., Brokemper, D., Alexiew, N. (2005): Beispiele für effiziente Oberflächenabdichtungen anhand der Deponien Koppelwald, Dillinger Hütte und Redlham: Aufbau, Statik, Verlegepläne, Bauausführung. Symposium Umweltgeotechnik des Ak 6.1 der DGGT e. V., TU Freiberg.

[8]   GDA E2-7 (2008): Nachweis der Gleitsicherheit von Abdichtungssystemen. Empfehlungen des Arbeitskreises 'Geotechnik der Deponiebauwerke' des Ak 6.1, Deutsche Gesellschaft für Geotechnik, Eds. Witt, K.-J. and Ramke, H.-G., Bautechnik, 85. Jahrgang, Ausgabe 9, Ernst & Sohn, 2008.

8.5 Example of Landfill Capping using Geosynthetic Reinforcement

Geosynthetic reinforcements are required in landfill capping systems using flexible membranes. Analyses are performed as described in Section 8.3.

8.5.1 Geometry, Soil Mechanics Parameters, Geosynthetic Properties and Data for a Selected Construction Vehicle

8.5.1.1 Geometry of the Liner System in the Slope

Length of slope: $\qquad$ $L = 30$ m

Slope angle (1 : n = 1 : 2): $\qquad$ $\beta = 26,6°$

Thickness of recultivation layer: $\qquad$ $d_1 = 1.0$ m

Thickness of drainage layer: $\qquad$ $d_2 = 0.3$ m

8.5.1.2 Characteristic Soil Mechanics Input Values

Drainage layer:

Gravel (16/32)   Angle of friction: $\varphi'_k = 32.5°$
                 Unit weight: $\gamma_k / \gamma'_k = 20/10$ kN/m^3
                 Cohesion: $c'_k = 0$ kN/m^2

Recultivation soil:

UL   Angle of friction: $\varphi'_k = 27.5°$
               Unit weight: $\gamma_k / \gamma'_k = 19/10$ kN/m^3
               Cohesion: $c'_k = 2.5$ kN/m^2

8.5.1.3 Geosynthetics

Types:

- flexible membranes structured on both sides,
- needle punched nonwoven as protective geotextile layer for flexible membrane,
- geogrid as reinforcement.

Friction properties [8]:

Flexible membrane/protective geotextile layer boundary from friction tests (governing):

Angle of friction: $\delta_k = 27°$

Adhesion: $a_k = 0$ kN/m^2

8.5.1.4 Data for a Selected Tracked Vehicle

Bulldozer, service weight: $G_R = 128$ kN on 2 tracks

Track width: $b_R = 0.6$ m per track

Track spacing: $e = 1.8$ m

Track length: $l_R = 2.6$ m per track

Dynamic coefficient based on DIN 1072: $\Phi = 1.4 - 0.1\,d_2 = 1.37$

Load distribution angle below tracks: $\alpha = 30°$

8.5.1.5 Construction State Definition

The system is designed without the recultivation layer, i.e. installation of drainage layer and trafficking.

8.5.2 Stability Analysis

Section 8.3.2 explains that adequate resistance against shear loading parallel to the layering is given for the following condition (also see Figure 8.5):

$$R_{t,d} + R_{B,d} - E_d \geq 0 \qquad \text{Eq. (8.16)}$$

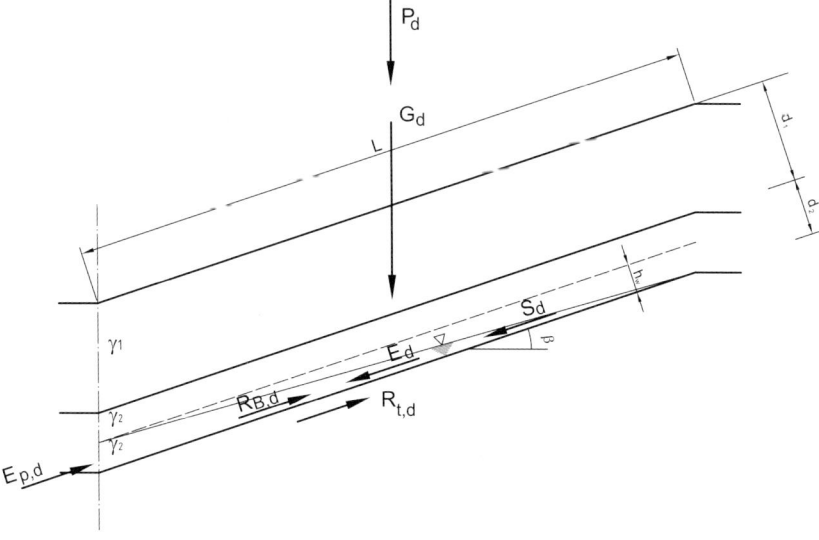

Figure 8.5 Actions and resistances in a parallel system

a) Design value of the actions E_d:

$$E_d = \gamma_G \cdot G_k \cdot \sin\beta + \gamma_Q \cdot P_k \cdot \sin\beta + P_{dt} \cdot \sin\beta + S_d + E_{a,d} \qquad \text{Eq. (8.17)}$$

Partial safety factors, see DIN 1054 for the GEO limit case, Table 2:

$\gamma_G = 1.0$ (final state) $\gamma_G = 1.0$ (construction state)
$\gamma_Q = 1.3$ (final state) $\gamma_Q = 1.2$ (construction state)

b) Design value of the friction resistance $R_{t,d}$:

$$R_{t,d} = (G_k + P_k) \cdot \cos\beta \cdot (\tan\delta_k) / \gamma_\delta + a_k / \gamma_a \cdot L + E_{p,d} \qquad \text{Eq. (8.18)}$$

Partial safety factors, see DIN 1054 for the GEO limit case, Table 3:

$\gamma_\varphi, \gamma_c = 1.25$ (final state) $\gamma_\varphi, \gamma_c = 1.15$ (construction state)
$\gamma_\delta, \gamma_a = 1.25$ (final state) $\gamma_\delta, \gamma_a = 1.15$ (construction state)

c) Characteristic value of the weight G_k:

$$G_k = [\gamma_k \cdot (d - 1/2 \cdot h_w) + \gamma_k' \cdot 1/2 \cdot h_w] \cdot L \qquad \text{Eq. (8.19)}$$

Precise determination of h_w compliant with GDA E2-20 using Lesaffre's method is dispensed with in this example. In simplification, $h_w = 1/2 \cdot d_2$ and percolation parallel to the slope are adopted.

$h_w = \frac{1}{2} \cdot 0.3 = 0.15$ m

For the construction state:

$G_k = [20 \cdot (0.3 - \frac{1}{2} \cdot 0.15) + 10 \cdot \frac{1}{2} \cdot 0.15] \cdot 30 = 157.5$ kN/m

For the final state:

$G_k = [20 \cdot (0.3 - \frac{1}{2} \cdot 0.15) + 10 \cdot \frac{1}{2} \cdot 0.15] \cdot 30 + 19 \cdot 1 \cdot 30 = 727.5$ kN/m

d) Characteristic value of the vertical force from live loads P_k (construction state):

$$P_k = G_R / b_i \qquad \text{Eq. (8.20)}$$

where:

$b_i = 2 \cdot (b_R + 2 \cdot d_i \cdot \tan\alpha) = 2 \cdot (0.6 + 2 \cdot 0.3 \cdot \tan 30°) = 1.89$ m
$P_k = 128 / 1.89 = 67.7$ kN/m

e) Design value of the braking and accelerating forces P_{dt} for the construction state:

Simplified analysis is performed using a dynamic coefficient Φ as described in Section 8.3.2.1 c).

$$P_{dt} = \gamma_Q \cdot (\Phi - 1) \cdot P_k \cdot \sin\beta$$
$$= 1.2 \cdot (1.37 - 1) \cdot 67.7 \cdot \sin 26.6° = 13.46 \text{ kN/m}$$

Eq. (8.21)

f) Design value of the seepage S_d:

$$S_d = \gamma_G \cdot 1/2 \cdot \gamma_w \cdot i \cdot h_w \cdot L$$

Eq. (8.22)

For the construction state:

$S_d = 1.0 \cdot \frac{1}{2} \cdot 10 \cdot \sin 26.6 \cdot 0.15 \cdot 30 = 10.06 \text{ kN/m}$

For the final state:

$S_d = 1.0 \cdot \frac{1}{2} \cdot 10 \cdot \sin 26.6 \cdot 0.15 \cdot 30 = 10.06 \text{ kN/m}$

g) Design value of the earth pressure $E_{a,d}$ at the top of the slope and the passive earth pressure $E_{p,d}$ at the bottom of the slope:

Because the layer is only thin the earth pressure and passive earth pressure do not have a significant impact and are not adopted below in line with Section 8.3.2.

h) Design strength of the geosynthetic reinforcement:

working $R_{B,d} \geq E_d - R_{t,d}$

For the construction state:

$$E_d = \gamma_G \cdot G_k \cdot \sin\beta + \gamma_Q \cdot P_k \cdot \sin\beta + P_{dt} + S_d + E_{a,d}$$
$$= 1.0 \cdot 157.7 \cdot \sin 26.6° + 1.2 \cdot 67.7 \cdot \sin 26.6° + 13.46 + 10.06 + 0$$
$$= 130.5 \text{ kN/m}$$

$$R_{t,d} = (G_k + P_k) \cdot \cos\beta \cdot (\tan\delta_k)/\gamma_\delta + a_k/\gamma_a \cdot L + 0$$
$$= (157.5 + 67.7) \cdot \cos 26.6° \cdot (\tan 27°)/1.15 + 0 + 0 = 89.2 \text{ kN/m}$$

work $R_{B,d} \geq 130.5 - 89.2 = 41.3 \text{ kN/m}$

For the final state:

$$E_d = \gamma_G \cdot G_k \cdot \sin\beta + S_d$$
$$= 1.0 \cdot 727.5 \cdot \sin 26.6° + 10.06 = 335.8 \text{ kN/m}$$

$$R_{t,d} = G_k \cdot \cos\beta \cdot (\tan\delta_k)/\gamma_\delta + a_k/\gamma_a \cdot 1 + 0$$
$$= 727.5 \cdot \cos 26.6° \cdot (\tan 27°)/1.25 + 0 + 0 = 265.2 \text{ kN/m}$$

working $R_{B,d} \geq 335.8 - 265.2$
$$\geq 70.6 \text{ kN/m} \textbf{ (governing)}$$

145

8.5.3 Analysing Reinforcement Failure

In line with the above analyses the governing condition of the cap in the final state is given by:

max. $R_{B,d}$ = 70.6 kN/m.

A polyester geogrid was used with a characteristic short-term strength of:

$R_{B,k}$ = 200 kN/m.

Design resistance of the reinforcement:

$R_{B,d} = R_{B,k} / (A_1 \cdot A_2 \cdot A_3 \cdot A_4 \cdot \gamma_M \cdot \eta_M)$

where:

A_1 reduction factor for the creep rupture strength (polymer creep),
A_2 reduction factor for transport, installation and compaction damage,
A_3 reduction factor for processing (joins, connections to structural elements, etc.),
A_4 reduction factor for environmental impacts (weather resistance, resistance against chemicals, microorganisms and animals),
γ_m structural resistance partial safety factor (here: LC1: 1.40),
η_m correction factor for the structural resistance as described in Section 8.3.1 (here: LC1: 1.1).

It is assumed that the reduction factors used in this example were verified by the manufacturers.

Design resistance of the reinforcement:

$R_{B,d}$ = 200 / (1.50 · 1.1 · 1.0 · 1.0 · 1.4 · 1.1) = 78.71 kN/m
    = 78.71 kN/m > working $R_{B,d}$ = 70.6 kN/m

8.5.4 Designing the Anchor Trench

8.5.4.1 Anchor Trench Geometry

Width of embankment top: L_1 = 2.0 m

Anchor trench slope angle: β_v = 26.6°

Selected depth of anchor trench: h_v = 0.5 m

Length of trench bottom: L_3 = 1.0 m

Anchor trench embankment length: l_v = L_2 = 1.10 m

Geogrid anchoring length: l_g = L_4 = 1.10 m

Slope angle (1 : n = 1 : 2): β = 26.6°

146

8.5.4.2 Friction Resistance Input Values

Drainage layer:

Gravel (16/32) Angle of friction: $\varphi'_k = 32.5°$
Unit weight: $\gamma_k = 20 \text{ kN/m}^3$
Cohesion: $c'_k = 0 \text{ kN/m}^2$

Recultivation soil:

UL Angle of friction: $\varphi'_k = 27.5°$
Unit weight: $\gamma_k = 19 \text{ kN/m}^3$
Cohesion: $c'_k = 2.5 \text{ kN/m}^2$

Anchor trench backfill:

SU Angle of friction: $\varphi'_k = 30°$
Unit weight: $\gamma_k = 19 \text{ kN/m}^3$

Ground:

SU Angle of friction: $\varphi'_k = 28°$
Unit weight: $\gamma_k = 19 \text{ kN/m}^3$
Cohesion: $c'_k = 2.5 \text{ kN/m}^2$

Liner system contact friction properties (governing):

Angle of friction: $\delta_k = 27°$
Adhesion: $a_k = 0 \text{ kN/m}^2$
Composite coefficient
between ground and geogrid: $\lambda = 0.9$

8.5.4.3 Safety Against Failure of the Anchor Trench

Boundary conditions:

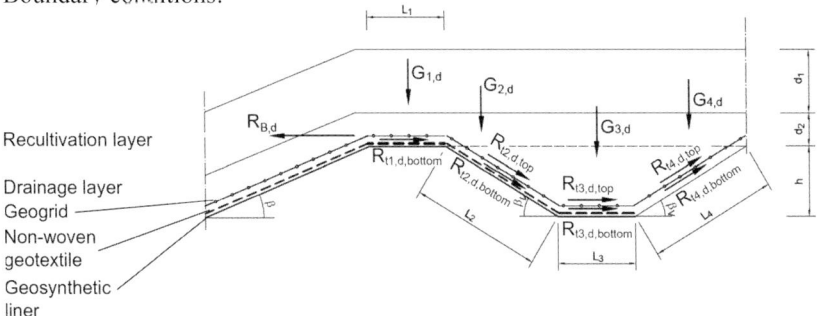

Figure 8.6 Safety against failure of the anchor trench

Limit state equation:

$$R_{t,d} - R_{B,d} - E_{a,d} \geq 0 \qquad \qquad \text{Eq. (8.23)}$$

147

where:

$$R_{t,d} = \Sigma\, R_{ti,d} = (\sigma_{vi,d} \cdot \tan \delta_{i,d} + a_{i,d}) \cdot L_i + (\sigma_{vi,d} \cdot \lambda \cdot \tan \varphi'_{i,d} + c'_{i,d}) \cdot L_i$$

$$= [\gamma_{k,i} \cdot d_i \cdot \cos \beta_i \cdot (\tan \delta_{i,k}) / \gamma_\delta + a_{i,k} / \gamma_a] \cdot L_i \qquad \text{Eq. (8.24)}$$

$$+ [\lambda_{k,i} \cdot d_i \cdot \cos \beta_i \cdot \lambda \cdot (\tan \varphi'_{i,k}) / \gamma_\varphi + c'_{i,k} / \gamma_c] \cdot L_i.$$

Note: To simplify the example cohesion and adhesion are not adopted below.

The final state is the governing state for the analysis. The active earth pressure $E_{a,d}$ is not taken into consideration.

Determining the design value of friction resistance:

$$R_{t,d} = \Sigma\, R_{ti,d} \qquad \text{Eq. (8.25)}$$

Friction resistance along L1:

$R_{t1,d,bottom}$ $= [(20 \cdot 0.3 + 19 \cdot 1.0) \cdot \tan 27° / 1.25 + 0] \cdot 2.0 = 20.38$ kN/m

Note: A resistance is only adopted on the bottom of the geogrid.

Friction resistance along L2:

$R_{t2,d,bottom}$ $= [(20 \cdot 0.3 + 19 \cdot 1.0 + 19 \cdot 0.5 \cdot 0.5)$
$\cdot \cos 26.6° \cdot \tan 27° / 1.25 + 0] \cdot 1.1$
$= 11.93$ kN/m

$R_{t2,d,top}$ $= [(20 \cdot 0.3 + 19 \cdot 1.0 + 19 \cdot 0.5 \cdot 0.5)$
$\cdot \cos 26.6° \cdot 0.9 \cdot \tan 30° / 1.25 + 0] \cdot 1.1$
$= 12.16$ kN/m

Friction resistance along L3:

$R_{t3,d,bottom}$ $= [(20 \cdot 0.3 + 19 \cdot 1.0 + 19 \cdot 0.5) \cdot (\tan 27°) / 1.25 + 0] \cdot 1.0$
$= 14.06$ kN/m

$R_{t3,d,top}$ $= [(20 \cdot 0.3 + 19 \cdot 1.0 + 19 \cdot 0.5) \cdot 0.9 \cdot (\tan 30°) / 1.25 + 0] \cdot 1.0$
$= 14.34$ kN/m

Friction resistance along L4:

$R_{t4,d,bottom}$ $= [(20 \cdot 0.3 + 19 \cdot 1.0 + 19 \cdot 0.5 \cdot 0.5) \cdot 0.9 \cdot (\tan 28°) / 1.25 + 0]$
$\cdot 1.10 \cdot \cos 26.6°$
$= 11.20$ kN/m

$R_{t4,d,top}$ $= [(20 \cdot 0.3 + 19 \cdot 1.0 + 19 \cdot 0.5 \cdot 0.5) \cdot 0.9 \cdot (\tan 30°) / 1.25 + 0]$
$\cdot 1.10 \cdot \cos 26.6°$
$= 12.16$ kN/m

Design value of the friction resistance:

$\Sigma\, R_{ti,d}$ $= 20.38 + 11.93 + 12.16 + 14.06 + 14.34 + 11.20 + 12.16$
$= 96.23$ kN/m > 70.6 kN/m

8.5.4.4 Analysing Failure of the Top of the Embankment

Boundary conditions:

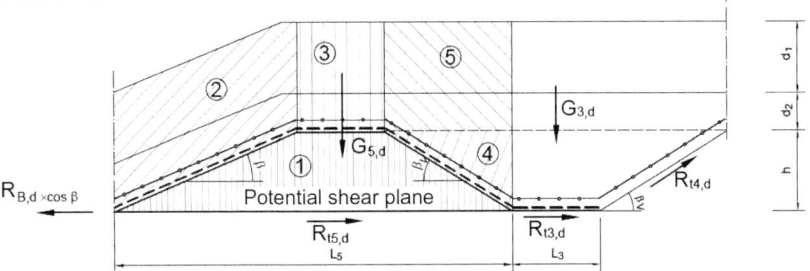

Figure 8.7 Safety against failure of the embankment top

Limit state equation:

$$R_{t,d} - R_{B,d} \cdot \cos\beta - E_{a,d} \geq 0 \qquad\qquad \text{Eq. (8.26)}$$

where:

$$R_{t,d} = \Sigma\, R_{ti,d} = (\sigma_{vi,d} \cdot \tan\varphi'_{i,d} + c'_{i,d}) \cdot L_i + (\sigma_{vi,d} \cdot \tan\delta_{i,d} + a_{i,d}) \cdot L_i \quad \text{Eq. (8.27)}$$

The final state is the governing state for the analysis. The length of the potential shear plane along L_5 and L_3 is defined.

Determining the design value of the resisting force:

$R_{t3,d,bottom}$ $= [(20 \cdot 0.3 + 19 \cdot 1.0 + 19 \cdot 0.5) \cdot \tan 27° / 1.25 + 0] \cdot 1.0$
$= 14.06$ kN/m

$R_{t5,d}$ $= [19 \cdot 0.5 \cdot 0.5 \cdot 1.0 + 19 \cdot 0.5 \cdot 2.0 + 19 \cdot 0.5 \cdot 0.5 \cdot 1.0$
$+ (20 \cdot 0.3 + 19 \cdot 1.0) \cdot 1.10$
$+ (20 \cdot 0.3 + 19 \cdot 1.0) \cdot 2.0 + 19 \cdot 0.5 \cdot 1.0 \cdot 0.5$
$+ (20 \cdot 0.3 + 19 \cdot 1.0) \cdot 1.0] \cdot \tan 28° / 1.25$
$= 57.7$ kN/m

Design value of the resisting force against failure of the anchor trench:

$\Sigma\, R_{ti,d} = 57.7 + 14.06 = 71.76 > 63.13 = R_{B,d} \cdot \cos\beta$

9 Reinforced Earth Structures over Point or Linear Bearing Elements

9.1 Definitions

A reinforced earth structure over point or linear, vertical bearing elements refers to a single- or multi-layer composite structure made of earth and geosynthetics, which rests on natural, soft ground and the bearing elements. The point or linear bearing elements are referred to below as bearing elements and the reinforced earth structure together with the bearing elements is referred to as the overall system or bearing structure.

The bearing elements are embedded in the natural ground down to deeper, stable soil strata and form a rigid, point or linear support for the reinforced earth structure relative to the soft ground. The terminology used is given in the embankment foundation example in Figure 9.1.

Reinforced earth structure is a single- or multi-layer, reinforced, composite structure made of earth and geosynthetics. It bridges the soft soil between bearing elements.

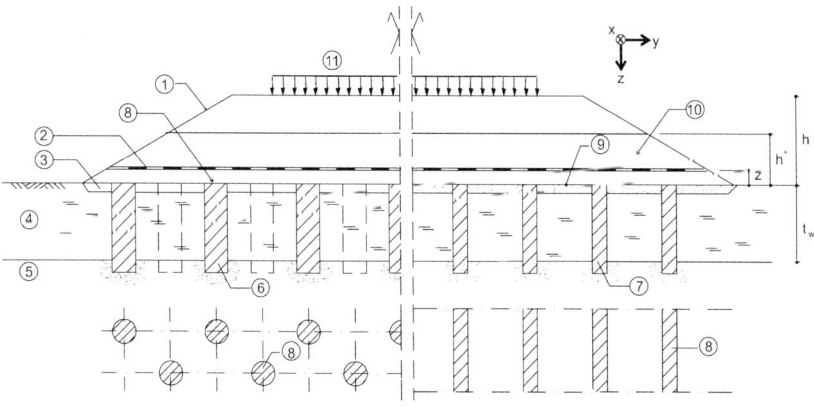

①	Reinforced earth structure	⑦	Linear bearing elements
②	Reinforcement plane	⑧	Support area of bearing elements
③	Working subgrade	⑨	Contact plane of reinforced earth structure
④	Soft ground	⑩	Region with granular soil
⑤	Deeper, stable soil strata	⑪	Surcharge
⑥	Point bearing elements		

Figure 9.1 Reinforced earth structure above point or linear bearing elements. Embankment foundation example

Recommendations for Design and Analysis of Earth Structures using Geosynthetic Reinforcements (EBGEO). German Geotechnical Society.
© 2011 Ernst & Sohn GmbH & Co. KG.
Published by Ernst & Sohn GmbH & Co. KG.

Point bearing elements are elements with predominantly round or square cross-sections, which may be arranged in a regular rectangular or triangular grid (square rotated through 45°) as shown in Figure 9.4.

Linear bearing elements are predominantly slab-like elements arranged in parallel (Figure 9.4).

Height h of the reinforced earth structure is the distance of the reinforced earth structure measured from the contact plane to the top of the reinforced earth structure as shown in Figure 9.1. It is assumed that the supporting surfaces of the bearing elements are approximately at the same elevation as the contact plane.

Height h*: Region in the reinforced earth structure in which granular soil is installed to DIN 1054 as shown in Figure 9.1.

Reinforcement plane: Plane as shown in Figure 9.1; for two-ply reinforcement the geometrical centre plane of the two layers (Figure 9.2).

Elevation z of the reinforcement plane is the vertical distance of the reinforcement plane from the contact plane of the reinforced earth structure as shown in Figure 9.1. For two-ply reinforcement the reinforcement plane lies central between the two layers (Figure 9.2).

Zone of influence A_E of the bearing elements is that part of the total area in the contact plane of the reinforced earth structure assigned to a point or linear bearing element as shown in Figure 9.3 and Figure 9.4.

Support area A_S is the surface area of a point or linear bearing element in the contact plane of the reinforced earth structure as shown in Figure 9.4. Typical support areas in the contact plane of the reinforced earth structure are shown in Figure 9.3.

Diameter d is the diameter of the support surface A_S of round point or linear bearing elements or their *caps* in the contact plane of the reinforced earth struc-

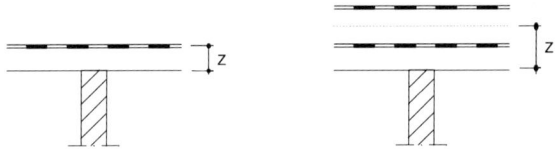

Figure 9.2 Location of reinforcement plane for one- and two-ply reinforcement

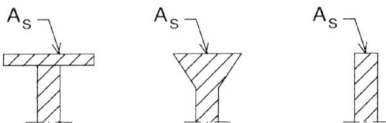

Figure 9.3 Typical support surfaces of point or linear bearing elements in elevation

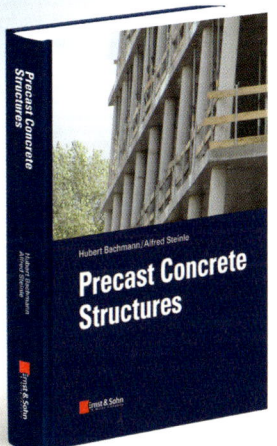

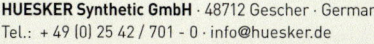

Das Kompendium der Geotechnik Karl Josef Witt (Hrsg.)

Teil 1: Geotechnische Grundlagen

7., überarb. u. aktualis. Auflage
2008. 838 S., 567 Abb., 86 Tab., Gb.
€ 179,– / sFr 283,–
ISBN: 978-3-433-01843-9

Inhalt:
– Sicherheitsnachweise im Erd- und Grundbau (Martin Ziegler)
– Baugrunderkundung im Feld (Klaus-Jürgen Melzer, Ulf Bergdahl, Edwin Fecker)
– Eigenschaften von Boden und Fels - ihre Ermittlung im Labor (Paul von Soos, Jens Engel)
– Charakterisierung von Schadstoffen im Baugrund und Grundwasser (Andreas Claussen)
– Stoffgesetze für Böden (Dimitrios Kolymbas, Ivo Herle)
– Erddruck (Achim Hettler)
– Stoffgesetze und Bemessungsverfahren für Festgestein (Erich Pimentel)
– Bodendynamik (Christos Vrettos)
– Numerische Verfahren der Geotechnik (Peter-Andreas von Wolffersdorff, Helmut Schweiger)
– Geodätische Überwachung von geotechnischen Bauwerken (Wilfried Schwarz, Klaus Linkwitz, Otto Heunecke)
– Geotechnische Messverfahren (Arno Thut)
– Massenbewegungen (Dieter Genske)

Teil 2: Geotechnische Verfahren

7., überarb. u. aktualis. Auflage
2009. 940 S., 500 Abb., Gb.
€ 179,– / sFr 283,–
ISBN: 978-3-433-01845-3

Inhalt:
– Erdbau (H.-H. Schmidt, T. Rumpelt)
– Baugrundverbesserung (W. Sondermann)
– Injektionen (St. Semprich, W. Hornich, G. Stadler)
– Unterfangungen und Nachgründungen (K. J. Witt)
– Bodenvereisung (W. Orth)
– Verpressanker (L. Wichter)
– Bohrtechnik (G. Ulrich, G. Ulrich)
– Horizontalbohrungen und Rohrvortrieb (H. Schad, T. Bräutigam, H.-J. Bayer)
– Rammen, Ziehen, Pressen, Rütteln (W. Paul, F. Berner)
– Grundwasserströmung - Grundwasserhaltung (B. Odenwald, U. Hekel, H. Hölscher, Henning Thormann)
– Abdichtungen und Fugen im Tiefbau (A. Haack)
– Geokunststoffe in Erd- und Grundbau (F. Saathoff, G. Bräu)
– Ingenieurbiologische Verfahren zur Böschungssicherung (E. Hacker, R. Johannsen)

Teil 3: Gründungen und geotechnische Bauwerke

7., überarb. u. aktualis. Auflage
2009. 940 S., 500 Abb., Gb.
€ 179,– / sFr 283,–
ISBN: 978-3-433-01846-0

Inhalt:
– Flachgründungen (U. Smoltczyk, N. Vogt)
– Pfahlgründungen (H.-G. Kempfert)
– Spundwände (W. Richwien, H.-U. Kalle, K.-H. Lambertz, Karl Morgen, H.-W. Vollstedt)
– Gründungen im offenen Wasser (J. G. de Gijt, Kerstin Lesny)
– Baugrubensicherung (A. Hettler, Anton Weißenbach)
– Pfahlwände, Schlitzwände, Dichtwände (M. Pulsfort, H.-G. Haugwitz)
– Gründung in Bergbaugebieten (D. Placzek)
– Erschütterungsschutz (Ch. Vrettos)
– Stützbauwerke und konstruktive Hangsicherungen (H. Brandl)

**Set-Preis
Grundbau-Taschenbuch
7. Auflage, Teile 1–3
zum Sonderpreis:
€ 483,–* / sFr 763,–**

ISBN: 978-3-433-01847-7

Ernst & Sohn
Verlag für Architektur und technische Wissenschaften GmbH & Co. KG

A Wiley Company
www.ernst-und-sohn.de

Für Bestellungen und Kundenservice:
Verlag Wiley-VCH
Boschstraße 12
69469 Weinheim
Telefon: +49(0) 6201 / 606-400
Telefax: +49(0) 6201 / 606-184
E-Mail: service@wiley-vch.de

* Der €-Preis gilt ausschließlich für Deutschland
006214106_my Irrtum und Änderungen vorbehalten.

Point bearing elements

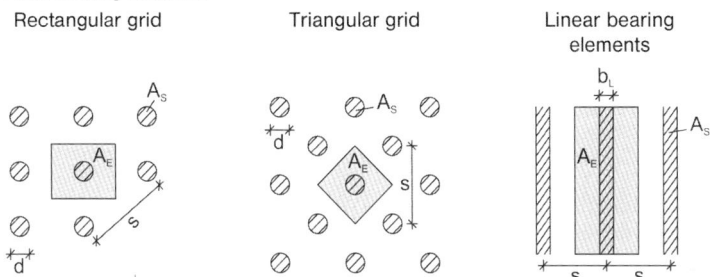

Figure 9.4 Spacing s of bearing elements, zone of influence A_E and support surface A_s (plan)

ture as shown in Figure 9.4. An equivalent diameter d_{Ers} may be derived from the support surface A_s according to Eq. (9.1) for different support surface shapes. Then:

$$d = d_{Ers.} = \sqrt{4 \cdot A_s / \pi}$$

Eq. (9.1)

Width b_L: width of linear bearing elements.

Spacing s of bearing elements: largest axial spacing of neighbouring bearing elements as shown in Figure 9.4.

Rectangular grid: Arrangement of point bearing elements in a rectangular plan along the longitudinal axis of the structure as shown in Figure 9.4.

The geosynthetic is rolled out and installed in the reinforcement plane along the longitudinal or transverse axis of the structure.

Triangular grid: Arrangement of point bearing elements in a square grid rotated by 45° relative to the longitudinal axis of the structure (Figure 9.4). Other triangular grid shapes (e.g. 60°) are not dealt with in these Recommendations.

The geosynthetics are arranged in the reinforcement plane as for the rectangular grid.

9.2 Applications and Modus Operandi

9.2.1 Applications

Geosynthetic-reinforced earth structures on point or linear bearing elements are suitable as systems for transmitting static and variable loads on soft soils to adequately load-bearing, deeper strata. Known applications include traffic embankments, reinforced soil replacement applications or tank foundations, for example. Recommendations for structural analysis, design, stability analysis and the execution of such systems are given below.

153

The ratio of the subgrade reaction moduli between the bearing element $k_{s,T}$ and the ground k_s in the contact plane of the reinforced earth structure should be greater than 75 if the analysis method described in Section 9.6 is adopted:

$$k_{s,T} / k_s > 75 \qquad\qquad \text{Eq. (9.2)}$$

The modulus of subgrade reaction of the ground k_s in the contact plane of the reinforced earth structure is determined as described in Section 9.6.3.5.

The modulus of subgrade reaction $k_{s,T}$ is a variable derived from the stiffness of the bearing elements. It is calculated in accordance with Eq. (9.3) from the characteristic value of the action $F_{s,k}$ as described in Section 9.6.3.3 and the ensuing anticipated settlement s_T of the bearing element in the plane of the support surface A_s:

$$k_{s,T} = \frac{F_{s,k}}{s_T \cdot A_s} \qquad\qquad \text{Eq. (9.3)}$$

The settlement s_T of the bearing element associated with the force $F_{s,k}$ can be determined from bearing element load testing based on DIN 1054 or on the basis of empirical data. Eq. (9.2) is generally fulfilled by all bearing elements stabilised by cements and installed in natural ground.

If poorer stiffness ratios are prevalent than that demanded by Eq. (9.2) the analysis method recommended in Section 9.6 for determining the tensile forces in the reinforcement may be adopted in approximation. The tensile forces thus determined are conservative. The assumptions that this method is based on are met increasingly badly as the stiffness ratio worsens. Refer to Section 10, in particular 10.6.3, Table 10.2, for details of the analysis of such systems.

Attention is drawn to the specific highway and railway engineering regulations. These EBGEO regulations do not apply without restriction or supplementary regulations for *floating foundations* or in cases where dynamic actions in soft soils considerably influence system behaviour. Nor do the regulations cover structures subjected to considerable horizontal forces (e.g. due to asymmetry or high lateral loads).

9.2.2   Modus Operandi

The reinforced earth structure is designed to ensure that the loads are transmitted to the bearing elements by redistributing the loads within the reinforced earth structure and prevent punching effects. The reinforcement *bridges* the soft soil between the bearing elements by membrane action. It is unloaded either partially or almost completely as a result, depending on the stiffness conditions between it, the reinforced earth structure and the bearing elements. In special cases the soft stratum can be entirely unloaded (e.g. groundwater table lowering once the system is complete or lateral excavation and loss of support). Active influence by the ground in the contact plane is then no longer given.

In embankments the reinforced earth structure can also accept spreading forces. The reinforcing effect of the geosynthetics in a reinforced earth structure can only unfold if the reinforcement is correctly anchored. In addition, the modus operandi of the system in terms of embankments requires that stability of the embankment slope is analysed.

From a structural perspective the modus operandi of the system in terms of load redistribution to the bearing elements can be modelled in a number of ways. Investigations using models dealt with in Section 9.6 can be found in [1], [2], [3], [4], [5], [6], [7], [8], [9], [10], [12] and [13].

The reinforcement is subjected to a load as a result of the vertical surcharge between the bearing elements, which is reduced as a result of arching, and unloaded due to the reaction force of the ground below the reinforcement. There is a relationship between the reinforcement sag and the reaction pressure. It is dependent on the relationship between the axial stiffness of the reinforcement and the rigidity of the ground. The ground reaction pressure increases with increasing reinforcement sag and the effect on the reinforcement decreases. Numerically, the effect on the reinforced earth structure can be determined on the basis of the arch model as described in [3]. Section 9.6 is also based on this.

Generally, the effectiveness of the system and unloading of the soft strata between the bearing elements increase with (Figure 9.1 to Figure 9.4):

– decreasing axial spacing s of the bearing elements,
– increasing height h of the earth structure,
– decreasing distance z of the reinforcement plane,
– increasing ratio d/s or b_L/s,
– increasing tensile force in the reinforcement, i.e. with increasing axial stiffness (short- and long-term stiffness) and tensile strength,
– increasing shear strength of the earth structure.

9.3   Design and Engineering Recommendations

Based on empiricism, and for practical reasons, a one- or two-ply reinforcement is recommended, where the following engineering options are common:

– for point bearing elements:
  · one- or two-ply biaxial,
  · two-ply orthogonal uniaxial.

– for linear bearing elements:
  · one- or two-ply uniaxial transverse to support surfaces.

If two reinforcement layers are used they are separated by a 15 cm to 30 cm thick layer of soil.

Note: If more than two layers of reinforcement are used the modus operandi of the system described in Section 9.2.2 generally cannot develop as described here. The design and analysis methods given cannot therefore be recommended. Among other things, widely deviating tensile forces in the individual layers of reinforcement shall be anticipated when using more than two layers of reinforcement (see [12] for more details).

Overlapping in biaxial reinforcements is only permitted over point bearing elements. In analogy, overlapping in uniaxial reinforcements are arranged above linear bearing elements. The overlap is at least as large as the equivalent diameter $d_{Ers.}$ or width b_L of the support surface and is analysed as described in Section 3.3.3.

Adequate anchorage of the reinforcement layers in both directions (parallel and perpendicular to the embankment axis) shall be demonstrated. Analysis is performed to Section 3.3.3.3, where only the maximum tensile force resulting from membrane action E_M is adopted. The anchoring length L_A required for analysis of the anchorage is shown in Figure 9.5, where the anchoring length in the perpendicular direction is given from the anchoring plane. Where the outer row of bearing elements lies within the anchoring plane as shown in Figure 9.5 the anchoring length is taken from the outer edge of these bearing elements.

Granular soils to DIN 1054 are used for the area of the reinforced earth structure with the height h^* above the contact plane as shown in Figure 9.1. The height h^* must meet the $h^* \geq (s-d)$ condition. Other defined embankment fill materials may be used above this region.

Recommended geometrical variables:

- $h/(s-d) \geq 0.8$ for predominantly static loads,

- A larger $h/(s-d)$ ratio is recommended for high variable loads. Current data indicates that the negative impact of the variable loads for $h/(s-d) \geq 2.0$ is negligible. More differentiated specifications can be defined in cooperation with the client.

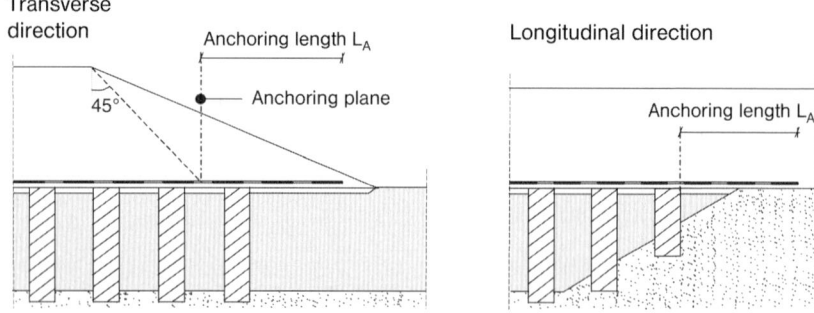

Transverse direction • Anchoring length L_A • 45° • Anchoring plane • Longitudinal direction • Anchoring length L_A

Figure 9.5 Anchoring length of reinforcement layers in longitudinal and transverse directions

156

- $d/s \geq 0.15$,
- $b_L/s \geq 0.15$,
- $z \leq 0.15$ m for one-ply reinforcement,
  $z \leq 0.30$ m for two-ply reinforcement,
- Based on previous experience the clear spacing between the support surfaces A_S of the bearing elements should be limited as follows:
  · $(s - d) \leq 3.0$ m or $(s - b_L) \leq 3.0$ m for predominantly static loads,
  · $(s - d) \leq 2.5$ m or $(s - b_L) \leq 2.5$ m for predominantly dynamic loads (Section 12),
  · $0.5 \leq s_x/s_y \leq 2.0$,

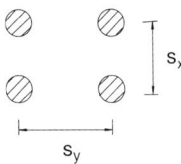

  · If this Recommendation is not followed and larger clear spacings $(s - d)$ or $(s - b_L)$ are planned, specific system serviceability analyses are recommended.
  · If a triangular grid is used, it should be as a square grid rotated through $45°$. Other forms of triangular grid are not dealt with by these Recommendations.

Recommended mechanical variables:

- characteristic effective friction angle $\varphi'_k \geq 30°$ or angle of total shear strength $\varphi'_{s,k} \geq 30°$ for the granular soils in the region of the reinforced earth structure with the height h^* as shown in Figure 9.1,
- design resistance of the geosynthetic reinforcement of each reinforcement layer $R_{B,d} \geq 30$ kN/m.

9.4   Actions and Resistances

Actions include permanent and variable loads to Section 12 and DIN 1054.

Resistances include:

- shear strength and stiffness of the soil in the reinforced earth structure,
- shear strength and stiffness of the ground,
- bearing capacity and deformability of the vertical bearing elements,
- tensile forces and axial stiffnesses of the geosynthetics,
- composite behaviour between geosynthetics and the surrounding ground.

9.5 Point and Linear Bearing Elements

Point bearing elements in the terms of these Recommendations are bored piles, micropiles, displacement piles, grouted vibrocolumns, columns manufactured using jet grouting and deep soil mixing, and other bearing elements displaying similar load-bearing behaviour.

Linear bearing elements in the terms of these recommendations are diaphragm walls, diaphragm wall elements, grouted walls, rectangular foundations manufactured using jet grouting and deep soil mixing, and rectangular or slab-like bearing elements displaying similar load-bearing behaviour.

See Section 9.2.1 for details of the use of geosynthetic-encased soil columns, non-grouted vibrocolumns or stabilising columns for ground improvement.

DIN standards, approvals, recommendations of the German Geotechnical Society (*Deutsche Gesellschaft für Geotechnik*) or similar regulations are available for the analysis and manufacture of the majority of bearing elements listed here. These regulations include minimum requirements for the shear strength of the soft strata and notes on any required buckling safety analyses. Due to the occasional low utilisation of the bearing elements lower embedment lengths of the bearing elements to those demanded for piles in DIN 1054 may be possible.

Observe Sections 9.7.1.3 and 9.7.2.3 as well as DIN 1054 for analysis of load-bearing capacity and serviceability of the bearing elements. Additionally, cyclic/dynamic actions (e.g. from live loads) and horizontal actions (e.g. from braking and centrifugal forces) should be analysed and evaluated to examine whether they have a governing impact on the load-bearing behaviour of the vertical bearing elements.

9.6 Analysing the Reinforced Earth Structure

9.6.1 General Recommendations

The characteristic effects from load redistribution within the reinforced earth structure and the membrane action of the geosynthetic reinforcement are determined in this section.

Load redistribution is described by ground arching. The principles are described in [2] and developed further in [3], [4] and [12]. Load redistribution is also described by what is known as the load redistribution factor E_L:

$$E_L = \frac{\sigma_{zs,k} \cdot A_s}{(\gamma_k \cdot h + p_k) \cdot A_E}$$

Eq. (9.4)

The load redistribution factor gives the proportion of the total load transferred directly into the bearing elements. The normal, characteristic ground stresses $\sigma_{zo,k}$ and $\sigma_{zs,k}$ in the reinforcement plane, assumed to be locally uniformly distributed,

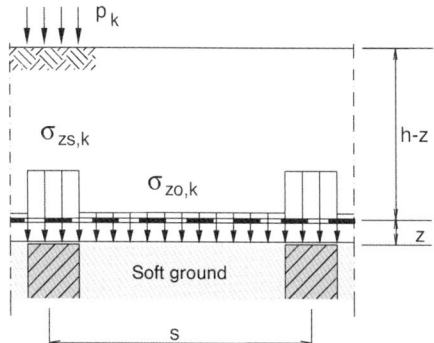

Figure 9.6 Stress redistribution in the contact plane as a result of arching

are determined as a function of the geometrical boundary conditions and the characteristic value of the earth structure's friction angle φ'_k or $\varphi'_{s,k}$. An external surcharge p_k at the height h may be taken into consideration on the top of the earth structure (Figure 9.6).

Note: Permanent (p_k) and variable surcharges (q_k) at not differentiated in the following, general description with a surcharge p_k. This shall be done when applying the partial safety factor concept. The notation p_Q was used for the variable surcharge below and in the example in Section 9.10.

The following characteristic effects shall be taken into consideration for the system:

- the normal stress $\sigma_{zo,k}$ on the plane between the support surfaces as shown in Figure 9.6,
- the normal stress $\sigma_{zs,k}$ on the support surfaces A_S in the contact plane of the reinforced earth structure as shown in Figure 9.6,
- outward directed shear and spreading forces in embankments,
- tensile stresses on the geosynthetics.

Here and below it is assumed in simplification that the stresses $\sigma_{zo,k}$ and $\sigma_{zs,k}$ are equal for the reinforcement plane and the contact plane. A requirement for this assumption is that the distance z is sufficiently small (also see Section 9.3).

The analysis method described here applies to the conditions recommended in Sections 9.2 and 9.3. If these conditions are deviated from supplementary investigations may be necessary.

9.6.2 Effect Situations

The characteristic effects are determined for all governing construction states and the system's final state. If the supporting subgrade reaction cannot be permanently guaranteed, no subgrade or a subgrade reduction shall be investigated

159

as a special case. A reduction or loss of subgrade may occur given inadequate consolidation of the soft strata, lateral excavation or lowering of the groundwater table, for example.

The governing construction states result from the varying heights h of the reinforced earth structure during it manufacture and compaction in layers. The final state of the system includes consideration of the completed structure with the planned actions for the design working life. The time-dependence of the reinforcement material and ground behaviour shall be taken into consideration where appropriate.

9.6.3   Characteristic Effects

9.6.3.1   Principles

The effects of permanent and of both permanent and variable actions shall be analysed to DIN 1054 in the STR limit state. Because the characteristic effects in the system used here are not proportional to the actions (non-applicability of the principle of superposition), the effects must first be separately determined for the permanent actions and for both the permanent and variable actions respectively, taking the impact of time into consideration where necessary.

9.6.3.2   Stress $\sigma_{zo,k}$ between the Bearing Elements

– Point bearing elements

For point bearing elements in a rectangular grid arrangement the normal stress $\sigma_{zo,G,k}$ from permanent actions or the normal stress $\sigma_{zo,G+Q,k}$ from permanent and variable actions as calculated in Eq. (9.5) or read off from Figure 9.7 pp., where $\sigma_{zo,k}$ is interpreted as $\sigma_{zo,G,k}$ or $\sigma_{zo,G+Q,k}$. Here, φ'_k is the characteristic value of the friction angle of the reinforced earth structure; $\varphi'_{s,k}$ can be used accordingly.

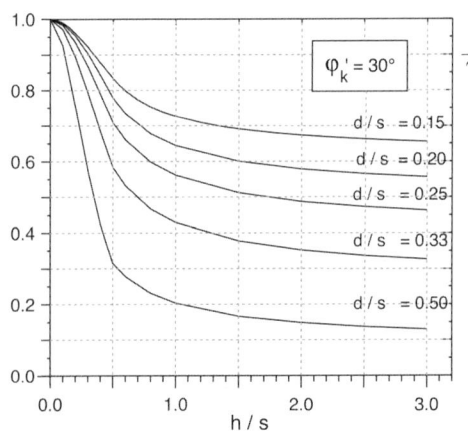

Figure 9.7 Normal stresses $\sigma_{zo,k}$ between the support surfaces in the contact plane of the reinforced earth structure for point bearing elements ($\varphi'_k = 30°$)

160

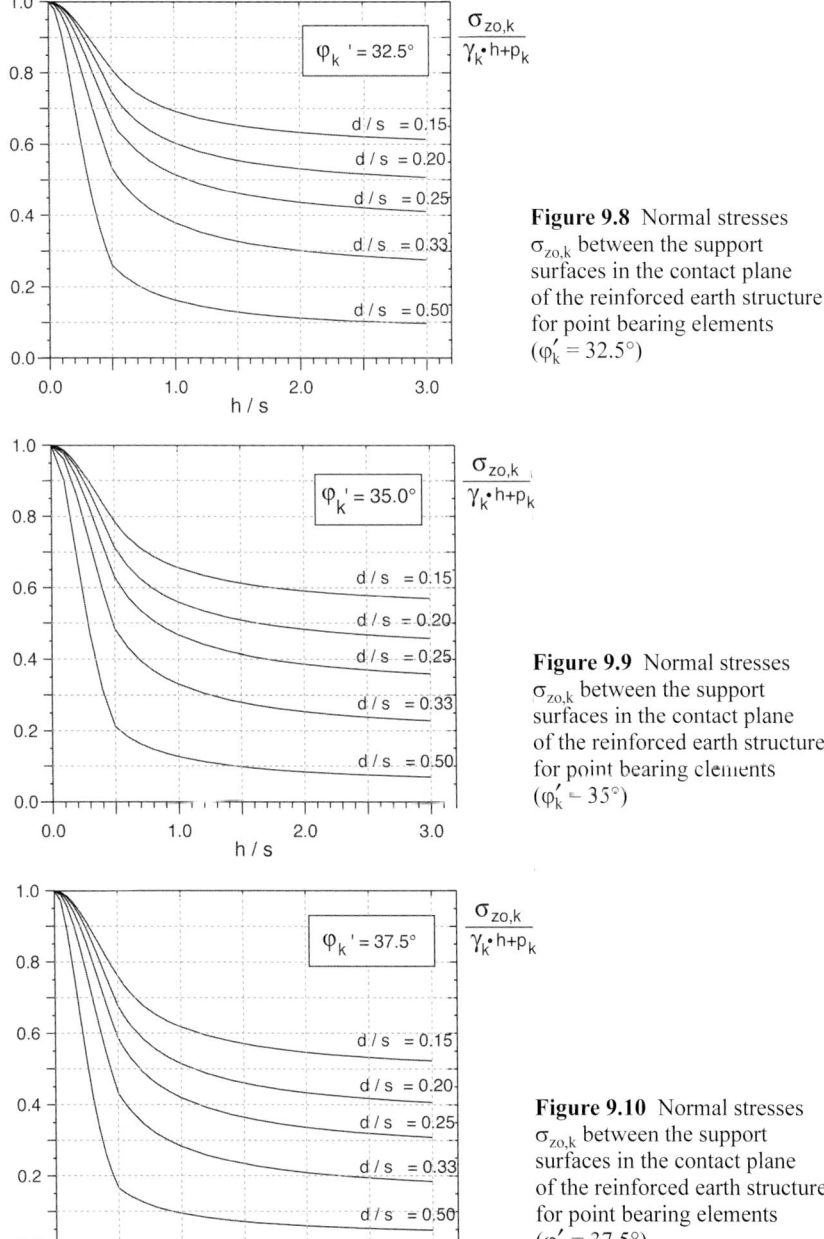

Figure 9.8 Normal stresses $\sigma_{zo,k}$ between the support surfaces in the contact plane of the reinforced earth structure for point bearing elements ($\varphi'_k = 32.5°$)

Figure 9.9 Normal stresses $\sigma_{zo,k}$ between the support surfaces in the contact plane of the reinforced earth structure for point bearing elements ($\varphi'_k = 35°$)

Figure 9.10 Normal stresses $\sigma_{zo,k}$ between the support surfaces in the contact plane of the reinforced earth structure for point bearing elements ($\varphi'_k = 37.5°$)

$$\sigma_{zo,G,k} = \lambda_1^{\chi} \cdot \left(\gamma_k + \frac{p_{G,k}}{h} \right)$$

$$\cdot \left\{ h \cdot (\lambda_1 + h_g^2 \cdot \lambda_2)^{-\chi} + h_g \cdot \left[\left(\lambda_1 + \frac{h_g^2 \cdot \lambda_2}{4} \right)^{-\chi} - (\lambda_1 + h_g^2 \cdot \lambda_2)^{-\chi} \right] \right\} \qquad \text{Gl. (9.5)}$$

$$\sigma_{zo,G+Q,k} = \lambda_1^{\chi} \cdot \left(\gamma_k + \frac{p_{G+Q,k}}{h} \right)$$

$$\cdot \left\{ h \cdot (\lambda_1 + h_g^2 \cdot \lambda_2)^{-\chi} + h_g \cdot \left[\left(\lambda_1 + \frac{h_g^2 \cdot \lambda_2}{4} \right)^{-\chi} - (\lambda_1 + h_g^2 \cdot \lambda_2)^{-\chi} \right] \right\} \qquad \text{Eq. (9.6)}$$

where:

γ_k characteristic unit weight of the soil in the reinforced earth structure in kN/m^3,

$p_{G,k}$ characteristic value of the permanent distributed load on the top of the reinforced earth structure in kN/m^2,

$p_{G+Q,k}$ characteristic value of the permanent and variable distributed load on the top of the reinforced earth structure in kN/m^2,

h_g arch height in m:

     $h_g = s/2$ for $h \geq s/2$,

     $h_g = h$ for $h < s/2$,

     h   see Figure 9.1,

     s and d   see Figure 9.4

K_{crit} critical principal stress ratio

$$K_{crit} = \tan^2 \left(45° + \frac{\varphi_k'}{2} \right) \qquad \text{Eq. (9.7)}$$

$$\chi = \frac{d \cdot (K_{crit} - 1)}{\lambda_2 \cdot s} \qquad \text{Eq. (9.8)}$$

$$\lambda_1 = \frac{1}{8} \cdot (s - d)^2 \qquad \text{Eq. (9.9)}$$

$$\lambda_2 = \frac{s^2 + 2 \cdot d \cdot s - d^2}{2 \cdot s^2} \qquad \text{Eq. (9.10)}$$

Proceed as follows to determine the stresses for a triangular grid arrangement of the bearing elements (square grid rotated through 45°):

- imaginary rotation of the triangular grid back to a square one parallel to the x- and y-directions,
- computation of stress $\sigma_{zo,k}$ in the imaginary rotated square grid.

Linear bearing elements

The normal stress $\sigma_{zo,k}$ for linear bearing elements can be read off Figure 9.11 pp., where b_L is the width of the linear bearing elements.

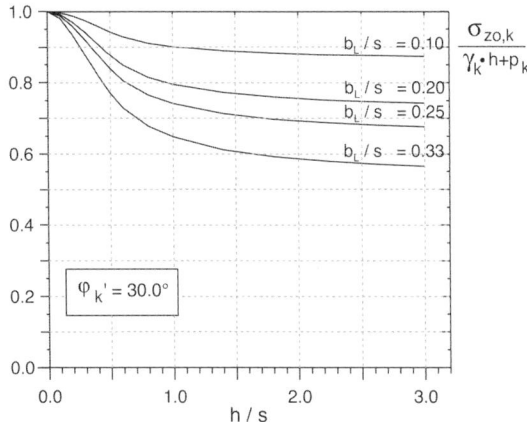

Figure 9.11 Normal stresses $\sigma_{zo,k}$ between the support surfaces in the contact plane of the reinforced earth structure for linear bearing elements ($\varphi'_k = 30°$)

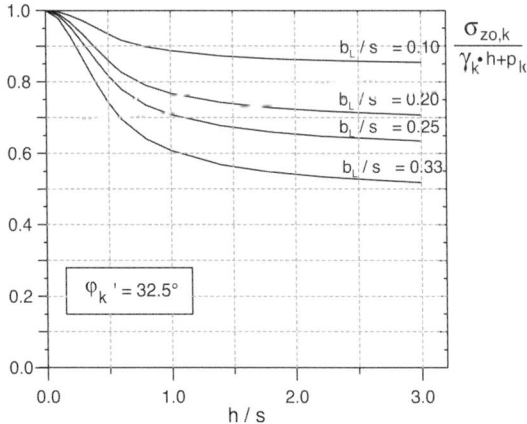

Figure 9.12 Normal stresses $\sigma_{zo,k}$ between the support surfaces in the contact plane of the reinforced earth structure for linear bearing elements ($\varphi'_k = 32.5°$)

When using Figure 9.7 pp. or Figure 9.11 pp. The permanent actions $p_{G,k}$ or the permanent and variable actions $p_{G+Q,k}$ respectively are adopted for p_k.

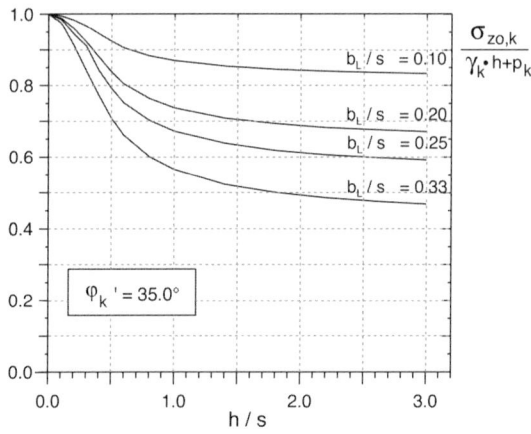

Figure 9.13 Normal stresses $\sigma_{zo,k}$ between the support surfaces in the contact plane of the reinforced earth structure for linear bearing elements ($\varphi_k' = 35°$)

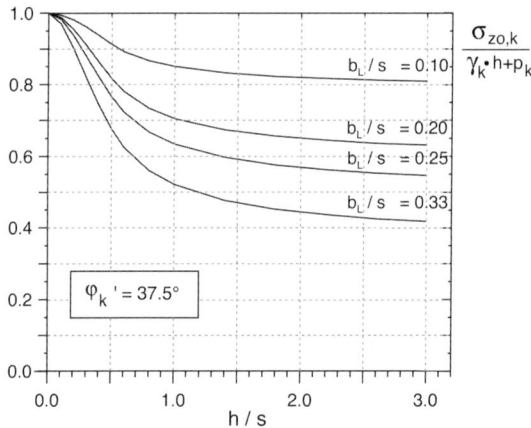

Figure 9.14 Normal stresses $\sigma_{zo,k}$ between the support surfaces in the contact plane of the reinforced earth structure for linear bearing elements ($\varphi_k' = 37.5°$)

9.6.3.3 Stress $\sigma_{zs,k}$ on the Bearing Elements

Using the designations as shown in Figure 9.4 the normal stress on the support surface A_S for point and linear bearing elements is given by Eqs. (9.11) and (9.12) for permanent, and for permanent and variable actions respectively.

$$\sigma_{zs,G,k} = [(\gamma_k \cdot h + p_{G,k}) - \sigma_{zo,G,k}] \cdot \frac{A_E}{A_S} + \sigma_{zo,G,k} \qquad \text{Eq. (9.11)}$$

$$\sigma_{zs,G+Q,k} = [(\gamma_k \cdot h + p_{G+Q,k}) - \sigma_{zo,G+Q,k}] \cdot \frac{A_E}{A_S} + \sigma_{zo,G+Q,k} \qquad \text{Eq. (9.12)}$$

164

The force on the bearing element resulting from $\sigma_{zs,k}$ is calculated using Eqs. (9.13) and (9.14). In addition, there are the normal forces on the support surface from the geosynthetic reinforcement, which are a function of subgrade reaction (see Section 9.6.3.5).

$$F_{S,G,k} = \sigma_{zs,G,k} \cdot A_S \qquad \text{Eq. (9.13)}$$

$$F_{S,G+Q,k} = \sigma_{zs,G+Q,k} \cdot A_S \qquad \text{Eq. (9.14)}$$

Generally, the resultant force on the bearing element is conservatively calculated using:

$$F_{S,G,k} = (\gamma_k \cdot h + p_{G,k}) \cdot A_E \qquad \text{Eq. (9.15)}$$

$$F_{S,G+Q,k} = (\gamma_k \cdot h + p_{G+Q,k}) \cdot A_E . \qquad \text{Eq. (9.16)}$$

This approach also comprises the loss of subgrade below the reinforcement plane effect situation.

9.6.3.4   Spreading Forces for Inclined Surface of Reinforced Earth Structure

Horizontal forces (*spreading forces*) occur in the contact plane below the slopes of embankments as a result of the absence of lateral supports in the reinforced earth structure. The spreading forces shall be accepted by the geosynthetic reinforcement and be transferred toward the embankment axis. Analysis of the characteristic effect on the geosynthetics as a result of spreading forces is dealt with in Section 9.6.3.5.

9.6.3.5   Effects on the Geosynthetic Reinforcement

The following notes apply to a geosynthetic reinforcement to which orthogonal bearing directions can be allocated. For analysis of a horizontal geosynthetic reinforcement installed above the bearing elements the normal stress σ_{zo} is assigned as an external action of the reinforcement as described in Section 9.6.3.2 or taking the load coverage area A_L into consideration. The resulting total load on the coverage area A_L is adopted approximately for point bearing elements as a triangular line road on a reinforcement strip of width b (see Figure 9.15), where b is the width of a pile head assumed to be rectangular. For round piles of diameter d an approximate equivalent width $b_{Ers.}$ is used as shown in Eq. (9.17).

$$b_{Ers.} = \frac{1}{2} \cdot d \cdot \sqrt{\pi} \qquad \text{Eq. (9.17)}$$

The equivalent width $b_{Ers.} = 1$ m is adopted below for linear bearing elements.

Two intersection directions occur for orthogonal reinforcement bearing directions. In addition, a coaxial x-y axis system with the associated load coverage areas A_{Lx} and A_{Ly} is introduced. The line loads $q_{z,W}$ [x] and $q_{z,W}$ [y] are imposed over the respective clear distance $l_{W,x} = (s_x - b_{Ers.})$ and $l_{W,y} = (s_y - b_{Ers.})$ (Figure 9.15).

Effects as a Result of Membrane Action

The resulting action F_k is determined as a characteristic value as shown in Figure 9.15 for a rectangular grid or a triangular grid of point bearing elements. The procedure is similar for an imaginary, rotated triangular grid as described in Section 9.6.3.2 (square rotated through 45°).

a)   *Rectangular grid:*

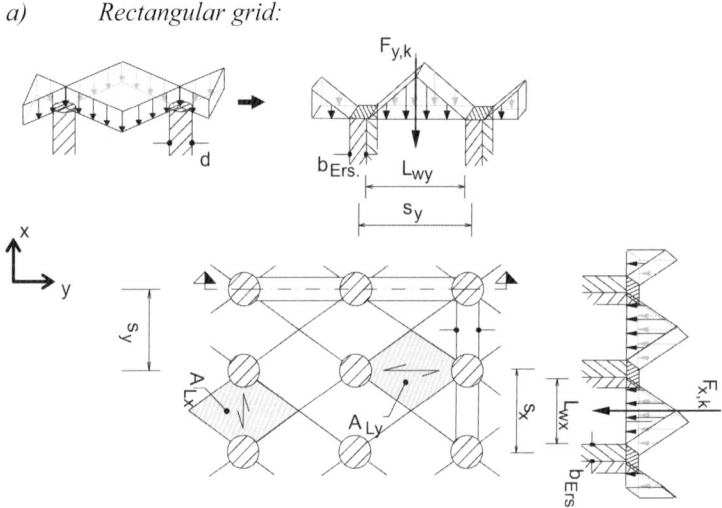

Figure 9.15 Resulting action F_k on the reinforcement plane (see Figure 9.2)

b) Load coverage areas:

$$A_{Lx} = \frac{1}{2} \cdot (s_x \cdot s_y) - \frac{d^2}{2} \cdot atn\left(\frac{s_y}{s_x}\right) \cdot \frac{\pi}{180} \qquad \text{Eq. (9.18)}$$

$$A_{Ly} = \frac{1}{2} \cdot (s_x \cdot s_y) - \frac{d^2}{2} \cdot atn\left(\frac{s_x}{s_y}\right) \cdot \frac{\pi}{180} \qquad \text{Eq. (9.19)}$$

c) Resulting load on a reinforcement strip of width $b_{Ers.}$:

$$F_{x,G,k} = A_{Lx} \cdot \sigma_{zo,G,k} \qquad \text{Eq. (9.20)}$$

$$F_{x,G+Q,k} = A_{Lx} \cdot \sigma_{zo,G+Q,k} \qquad \text{Eq. (9.21)}$$

$$F_{y,G,k} = A_{Ly} \cdot \sigma_{zo,G,k} \qquad \text{Eq. (9.22)}$$

$$F_{y,G+Q,k} = A_{Ly} \cdot \sigma_{zo,G+Q,k} \qquad \text{Eq. (9.23)}$$

Figure 9.15 is applied accordingly for linear bearing elements, where F_k is adopted in the perpendicular direction for a 1 m wide strip.

166

J_k is the characteristic value of the axial stiffness ('modulus of tension') of the geosynthetic reinforcement in kN/m. The axial stiffness is a function of the load and decreases with time due to creep strain. This is taken into consideration for more detailed investigations depending on the analysis case. The reinforcement isochrones representing the relationship between tensile force, time and strain are used for this purpose.

The value of the maximum strain in a geosynthetic reinforcement can be taken from Figure 9.16 pp. The effect E (tensile force in the reinforcement) as a result of membrane action is given by Eqs. (9.24) and (9.25)

$$E_{M,G,k} = \varepsilon_{G,k} \cdot J_k \qquad\qquad \text{Eq. (9.24)}$$

$$E_{M,G+Q,k} = \varepsilon_{G+Q,k} \cdot J_k \qquad\qquad \text{Eq. (9.25)}$$

Variable	Unit	Designation
$k_{s,k}$	[kN/m³]	Modulus of subgrade reaction of the soft soil between the bearing elements as described in Section 9.6.3.5
L_W	[m]	Clear spacing between the support surfaces (L_{Wx}, L_{Wy} see Figure 9.15)
J_k	[kN/m]	Axial stiffness of the geosynthetic reinforcement
F_k	[kN]	Resulting load $F_{G,k}$ or $F_{G+Q,k}$
max. ε_k	[%]	Maximum strain $\varepsilon_{G,k}$ or $\varepsilon_{G+Q,k}$
f	[m]	Geosynthetic reinforcement sag

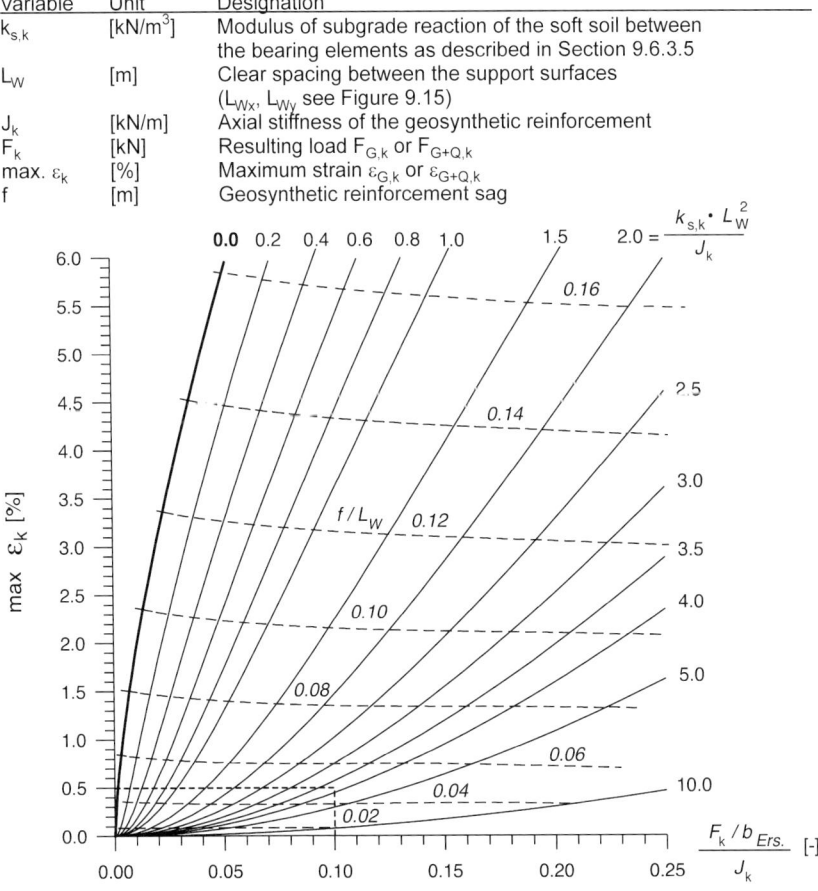

Figure 9.16 Maximum strain ε_k of reinforcement between support surfaces

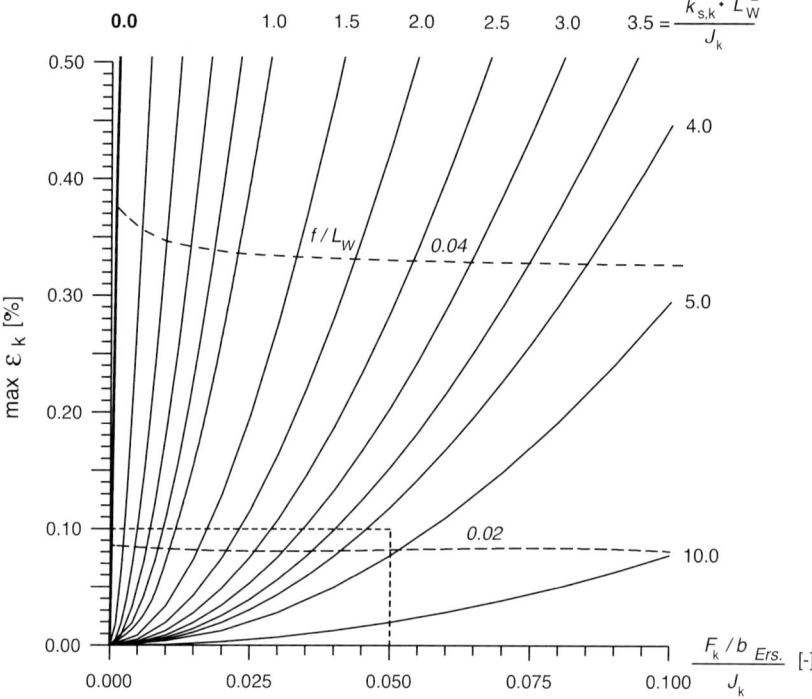

Figure 9.17 Maximum strain ε_k of reinforcement between the support surfaces (excerpt 1)

The effects in two-ply reinforcements may be divided according to the ratio of the characteristic reinforcement stiffnesses J_k inasmuch as the vertical distance z of the reinforcement plane from the contact plane meets the demands of Section 9.3.

When using Figure 9.16 the permanent actions or both the permanent and variable actions respectively are adopted for F_k as shown in Figure 9.15. Figure 9.16 applies to sufficiently vertically inflexible bearing elements or support surfaces (Eq. (9.2)).

Note: Peculiarities of the triangular grid:
The characteristic effect of the geosynthetic reinforcement when using a triangular grid (square bearing element grid rotated through 45°) is determined as described above using an imaginary, rotated square grid. The resulting geosynthetic reinforcement is once again installed over the grid diagonals parallel to the x- and y-axes.

168

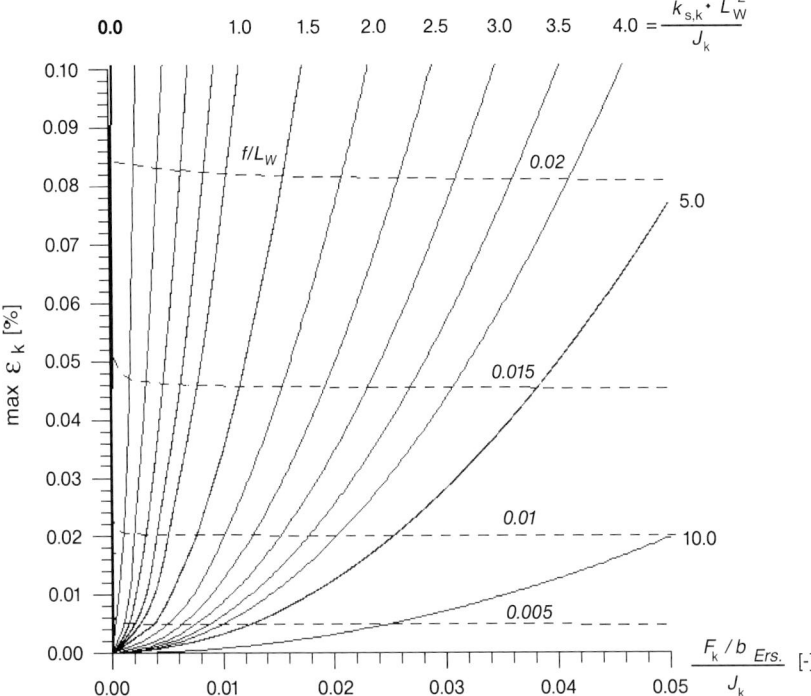

Figure 9.18 Maximum strain ε_k of reinforcement between the support surfaces (excerpt 2)

Estimating the Subgrade Reaction Moduli of the Soft Strata

For a homogeneous soft stratum the modulus of subgrade reaction of the stratum can be estimated from the constrained modulus of the stratum E_s and its thickness t_W using Eq. (9.26), disregarding stress reduction with depth (infinite stress surface).

$$k_s = \frac{E_{s,k}}{t_W}$$
Eq. (9.26)

Several soil strata i below the reinforcement can be approximately modelled using a mean modulus of subgrade reaction k_s weighted proportional to the layer thicknesses t_{Wi} as used in Eq. (9.27) (Figure 9.19).

169

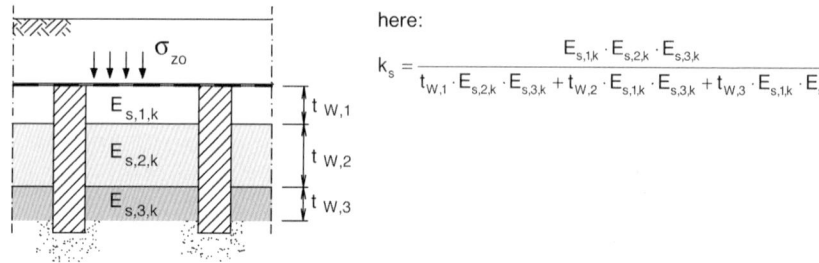

Figure 9.19 Estimate of modulus of subgrade reaction for stratified ground below the reinforcement plane

$$k_s = \frac{\prod\limits_{n=1}^{i} E_{s,n}}{\sum\limits_{n=1}^{i} t_{W,n} \cdot \prod\limits_{m=1}^{i} E_{s,m}}, \quad m \neq n \qquad \text{Eq. (9.27)}$$

Note 1: *A weighted modulus of subgrade reaction as used in Eq. (9.27) may lead to a considerable over- or under-estimation of the subgrade reaction for highly variable stiffnesses of the prevalent strata.*

Note 2: *The design method is highly sensitive in terms of the subgrade approach used. More precise investigations of the subgrade reaction may be necessary.*

When specifying the modulus of subgrade reaction the possibility of further settlement of the soft strata (e.g. as a result of groundwater fluctuations) leading to a reduction in the subgrade reaction in the contact plane shall be investigated.

Effects as a Result of Spreading

Two procedures are differentiated for determining the effects on the geosynthetic reinforcement resulting from spreading.

Procedure 1

The spreading force is derived from an assumed active earth pressure, accumulated between the top of the reinforced earth structure and the reinforcement, and allocated to the reinforcement as an effect ΔE_k. The active earth pressure is determined to DIN 4085, taking the existing surcharges p_k into consideration.

If the soft stratum extends to the contact plane of the earth structure (Figure 9.20) or only a thin, granular cover is present, the spreading force is given by Eqs. (9.28) and (9.29).

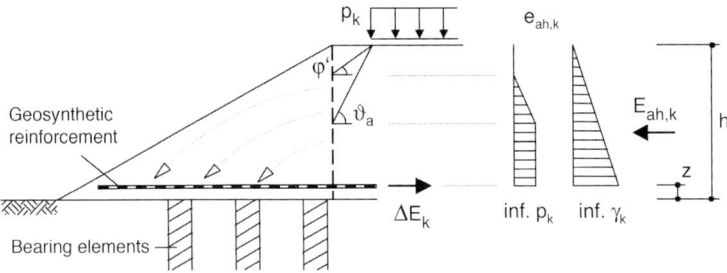

Figure 9.20 Additional effects in the geosynthetic reinforcement on an embankment slope using Procedure 1 without or with only a thin cover

$$\Delta E_{G,k} = E_{ah,G,k} \qquad \text{Eq. (9.28)}$$

$$\Delta E_{G+Q,k} = E_{ah,G+Q,k} \qquad \text{Eq. (9.29)}$$

If the soft stratum as shown in Figure 9.21 is overlain by a granular cover, the passive earth pressure $E_{ph,k}$ of the cover may be adopted when determining the effect on the reinforcement as a result of spreading as used in Eqs. (9.30) and (9.31). However, the passive earth pressure may only be adopted if the following conditions are met:

- the blanket layer shall be of at least medium-dense compaction,
- the characteristic passive earth pressure of the blanket may only be adopted at 50% of the passive earth pressure calculated to DIN 4085.
- the blanket shall be at least 20% of the embankment height in thickness, but at least 1 m.

$$\Delta E_{G,k} = E_{ah,G,k} - 0.5 \cdot E_{ph,k} \qquad \text{Eq. (9.30)}$$

$$\Delta E_{G+Q,k} = E_{ah,G+Q,k} - 0.5 \cdot E_{ph,k} \qquad \text{Eq. (9.31)}$$

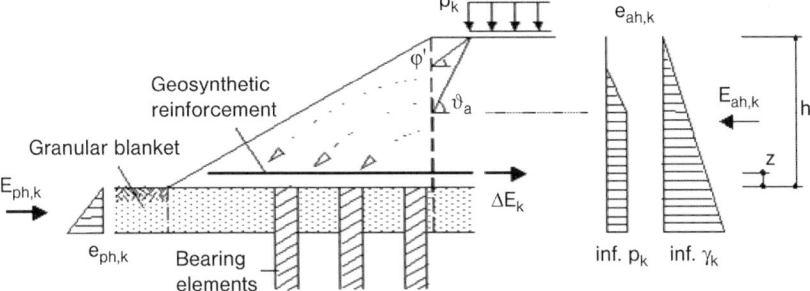

Figure 9.21 Additional effects in the geosynthetic reinforcement on an embankment slope using Procedure 1 taking the load-bearing action of a granular blanket into consideration

171

Using the analyses in line with Procedure 1 it may be assumed that the geosynthetic reinforcement is adequately designed and the vertical bearing elements show no inadmissible deformations. It is generally unnecessary to analyse the deformation of bearing elements.

Procedure 2

Alternatively to Procedure 1 the effects on the geosynthetic reinforcement resulting from spreading may also be determined using Figure 9.22 and Eqs. (9.32) and (9.33). In accordance with [11] the geosynthetic reinforcement is allocated either only the effects as a result of membrane action or as a result of spreading action from adopting earth pressure using Procedure 1, where the maximum is the governing value.

In Procedure 2 it can no longer be assumed that the vertical bearing elements display no unacceptable deformation. Analysis of deformation of the vertical bearing elements, e.g. using numerical methods, is therefore recommended.

$$E_{G,k} = \max \begin{cases} E_{ah,G,k} \\ E_{M,G,k} \end{cases} \qquad \text{Eq. (9.32)}$$

$$E_{G+Q,k} = \max \begin{cases} E_{ah,G+Q,k} \\ E_{M,G+Q,k} \end{cases} \qquad \text{Eq. (9.33)}$$

where:

$E_{ah,G,k}$ and
$E_{ah,G+Q,k}$ resulting active earth pressure, accumulated between the top of the reinforced earth structure and the reinforcement, taking the existing surcharges p_k to DIN 4085 into consideration, also see Figure 9.22,
$E_{M,G,k}$ and
$E_{M,G+Q,k}$ as used in Eqs. (9.24) and (9.25).

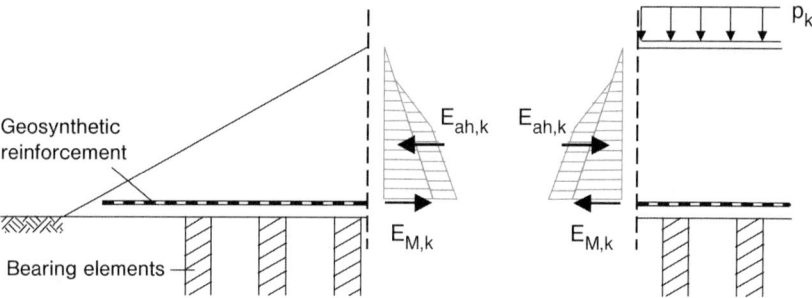

Figure 9.22 Governing effects in the geosynthetic reinforcement in embankment slopes using Procedure 2

172

In Figure 9.22 and Eqs. (9.32) and (9.33) the procedure is given for a case assuming the soft stratum reaches the earth structure's contact plane or only a thin blanket is present.

If the soft stratum is overlain by a granular blanket layer, proceed as for Procedure 1. The effect on the geosynthetic reinforcement as a result of spreading action is given by Eqs. (9.34) and (9.35).

$$E_{G,k} = \max \begin{cases} E_{ah,G,k} - 0.5 \cdot E_{ph,k} \\ E_{M,G,k} \end{cases} \qquad \text{Eq. (9.34)}$$

$$E_{G+Q,k} = \max \begin{cases} E_{ah,G+Q,k} - 0.5 \cdot E_{ph,k} \\ E_{M,G+Q,k} \end{cases} \qquad \text{Eq. (9.35)}$$

where:

$E_{ah,G,k}$ and
$E_{ah,G+Q,k}$ resulting active earth pressure, accumulated between the top of the reinforced earth structure and the base of the granular blanket layer, taking the existing surcharges p_k to DIN 4085 into consideration, also see Figure 9.21,

$E_{ph,k}$ resulting passive earth pressure from the granular blanket layer, also see Figure 9.21,

$E_{M,G,k}$ and
$E_{M,G+Q,k}$ as used in Eqs. (9.24) and (9.25).

9.6.4 Analysing Effect on the Geosynthetic Reinforcement using Numerical Methods

Comparative investigations of test and analysis results have shown that numerical analyses, e.g. using the finite element method, result in substantially reduced characteristic tensile stresses in the geosynthetic reinforcement for a reinforced earth structure on point bearing elements and using a common reinforcement element model.

Because cause and effect and their interrelationships are not yet sufficiently clarified, the effect on the geosynthetic reinforcement and the system design may not be based on numerical analyses.

On the other hand, assuming correct models are used numerical analyses provide realistic deformation results for the bearing elements (Section 9.7.2.3) and in terms of overall deformation (Section 9.7.2.4).

9.6.5　Analysing Effects on Geosynthetic Reinforcement for Dynamic Actions

High variable loads can negatively impact on arching above the bearing elements. Geometrical recommendations for adopting variable loads, such as limit values for the geometrical ratio $h/(s - d)$, are given in Section 9.3. If these limit values are adhered to it may be assumed that the arch remains intact under dynamic actions. If the limit values are not achieved, arch deterioration and simultaneously increased effects in the geosynthetic reinforcement may occur as a function of the geometrical ratio $h/(s - d)$, the magnitude of the dynamic actions (load amplitude and load frequency) and their frequency of occurrence. Arch deterioration and the surcharge on the geosynthetics can be approximately described with the aid of a simplified approach (arching reduction factor method). Additional notes can be found in [12].

9.7　Analyses and Design

9.7.1　Analysing Bearing Capacity

9.7.1.1　General Recommendations

Ultimate limit state analyses for the STR and GEO limit states are performed to DIN 1054.

The effects described in Section 6.3, here with the general notation E, are adopted as characteristic values E_k for analysis of the STR limit state and the design values for effects E_d determined using Eq. (9.36) by multiplying by the partial safety factors γ for actions to DIN 1054, Table 2.

$$E_d = E_{G,k} \cdot \gamma_G + (E_{G+Q,k} - E_{G,k}) \cdot \gamma_Q \qquad \text{Eq. (9.36)}$$

The ultimate limit state analyses for the individual structural elements are carried out for the STR limit state using the limit state equation (9.37), where R_d represents the design values of the resistances.

$$E_d \leq R_d \qquad \text{Eq. (9.37)}$$

Analysis of the GEO limit state comprises an investigation of the overall system, where the actions and resistances are introduced into the analysis from the outset as design values. Details can be taken from DIN 1054.

9.7.1.2　Analysing the Geosynthetic Reinforcement

The ultimate limit state of the geosynthetic reinforcement in the STR limit state is analysed as described in Section 9.6.3.5 using Eq. (9.38) for Procedure 1 and Eq. (9.39) for Procedure 2:

$$\Delta E_d + E_{M,d} \leq R_{B,d} \qquad \text{Eq. (9.38)}$$

$$\max\begin{cases}E_{ah,d}\\E_{M,d}\end{cases}\le R_{B,d}\quad\text{or}\quad\max\begin{cases}E_{ah,d}-0.5\cdot E_{ph,d}\\E_{M,d}\end{cases}\le R_{B,d}\qquad\text{Eq. (9.39)}$$

where:

ΔE_d design value of the effect as a result of spreading from Procedure 1 as described in Section 9.6.3.5. For linear structures (traffic embankments, etc.) ΔE_d can be adopted as zero longitudinal to the axis. This also applies perpendicular to the axis if the reinforced earth structure is predominantly below grade and adequate lateral support can be assumed.

$E_{M,d}$ design value of the effect as a result of membrane action as used in Eq. (9.36) and described in Section 9.6.3.5,

$E_{ah,d}$ design value of active earth pressure from Procedure 2 as described in Section 9.6.3.5,

$E_{ph,d}$ design value of passive earth pressure from Procedure 2 as described in Section 9.6.3.5,

$R_{B,d}$ design resistance of the geosynthetic reinforcement.

The design resistance of the geosynthetic reinforcement is determined as described in Section 3.3 and modified as follows for the case at hand:

$$R_{B,d}=\frac{R_{B,k,5\%}}{\gamma_M}\cdot\eta_M\qquad\text{Eq. (9.40)}$$

Where $\eta_M = 1.1$ is the calibration factor for modifying the level of safety in the STR limit state.

9.7.1.3   Analysing Bearing Elements

Analysis of the ultimate limit state of the bearing elements for the STR limit state is carried out using Eq. (9.41):

$$E_{S,d}\le R_d\qquad\text{Eq. (9.41)}$$

where:

R_d design value of the ultimate limit state of the bearing elements as given in the respective standards/approvals,

$E_{S,d}$ design value of the effect as used in Eq. (9.36) and described in Section 9.6.3.3.

If the ultimate limit state of the geosynthetic reinforcement is analysed using Eqs. (9.28), (9.29), (9.30), (9.31) and (9.38) (Procedure 1 as described in Section 9.3.6.5), empiricism indicates that analysis of the horizontal bearing capacity of the bearing elements can be dispensed with. If this analysis method is not employed the horizontal effects on the bearing elements shall be analysed separately.

175

9.7.1.4 Analysing Overall Stability

The stability of the overall system in the GEO limit state is analysed to DIN 1054 and Section 3. Any slip planes intersecting the reinforcement layers and the bearing elements, or local failure mechanisms in the slope and toe zones of embankment structures, are investigated. Where slip planes intersect bearing elements or the geosynthetic reinforcement the resistances of the geosynthetic reinforcement may be adopted as restraining forces. If no more detailed analyses of the transfer of any horizontal actions by the bearing elements are carried out, the vertical bearing elements may not be adopted for this purpose. The effects are determined with the aid of the design values of the actions and compared to the design values of the resistances (DIN 1054 and Section 3).

9.7.2 Serviceability Limit State Analysis

9.7.2.1 General Recommendations

The serviceability limit states refer to the tolerable system deformations. The overall system deformations comprise:

- deformations in the reinforced earth structure,
- deformations in the bearing elements and the ground.

These deformations can be considered singly or as a whole, or analyses performed based on them.

Analysis of the serviceability limit state is particularly important if the geometrical recommendations in Section 9.3 are not adopted.

9.7.2.2 Deformations in the Reinforced Earth Structure

The reinforcement is strained during manufacture of the reinforced earth structure in layers and in particular during installation and compaction of the initial soil layers. These strains can lead to settlement depressions between the contact planes. However, the layered earth structure compensates for the settlement depressions in each subsequent soil layer, such that the upper surface of the earth structure is manufactured as a flat plane. The geosynthetic reinforcement retains a permanent strain, which is not, however, critical to serviceability considerations. Generally, the serviceability analysis is governed only by the additional deformations in the reinforced earth structure occurring **after** its manufacture, due to the actions described in Section 9.4.

These additional strains, occurring after the reinforced earth structure is manufactured until the end of its operational life (generally 100 years), should meet the conditions of Eq. (9.42):

$$\Delta\varepsilon_{kr} \le 2\% \qquad \text{Eq. (9.42)}$$

where $\Delta\varepsilon_{kr}$ is the additional creep strain of the reinforcement for the period considered. In certain applications (e.g. transport structures sensitive to deformation or structures with small cover depths h) it may be necessary to further limit the additional strains $\Delta\varepsilon_{kr}$ in terms of the sag Δf.

9.7.2.3 Deformation of Bearing Elements

The deformations in the bearing elements can be estimated from the results of load testing, data recorded during manufacture or empirical data from similar ground conditions and actions. In addition, bearing element-specific analytical methods are available for estimating deformations. Reliable data can also be collected using:

- the observational method to DIN 1054 or
- numerical methods.

9.7.2.4 Analysing Overall Deformations

Reliable analytical methods for forecasting the overall system deformations are not currently available. The overall deformations can be approximately determined using numerical methods. Use of the observational method to DIN 1054 is recommended for ensuring serviceability.

9.8 Notes on Execution

9.8.1 Enabling Works

The site shall be provided with a load-bearing working subgrade to facilitate subsequent works. If the natural ground is too weak to support the loads from rigs and site traffic, special measures for manufacturing a load-bearing working subgrade shall be specified and executed.

The planned locations of point and linear bearing elements shall be surveyed using suitable methods.

The impacts of construction projects immediately adjacent to existing structures with high safety requirements (railways, roads) and which impact their bearing capacity and deformation behaviour, shall be limited such that no damage is caused to the existing structures.

9.8.2 Point and Linear Bearing Elements

All bearing elements shall be installed in load-bearing ground. Sufficient embedment of the bearing elements in the load-bearing ground shall be demonstrated using suitable methods.

The bearing element support surfaces are manufactured with their planned shape, dimensions and elevations. The installed bearing elements shall be protected against damage from actions occurring during site operations.

9.8.3   Reinforced Earth Structures

Once the bearing elements are manufactured and their elevations calibrated a 10 cm to 15 cm thick levelling course of compacted, granular soil or a similar protective layer (e.g. a nonwoven) should be installed above the reinforced earth structure's contact plane. The levelling course is graded.

The reinforcement geosynthetics are installed on the levelling course surface according to a diagram compiled by the planner. The geosynthetics are laid flat on the subgrade, free from folds and creases, tensioned and secured as required.

Overlapping in the direction of the principal tension is avoided. If overlapping cannot be avoided in exceptional cases, the planner provides a numerical analysis of adequate transmission of tensile forces. The vertical spacing of the geosynthetic reinforcement layers is ≥ 15 cm for multi-layer installations.

The laid out and uncovered geosynthetics may not be traversed by vehicles. The cover fill is emplaced such that the effects on the reinforcement are minor and the position of the reinforcement is not altered.

The geosynthetics may only be traversed by vehicles when the cover fill is at least 15 cm thick.

If no higher demands are made on the reinforced earth structure, a relative compaction $D_{Pr} \geq 0.97$ shall be achieved.

Empiricism indicates that careful compaction of the initial soil layers above the lowest reinforcement layer proves favourable for the reinforced earth structure's subsequent behaviour (reinforcement 'pre-straining', low deformability and greater shear resistance of the embankment soil). The effects from compaction equipment are adopted when designing the geosynthetics. Care must also be taken that the bearing elements are not damaged.

9.9   Bibliography

[1]   Hewlett, W. J., Randolph, M. F., Aust, M. I. E. (1988): Analysis of piled embankments. Ground Engineering Vol. 21, pp. 12–17.

[2]   Kempfert, H.-G., Stadel, M., Zaeske, D. (1997): Berechnung von geokunststoffbewehrten Tragschichten über Pfahlelementen. Bautechnik Jahrgang 75, Heft 12, pp. 818–825.

[3]   Zaeske, D. (2001): Zur Wirkungsweise von unbewehrten und bewehrten mineralischen Tragschichten über pfahlartigen Gründungselementen. Schriftenreihe Geotechnik, Universität Kassel, Heft 10.

[4]   Zaeske, D., Kempfert, H.-G. (2002): Berechnung und Wirkungsweise von unbewehrten und bewehrten mineralischen Tragschichten auf punkt- und linienförmigen Traggliedern. Bauingenieur, Volume 77, February 2002.

[5] Kempfert, H.-G., Zaeske, D., Alexiew, D. (1999): Interactions in reinforced bearing layers over partial supported underground. Proc. of the 12th ECSMGE, Amsterdam, 1999. Balkema, Rotterdam, 1999. pp. 1527–1532.

[6] Alexiew, D., Gartung, E. (1999): Geogrid reinforced railway embankment on piles – performance monitoring 1994-1998. Proc. 1st South American Symposium on Geosynthetics, Rio de Janeiro, 1999. pp. 403–411.

[7] Alexiew, D., Vogel, W. (2001): Railroads on piled embankments in Germany: Milestone projects, in: Landmarks in Earth Reinforcement, Swets & Zeitlinger, pp. 185–190.

[8] Alexiew, D. (2004): Geogitterbewehrte Dämme auf pfahlähnlichen Elementen: Grundlagen und Projekte. Die Bautechnik 81, Heft 9/2004.

[9] Kempfert, H.-G., Göbel, C., Alexiew, D., Heitz, C. (2004): German Recommendations for Soil Reinforcement above Pile-Elements. EuroGeo3, Third Geosynthetics Conference, Munich, Volume I, pp. 279–283.

[10] Geduhn, M., Vollmert, L. (2005): Verformungsabhängige Spannungszustände bei horizontalen Geokunststoffbewehrungen über Pfahlelementen in der Dammbasis. Bautechnik 82, Heft 9/2005.

[11] Love, J., Milligan, G. (2003): Design methods for basally reinforced pile-supported embankments over soft ground. Ground engineering, March 2003, pp. 39–43.

[12] Heitz, C. (2006): Bodengewölbe unter ruhender und nichtruhender Belastung bei Berücksichtigung von Bewehrungseinlagen aus Geogittern. Schriftenreihe Geotechnik, University of Kassel, Heft 19.

[13] Kempfert, H.-G., Lüking, J., Gebreselassie, B. (2009): Dimensionierung von bewehrten Erdkörpern auf punktförmigen Traggliedern bei besonderen Randbedingungen. 11. Informations- und Vortragstagung über 'Kunststoffe in der Geotechnik', Munich February 2009, Geotechnik Sonderheft 2009, pp. 35–43.

9.10   Design Example: Reinforced Earth Structures over Point or Linear Bearing Elements

9.10.1   Geometry, Loads, Soil Mechanics Parameters, Reinforcement and Effect Situation Parameters

Geometry and loads:

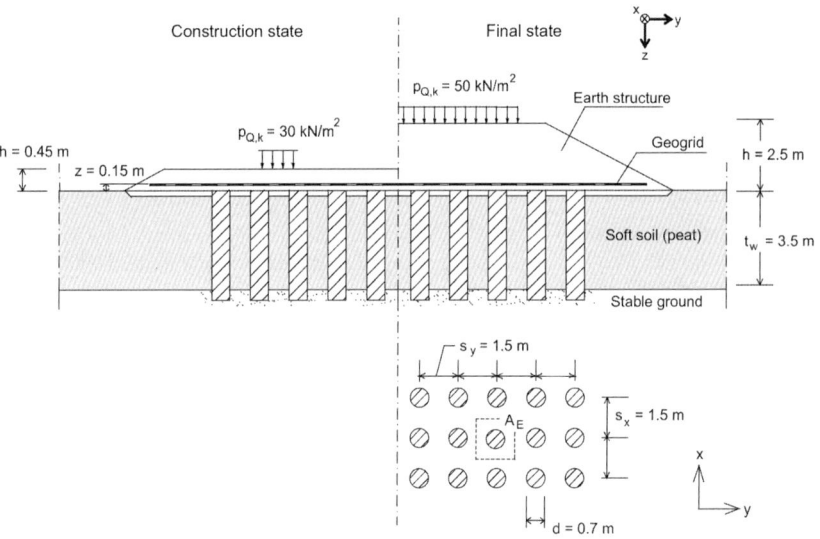

Figure 9.23 Geometry and loads

Soil mechanics parameters:

		Soft layer	Earth structure
E_s	$[kN/m^2]$	500	–
γ_k	$[kN/m^3]$	–	18
φ'_k	$[°]$	–	35

Reinforcement parameters:

The short-term load-extension curve and the isochrones for both directions shall be known for the analysis. The transient axial stiffnesses $J_{x,t}$ and $J_{y,t}$ are determined from these curves. A simplifying curve linearisation may be performed (*cf.* Section 9.10.2.2). It is recommended to relate the linearisation to a strain of $\varepsilon = 2.5\%$ for this example (Figure 9.24). Additionally, the reduction factors A_1 to A_5 shall be known.

Dynamic Design Case:

As an approximation, Dynamic Design Case 1 as described in Section 12 is assumed. Therefore $A_5 = 1.0$.

A geogrid with a short-term strength of 400 kN/m parallel to the roll-out axis and 200 kN/m perpendicular to the roll-out axis is used. The short-term load-extension curve and the isochrones for the geogrid are given in Figure 9.24.

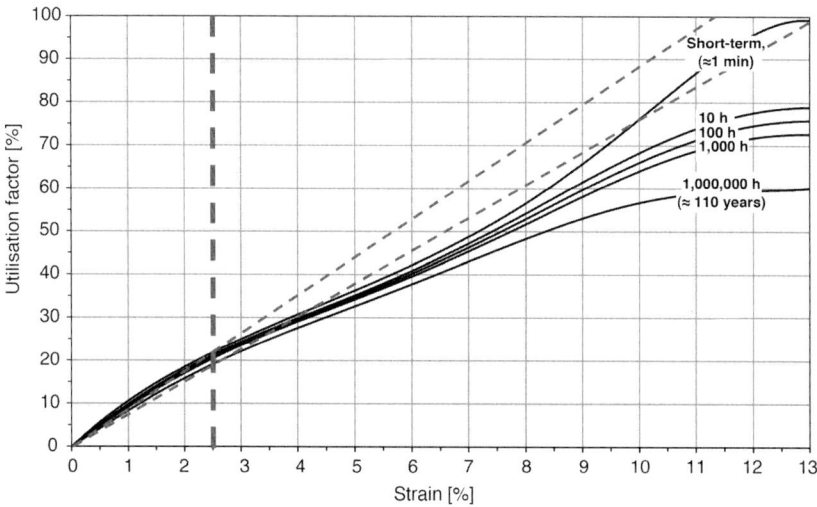

Figure 9.24 Short-term load-extension curve with isochrones

The reduction factors are:

$A_1 = 1.26$ $t = 10$ h, $\varepsilon_{Failure} = 13\%$),
$A_1 = 1.34$ $t = 500$ h, $\varepsilon_{Failure} = 13\%$),
$A_1 = 1.36$ $t = 1{,}000$ h, $\varepsilon_{Failure} = 13\%$),
$A_1 = 1.65$ $t = 1{,}000{,}000$ h, $\varepsilon_{Failure} = 13\%$),
$A_2 = 1.10$ $A_3 = 1.00$ $A_4 = 1.00$ $A_5 = 1.00$.

The geogrid is rolled out perpendicular to the x-axis (embankment axis). This means a short-term strength in the x-axis of 200 kN/m and in the y-axis (perpendicular to the embankment axis) of 400 kN/m, together with an elongation at failure of 13% in both directions.

181

Effect situations:

Three effect situations are considered for analysis:

Situation 1: construction state with an embankment height of h = 0.45 m (z = 0.15 m, i.e. the top of the embankment is 0.3 m above the geosynthetic reinforcement) and a load p_k = 30 kN/m² from compaction and traffic ,
('temporary design situation');
estimated time: t_1 = 10 h.

Situation 2: construction state with an embankment height of h = 2.5 m and a load $p_{Q,k}$ = 30 kN/m² from compaction and traffic,
('temporary design situation');
estimated time: t_2 = 500 h.

Situation 3: final state with an embankment height of h = 2.5 m and a live load $p_{Q,k}$ = 50 kN/m²,
('permanent design situation');
estimated time: t_3 = 1,000,000 h (~ 110 a).

In this case the site investigation report states that groundwater lowering is planned for a later date. It is therefore necessary to additionally investigate the loss or reduction of the supporting subgrade reaction of the ground as described in Section 9.6.2 as a special case.

Special case: loss of balancing ground reaction stress,
('accidental design situation');
estimated time: T_4 = 1,000,000 h (~ 110 a).

General notes:

- Conservatively, live load distribution with depth was not adopted. If a distribution is adopted, the distributed live load is adopted at the level of the arch apex, i.e. at h_g. However, this distribution does **not** apply to any adopted spreading force!
- Any load increase factors required for adopting live loads as described in Section 12 should already be included in the characteristic variable actions $p_{Q,k}$ for the case discussed here.
- In the final state: h / (s – d) = 2.5 / (2.12 – 0.7) = 1.75. In this example the h / (s – d) ≥ 2.0 condition as described in Section 9.3 for disregarding a negative impact as a result of the variable load for the final state is disregarded in agreement with the client.
- It shall be demonstrated that the calculated strain does not exceed the elongation at failure of the selected reinforcement. If this does occur a different reinforcement shall be selected until the calculated strain is lower than the elongation at failure.

In this design example the design values of the effects on the geosynthetic reinforcement are determined as described in Section 9.7.1.2.

9.10.2 Effect Situation 1: Construction State (t_1 = 10 h)

9.10.2.1 Load Redistribution in the Reinforced Earth Structure

Input values for analysis using Eq. (9.5) pp.:

Governing bearing element axis centres:

$$s = \sqrt{1.50^2 + 1.50^2} = \sqrt{4.50} = 2.1213 \text{ m}$$

Arch height: $h = 0.45 \text{ m} < s/2 = 1.06 \text{ m} \rightarrow h_g = h = 0.45 \text{ m}$

Base course material: $K_{crit} = 3.69$ ($\varphi'_k = 35°$)

Earth pressure coefficient: $K_{ah} = 0.271$

$$\lambda_1 = \frac{1}{8} \cdot (s-d)^2 = \frac{1}{8} \cdot (2.1213 - 0.70)^2 = 0.2525$$

$$\lambda_2 = \frac{s^2 + 2 \cdot d \cdot s - d^2}{2 \cdot s^2} = \frac{4.5 + 2 \cdot 0.70 \cdot 2.1213 - 0.70^2}{2 \cdot 4.5} = 0.7755$$

$$\chi = \frac{d \cdot (K_{crit} - 1)}{\lambda_2 \cdot s} = \frac{0.70 \cdot (3.690 - 1)}{0.7755 \cdot 2.1213} = 1.1447$$

Stress $\sigma_{zo,k}$ between the vertical bearing elements:

Characteristic stress $\sigma_{zo,G,k}$ as a result of permanent loads:

$$\sigma_{zo,G,k} = 0.2525^{1.1447} \cdot 18 \cdot \left\{ 0.45 \cdot (0.2525 + 0.45^2 \cdot 0.7755)^{-1.1447} + 0.45 \right.$$

$$\cdot \left[\left(0.2525 + \frac{0.45^2 \cdot 0.7755}{4} \right)^{-1.1447} - (0.2525 + 0.45^2 \cdot 0.7755)^{-1.1447} \right] \right\}$$

$$= 6.87 \text{ kN/m}^2$$

or from Figure 9.12 for $\varphi'_k = 35°$ and $h/s = 0.45/2.1213 = 0.212$ and $d/s = 0.70/2.1213 = 0.33$.

$$\frac{\sigma_{zo,G,k}}{\gamma_k \cdot h} = 0.85 \quad \rightarrow \quad \sigma_{zo,G,k} = 6.89 \text{ kN/m}^2$$

Characteristic stress $\sigma_{zo,G+Q,k}$ as a result of permanent and variable loads:

$$\sigma_{zo,G+Q,k} = 0.2525^{1.1447} \cdot \left(18 + \frac{30}{0,45}\right) \cdot \left\{ 0.45 \cdot (0.2525 + 0.45^2 \cdot 0.7755)^{-1.1447} + 0.45 \right.$$

$$\left. \cdot \left[\left(0.2525 + \frac{0.45^2 \cdot 0.7755}{4}\right)^{-1.1447} - (0.2525 + 0.45^2 \cdot 0.7755)^{-1.1447} \right] \right\}$$

$$= 32.29 \text{ kN/m}^2$$

or from Figure 9.11 pp.:

$$\frac{\sigma_{zo,G+Q,k}}{\gamma_k \cdot h + p_{Q,k}} = 0.85 \quad \rightarrow \quad \sigma_{zo,G+Q,k} = 32.39 \text{ kN/m}^2 \, .$$

Analysis of stress $\sigma_{zs,k}$ on the vertical bearing elements:

Characteristic stress $\sigma_{zs,G,k}$ as a result of permanent loads:

$$A_E = s_x \cdot s_y = 1.5 \cdot 1.5 = 2.25 \text{ m}^2$$

$$A_S = \pi \cdot \frac{d^2}{4} = \pi \cdot \frac{0.7^2}{4} = 0.385 \text{ m}^2$$

$$\sigma_{zs,G,k} = [(\gamma_k \cdot h) - \sigma_{zo,G,k}] \cdot \frac{A_E}{A_S} + \sigma_{zo,G,k}$$

$$= [(18 \cdot 0.45) - 6.87] \cdot \frac{2.25}{0.385} + 6.87 = 14.06 \text{ kN/m}^2$$

Load redistribution factor:

$$E_L = \frac{\sigma_{zs,G,k} \cdot A_s}{\gamma_k \cdot h \cdot A_E} = \frac{14.06 \cdot 0.385}{18 \cdot 0.45 \cdot 2.25} = 0.297$$

This means that 29.7% of the total load is transmitted directly to the bearing elements.

Characteristic stress $\sigma_{zs,G+Q,k}$ as a result of permanent and variable loads:

$$\sigma_{zs,G+Q,k} = [(\gamma_k \cdot h + p_{Q,k}) - \sigma_{zo,G+Q,k}] \cdot \frac{A_E}{A_S} + \sigma_{zo,G+Q,k}$$

$$= [(18 \cdot 0.45 + 30) - 32.29] \cdot \frac{2.25}{0.385} + 32.29 = 66.24 \text{ kN/m}^2$$

(Load redistribution factor:

$$E_L = \frac{\sigma_{zs,G+Q,k} \cdot A_s}{(\gamma_k \cdot h + p_{Q,k}) \cdot A_E} = 0.297 \,)$$

9.10.2.2 Characteristic Effects in the Geosynthetic Reinforcement

The subgrade reaction of the ground is described in Eq. (9.26) by a modulus of subgrade reaction $k_{s,k}$.

$$k_{s,k} = \frac{E_s}{t_w} = \frac{500}{3.5} = 143 \text{ kN/m}^3$$

The geogrid used has a short-term strength in the x-axis (embankment axis) of 200 kN/m and in the y-axis (perpendicular to the embankment axis) of 400 kN/m.

Simplifying linearisation of the curve is first carried out at $\varepsilon = 2.5\%$. This gives the following axial stiffnesses:

Time t	Axial stiffness	
	$J_{x,t}$ [kN/m]	$J_{y,t}$ [kN/m]
Short term	1,756	3,512
10 h	1,688	3,376
100 h	1,660	3,320
500 h	1,648	3,296
1,000 h	1,632	3,264
1,000,000 h	1,520	3,040

Description: the axial stiffnesses $J_{x,10h}$ and $J_{y,10h}$ in the failure state are given for $\varepsilon = 2.5\%$ in Figure 9.25 as:

$$J_{x,10h} = \frac{21.1}{2.5} \cdot 200 = 1,688 \text{ kN/m} \quad \text{and} \quad J_{y,10h} = \frac{21.1}{2.5} \cdot 400 = 3,376 \text{ kN/m}$$

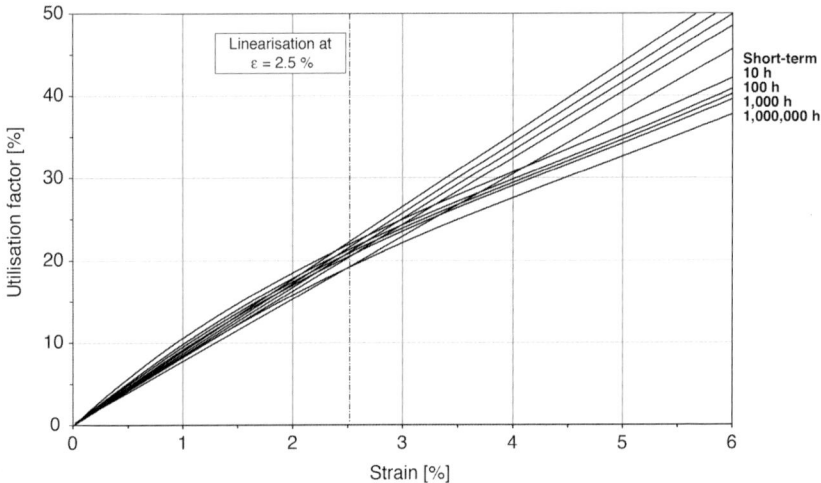

Figure 9.25 Linearisation for ε = 2.5%

The analysis is shown below for the x- and y-axes.

x-axis	y-axis
Equivalent width and clear spacing:	
$b_{Ers.} = \dfrac{1}{2} \cdot d \cdot \sqrt{\pi} = \dfrac{1}{2} \cdot 0.7 \cdot \sqrt{\pi} = 0.62\ m$	
$L_{w,x} = s_x - b_{Ers.} = 1.50 - 0.62 = 0.88\ m$	$L_{w,y} = s_y - b_{Ers.} = 1.50 - 0.62 = 0.88\ m$
Load coverage area:	
$A_{Lx} = \dfrac{1}{2} \cdot s_x \cdot s_y - \dfrac{d^2}{2} \cdot atn\left(\dfrac{s_y}{s_x}\right) \cdot \dfrac{\pi}{180}$ $= \dfrac{1}{2} \cdot 1.5 \cdot 1.5 - \dfrac{0.7^2}{2} \cdot atn\left(\dfrac{1.5}{1.5}\right) \cdot \dfrac{\pi}{180}$ $= 0.933\ m^2$	$A_{Ly} = \dfrac{1}{2} \cdot s_x \cdot s_y - \dfrac{d^2}{2} \cdot atn\left(\dfrac{s_x}{s_y}\right) \cdot \dfrac{\pi}{180}$ $= \dfrac{1}{2} \cdot 1.5 \cdot 1.5 - \dfrac{0.7^2}{2} \cdot atn\left(\dfrac{1.5}{1.5}\right) \cdot \dfrac{\pi}{180}$ $= 0.933\ m^2$

x-axis	y-axis
Effects as a result of permanent loads	
Resulting load on a reinforcement strip of width $b_{Ers.}$:	
$F_{x,G,k} = A_{Lx} \cdot \sigma_{zo,G,k} = 0.933 \cdot 6.87$ $\quad\quad = 6.41 \text{ kN/m}$	$F_{y,G,k} = A_{Ly} \cdot \sigma_{zo,G,k} = 0.933 \cdot 6.87$ $\quad\quad = 6.41 \text{ kN/m}$
Determining maximum strains:	
$\dfrac{k_{s,k} \cdot L_{w,x}^2}{J_{x,10h}} = \dfrac{143 \cdot 0.88^2}{1,688} = 0.0656$ $\dfrac{F_{x,G,k} / b_{Ers.}}{J_{x,10h}} = \dfrac{6.41 / 0.62}{1,688} = 0.0061$ From Figure 9.16: max $\varepsilon_{x,G,k} = 0.96\%$	$\dfrac{k_{s,k} \cdot L_{w,y}^2}{J_{y,10h}} = \dfrac{143 \cdot 0.88^2}{3,376} = 0.0328$ $\dfrac{F_{y,G,k} / b_{Ers.}}{J_{y,10h}} = \dfrac{6.41 / 0.62}{3,376} = 0.0031$ From Figure 9.16: max $\varepsilon_{y,G,k} = 0.65\%$
Tensile forces from membrane action:	
$E_{M,x,G,k} = $ max $\varepsilon_{x,G,k} \cdot J_{x,10h}$ $\quad = \dfrac{0.96}{100} \cdot 1,688 = 16.20 \text{ kN/m}$	$E_{M,y,G,k} = $ max $\varepsilon_{y,G,k} \cdot J_{y,10h}$ $\quad = \dfrac{0.65}{100} \cdot 3,376 = 21.94 \text{ kN/m}$
Effects from spreading forces:	
–	$\Delta E_{y,G,k} = \dfrac{1}{2} \cdot \gamma_k \cdot (h - z)^2 \cdot K_{agh}$ $\quad = \dfrac{1}{2} \cdot 18 \cdot 0.30^2 \cdot 0.271$ $\quad = 0.22 \text{ kN/m}$
Total effect in the geosynthetics:	
max $E_{x,G,k} = E_{M,x,G,k}$ $\quad = 16.20 \text{ kN/m}$	max $E_{y,G,k} = E_{M,y,G,k} + \Delta E_{y,G,k}$ $\quad = 22.16 \text{ kN/m}$

x-axis	y-axis
Effects as a result of permanent and variable loads	

Resulting load on a reinforcement strip of width $b_{Ers.}$:

$F_{x,G+Q,k} = A_{Lx} \cdot \sigma_{zo,G+Q,k}$ $= 0.933 \cdot 32.29 = 30.13 \text{ kN/m}$	$F_{y,G+Q,k} = A_{Ly} \cdot \sigma_{zo,G+Q,k}$ $= 0.933 \cdot 32.29 = 30.13 \text{ kN/m}$

Determining maximum strains:

$\dfrac{k_{s,k} \cdot L_{w,x}^2}{J_{x,10h}} = \dfrac{143 \cdot 0.88^2}{1,688} = 0.0656$	$\dfrac{k_{s,k} \cdot L_{w,y}^2}{J_{y,10h}} = \dfrac{143 \cdot 0.88^2}{3,376} = 0.0328$
$\dfrac{F_{x,G+Q,k} / b_{Ers.}}{J_{x,10h}} = \dfrac{30.13 / 0.62}{1,688} = 0.0288$	$\dfrac{F_{y,G+Q,k} / b_{Ers.}}{J_{y,10h}} = \dfrac{30.13 / 0.62}{3,376} = 0.0144$
From Figure 9.16: max $\varepsilon_{x,G+Q,k} = 3.47 \%$	From Figure 9.16: max $\varepsilon_{y,G+Q,k} = 2.22 \%$

Tensile forces from membrane action:

$E_{M,x,G+Q,k} = \max \varepsilon_{x,G+Q,k} \cdot J_{x,10h}$ $= \dfrac{3.47}{100} \cdot 1,688 = 58.57 \text{ kN/m}$	$E_{M,y,G+Q,k} = \max \varepsilon_{y,G+Q,k} \cdot J_{y,10h}$ $= \dfrac{2.22}{100} \cdot 3,376 = 74.95 \text{ kN/m}$

Effects from spreading forces:

–	$\Delta E_{y,G+Q,k} = \dfrac{1}{2} \cdot \gamma_k \cdot (h-z)^2 \cdot K_{agh} +$ $+ p_{Q,k} \cdot (h-z) \cdot K_{aph}$ $= \dfrac{1}{2} \cdot 18 \cdot 0.30^2 \cdot 0.271 +$ $+ 30 \cdot 0.30 \cdot 0.271$ $= 2.66 \text{ kN/m}$

Total effect in the geosynthetics:

$\max E_{x,G+Q,k} = E_{M,x,G+Q,k}$ $= 58.57 \text{ kN/m}$	$\max E_{y,G+Q,k} = E_{M,y,G+Q,k} + \Delta E_{y,G+Q,k}$ $= 77.61 \text{ kN/m}$

9.10.3 Effect Situation 2: Construction State (t_2 = 500 h)

9.10.3.1 Load Redistribution in the Reinforced Earth Structure

Arch height: h = 2.50 m > s/2 = 2.12/2 = 1.06 m $\Rightarrow$ h_g = s/2 = 1.06 m

Stress $\sigma_{zo,k}$ between the vertical bearing elements

Characteristic stress $\sigma_{zo,G,k}$ as a result of permanent loads:

$$\sigma_{zo,G,k} = 0.2525^{1.1447} \cdot 18 \cdot \left\{ 2.50 \cdot (0.2525 + 1.06^2 \cdot 0.7755)^{-1.1447} + 1.06 \right.$$

$$\left. \cdot \left[\left(0.2525 + \frac{1.06^2 \cdot 0.7755}{4} \right)^{-1.1447} - (0.2525 + 1.06^2 \cdot 0.7755)^{-1.1447} \right] \right\}$$

$$= 14.05 \text{ kN/m}^2$$

or from Figure 9.9 for φ'_k = 35° and h/s = 2.50/2.1213 = 1.179 and
d/s = 0.70/2.1213 = 0.33

$$\frac{\sigma_{zo,G,k}}{\gamma_k \cdot h} = 0.31 \quad \rightarrow \quad \sigma_{zo,G,k} = 13.95 \text{ kN/m}^2 .$$

Characteristic stress $\sigma_{zo,G+Q,k}$ as a result of permanent and variable loads:

$$\sigma_{zo,G+Q,k} = 0.2525^{1.1447} \cdot \left(18 + \frac{30}{2.50} \right) \cdot \left\{ 2.50 \cdot (0.2525 + 1.06^2 \cdot 0.7755)^{-1.1447} + 1.06 \right.$$

$$\left. \cdot \left[\left(0.2525 + \frac{1.06^2 \cdot 0.7755}{4} \right)^{-1.1447} - (0.2525 + 1.06^2 \cdot 0.7755)^{-1.1447} \right] \right\}$$

$$= 23.41 \text{ kN/m}^2$$

or from Figure 9.9:

$$\frac{\sigma_{zo,G+Q,k}}{\gamma_k \cdot h + p_{Q,k}} = 0.31 \quad \rightarrow \quad \sigma_{zo,G+Q,k} = 23.25 \text{ kN/m}^2 .$$

189

Analysis of stress $\sigma_{zs,k}$ on the vertical bearing elements

Characteristic stress $\sigma_{zs,G,k}$ as a result of permanent loads:

$$\sigma_{zs,G,k} = [(\gamma_k \cdot h) - \sigma_{zo,G,k}] \cdot \frac{A_E}{A_S} + \sigma_{zo,G,k}$$

$$= [(18 \cdot 2.50) - 14.05] \cdot \frac{2.25}{0.385} + 14.05 = 194.93 \text{ kN/m}^2$$

Load redistribution factor:

$$E_L = \frac{\sigma_{zs,G,k} \cdot A_s}{\gamma_k \cdot h \cdot A_E} = \frac{194.93 \cdot 0.385}{18 \cdot 2.50 \cdot 2.25} = 0.741$$

This means that 74.1% of the total load is transmitted directly to the bearing elements.

Characteristic stress $\sigma_{zs,G+Q,k}$ as a result of permanent and variable loads:

$$\sigma_{zs,G+Q,k} = [(\gamma_k \cdot h + p_{Q,k}) - \sigma_{zo,G+Q,k}] \cdot \frac{A_E}{A_S} + \sigma_{zo,G+Q,k}$$

$$= [(18 \cdot 2.50 + 30) - 23.41] \cdot \frac{2.25}{0.385} + 23.41 = 324.91 \text{ kN/m}^2$$

(Load redistribution factor:

$$E_L = \frac{\sigma_{zs,G+Q,k} \cdot A_s}{(\gamma_k \cdot h + p_{Q,k}) \cdot A_E} = 0.741 \text{)}$$

9.10.3.2 Characteristic Effects in the Geosynthetic Reinforcement

Modulus of subgrade reaction $k_{s,k}$: $k_{s,k} = 143$ kN/m³

The analysis is shown below for the x- and y-axes.

x-axis	y-axis
Effects as a result of permanent loads	
Resulting load on a reinforcement strip of width $b_{Ers.}$:	
$F_{x,G,k} = A_{Lx} \cdot \sigma_{zo,G,k} = 0.933 \cdot 14.05$ $= 13.11$ kN/m	$F_{y,G,k} = A_{Ly} \cdot \sigma_{zo,G,k} = 0.933 \cdot 14.05$ $= 13.11$ kN/m
Determining maximum strains:	
$\dfrac{k_{s,k} \cdot L_{w,x}^2}{J_{x,500h}} = \dfrac{143 \cdot 0.88^2}{1,648} = 0.0672$ $\dfrac{F_{x,G,k} / b_{Ers.}}{J_{x,500h}} = \dfrac{13.11 / 0.62}{1,648} = 0.0128$ From Figure 9.16: max $\varepsilon_{x,G,k} = 1.82$ %	$\dfrac{k_{s,k} \cdot L_{w,y}^2}{J_{y,500h}} = \dfrac{143 \cdot 0.88^2}{3,296} = 0.0336$ $\dfrac{F_{y,G,k} / b_{Ers.}}{J_{y,500h}} = \dfrac{13.11 / 0.62}{3,296} = 0.0064$ From Figure 9.16: max $\varepsilon_{y,G,k} = 1.19$ %
Tensile forces from membrane action:	
$E_{M,x,G,k} = \dfrac{1.82}{100} \cdot 1,648 = 29.99$ kN/m	$E_{M,y,G,k} = \dfrac{1.19}{100} \cdot 3,296 = 39.22$ kN/m
Effects from spreading forces	
–	$\Delta E_{y,G,k} = \dfrac{1}{2} \cdot 18 \cdot 2.35^2 \cdot 0.271$ $= 13.47$ kN/m
Total effect in the geosynthetics:	
max $E_{x,G,k} = E_{M,x,G,k}$ $= 29.99$ kN/m	max $E_{y,G,k} = E_{M,y,G,k} + \Delta E_{y,G,k}$ $= 52.69$ kN/m

x-axis	y-axis
Effects as a result of permanent and variable loads	
Resulting load on a reinforcement strip of width $b_{Ers.}$:	
$F_{x,G+Q,k} = 0.933 \cdot 23.41 = 21.84\ \text{kN/m}$	$F_{y,G+Q,k} = 0.933 \cdot 23.41 = 21.84\ \text{kN/m}$
Determining maximum strains	
$\dfrac{k_{s,k} \cdot L^2_{w,x}}{J_{x,500h}} = \dfrac{143 \cdot 0.88^2}{1,648} = 0.0672$ $\dfrac{F_{x,G+Q,k} / b_{Ers.}}{J_{x,500h}} = \dfrac{21.84 / 0.62}{1,648} = 0.0214$ From Figure 9.16: max $\varepsilon_{x,G+Q,k} = 2.74\%$	$\dfrac{k_{s,k} \cdot L^2_{w,y}}{J_{y,500h}} = \dfrac{143 \cdot 0.88^2}{3,296} = 0.0336$ $\dfrac{F_{y,G+Q,k} / b_{Ers.}}{J_{y,500h}} = \dfrac{21.84 / 0.62}{3,296} = 0.0107$ From Figure 9.16: max $\varepsilon_{y,G+Q,k} = 1.77\%$
Tensile forces from membrane action:	
$E_{M,x,G+Q,k} = \dfrac{2.74}{100} \cdot 1,648 = 45.16\ \text{kN/m}$	$E_{M,y,G+Q,k} = \dfrac{1.77}{100} \cdot 3,296 = 58.34\ \text{kN/m}$
Effects from spreading forces:	
—	$\begin{aligned} \Delta E_{y,G+Q,k} &= \frac{1}{2} \cdot 18 \cdot 2.35^2 \cdot 0.271 + \\ &\quad + 30 \cdot 2.35 \cdot 0.271 \\ &= 32.57\ \text{kN/m} \end{aligned}$
Total effect in the geosynthetics:	
max $E_{x,G+Q,k} = 45.16\ \text{kN/m}$	$\begin{aligned} \text{max } E_{y,G+Q,k} &= 58.34 + 32.57 \\ &= 90.91\ \text{kN/m} \end{aligned}$

9.10.4 Effect Situation 3: Final State (t_3 = 1,000,000 h)

9.10.4.1 Load Redistribution in the Reinforced Earth Structure

Arch height: $h = 2.50$ m $> s/2 = 2.12/2 = 1.06$ m $\Rightarrow h_g = s/2 = 1.06$ m

Stress $\sigma_{zo,k}$ between the vertical bearing elements

Characteristic stress $\sigma_{zo,G,k}$ as a result of permanent loads:

see Section 9.10.3.1: $\sigma_{zo,G,k} = 14.05$ kN/m^2

Characteristic stress $\sigma_{zo,G+Q,k}$ as a result of permanent and variable loads:

$$\sigma_{zo,G+Q,k} = 0.2525^{1.1447} \cdot \left(18 + \frac{50}{2.50}\right) \cdot \left\{ 2.50 \cdot (0.2525 + 1.06^2 \cdot 0.7755)^{-1.1447} + 1.06 \right.$$

$$\left. \cdot \left[\left(0.2525 + \frac{1.06^2 \cdot 0.7755}{4}\right)^{-1.1447} - (0.2525 + 1.06^2 \cdot 0.7755)^{-1.1447} \right] \right\}$$

$$= 29.65 \text{ kN/m}^2$$

or from Figure 9.8:

$$\frac{\sigma_{zo,G+Q,k}}{\gamma_k \cdot h + p_{Q,k}} = 0.31 \quad \rightarrow \quad \sigma_{zo,G+Q,k} = 29.45 \text{ kN/m}^2$$

Analysis of stress $\sigma_{zs,k}$ on the vertical bearing elements

Characteristic stress $\sigma_{zs,G,k}$ as a result of permanent loads:

see Section 9.10.3.1: $\sigma_{zs,G,k} = 194.93$ kN/m^2

Load redistribution factor $E_L = 0.741$

This means that 74.1% of the total load is transmitted directly to the bearing elements.

Characteristic stress $\sigma_{zs,G+Q,k}$ as a result of permanent and variable loads:

$$\sigma_{zs,G+Q,k} = [(\gamma_k \cdot h + p_{Q,k}) - \sigma_{zo,G+Q,k}] \cdot \frac{A_E}{A_S} + \sigma_{zo,G+Q,k}$$

$$= [(18 \cdot 2.50 + 50) - 29.65] \cdot \frac{2.25}{0.385} + 29.65 = 411.57 \text{ kN/m}^2$$

(Load redistribution factor:

$$E_L = \frac{\sigma_{zs,k} \cdot A_s}{(\gamma_k \cdot h + p_{Q,k}) \cdot A_E} = 0.741)$$

9.10.4.2 Characteristic Effects in the Geosynthetic Reinforcement

Modulus of subgrade reaction $k_{s,k}$: $k_{s,k}$ = 143 kN/m^3

The analysis is shown below for the x- and y-axes.

x-axis	y-axis
Effects as a result of permanent loads	
Resulting load on a reinforcement strip of width $b_{Ers.}$:	
$F_{x,G,k} = A_{Lx} \cdot \sigma_{zo,G,k} = 0.933 \cdot 14.05$ $= 13.11 \, \text{kN/m}$	$F_{y,G,k} = A_{Ly} \cdot \sigma_{zo,G,k} = 0.933 \cdot 14.05$ $= 13.11 \, \text{kN/m}$
Determining maximum strains:	
$\dfrac{k_{s,k} \cdot L_{w,x}^2}{J_{x,110a}} = \dfrac{143 \cdot 0.88^2}{1,520} = 0.0729$ $\dfrac{F_{x,G,k} / b_{Ers.}}{J_{x,110a}} = \dfrac{13.11 / 0.62}{1,520} = 0.014$ From Figure 9.16: max $\varepsilon_{x,G,k}$ = 1.91 %	$\dfrac{k_{s,k} \cdot L_{w,y}^2}{J_{y,110a}} = \dfrac{143 \cdot 0.88^2}{3,040} = 0.0364$ $\dfrac{F_{y,G,k} / b_{Ers.}}{J_{y,110a}} = \dfrac{13.11 / 0.62}{3,040} = 0.007$ From Figure 9.16: max $\varepsilon_{y,G,k}$ = 1.25 %
Tensile forces from membrane action:	
$E_{M,x,G,k} = \dfrac{1.91}{100} \cdot 1,520 = 29.03 \, \text{kN/m}$	$E_{M,y,G,k} = \dfrac{1.25}{100} \cdot 3,040 = 38.00 \, \text{kN/m}$
Effects from spreading forces:	
—	$\Delta E_{y,G,k} = \dfrac{1}{2} \cdot 18 \cdot 2.35^2 \cdot 0.271$ $= 13.47 \, \text{kN/m}$
Total effect in the geosynthetics:	
max $E_{x,G,k} = E_{M,x,G,k} = 29.03 \, \text{kN/m}$	max $E_{y,G,k} = 38.00 + 13.47 = 51.47 \, \text{kN/m}$

x-axis	y-axis
Effects as a result of permanent and variable loads	
Resulting load on a reinforcement strip of width $b_{Ers.}$:	
$F_{x,G+Q,k} = 0.933 \cdot 29.65 = 27.66$ kN/m	$F_{y,G+Q,k} = 0.933 \cdot 29.65 = 27.66$ kN/m
Determining maximum strains:	
$\dfrac{k_{s,k} \cdot L^2_{w,x}}{J_{x,110a}} = \dfrac{143 \cdot 0.88^2}{1,520} = 0.0729$ $\dfrac{F_{x,G+Q,k} / b_{Ers.}}{J_{x,110a}} = \dfrac{27.66 / 0.62}{1,520} = 0.0294$ From Figure 9.16: max $\varepsilon_{x,G+Q,k} = 3.47\%$	$\dfrac{k_{s,k} \cdot L^2_{w,y}}{J_{y,110a}} = \dfrac{143 \cdot 0.88^2}{3,040} = 0.0364$ $\dfrac{F_{y,G+Q,k} / b_{Ers.}}{J_{y,110a}} = \dfrac{27.66 / 0.62}{3,040} = 0.0147$ From Figure 9.16: max $\varepsilon_{y,G+Q,k} = 2.23\%$
Tensile forces from membrane action:	
$E_{M,x,G+Q,k} = \dfrac{3.47}{100} \cdot 1,520 = 52.74$ kN/m	$E_{M,y,G+Q,k} = \dfrac{2.23}{100} \cdot 3,040 = 67.79$ kN/m
Effects from spreading forces:	
–	$\begin{aligned} \Delta E_{y,G+Q,k} &= \dfrac{1}{2} \cdot 18 \cdot 2.35^2 \cdot 0.271 \\ &\quad + 50 \cdot 2.35 \cdot 0.271 \\ &= 45.31 \text{ kN/m} \end{aligned}$
Total effect in the geosynthetics:	
max $E_{x,G+Q,k} = 52.74$ kN/m	$\begin{aligned} \text{max } E_{y,G+Q,k} &= 67.79 + 45.31 \\ &= 113.10 \text{ kN/m} \end{aligned}$

9.10.5 Special Case: Loss of Subgrade Reaction ($t_4 = 1,000,000$ h)

9.10.5.1 Load Redistribution in the Base Course

Load redistribution in the base course and the stresses $\sigma_{zo,k}$ and $\sigma_{zs,k}$ correspond to the values in Section 9.10.4.1.

9.10.5.2 Characteristic Effects in the Geosynthetic Reinforcement

Modulus of subgrade reaction $k_{s,k}$: $k_{s,k} = 0$ kN/m^3

Time: $T_4 = 1,000,000$ h (~ 110 a).

The analysis is shown below for the x- and y-axes.

x-axis	y-axis
Effects as a result of permanent loads	
Resulting load on a reinforcement strip of width $b_{Ers.}$:	
$F_{x,G,k} = 13.11$ kN/m	$F_{y,G,k} = 13.11$ kN/m
Determining maximum strains:	
$\dfrac{k_{s,k} \cdot L_{w,x}^2}{J_{x,110a}} = 0$ $\dfrac{F_{x,G,k} / b_{Ers.}}{J_{x,110a}} = \dfrac{13.11 / 0.62}{1,520} = 0.014$ From Figure 9.16: max $\varepsilon_{x,G,k} = 2.4\%$	$\dfrac{k_{s,k} \cdot L_{w,y}^2}{J_{y,110a}} = 0$ $\dfrac{F_{y,G,k} / b_{Ers.}}{J_{y,110a}} = \dfrac{13.11 / 0.62}{3,040} = 0.007$ From Figure 9.16: max $\varepsilon_{y,G,k} = 1.50\%$
Tensile forces from membrane action:	
$E_{M,x,G,k} = \dfrac{2.4}{100} \cdot 1,520 = 36.48$ kN/m	$E_{M,y,G,k} = \dfrac{1.5}{100} \cdot 3,040 = 45.60$ kN/m
Effects from spreading forces:	
–	$\Delta E_{y,G,k} = 13.47$ kN/m
Total effect in the geosynthetics:	
max $E_{x,G,k} = 36.48$ kN/m	max $E_{y,G,k} = 45.60 + 13.47$ $= 59.07$ kN/m

x-axis	y-axis
Effects as a result of permanent and variable loads	
Resulting load on a reinforcement strip of width $b_{Ers.}$:	
$F_{x,G+Q,k} = 27.66 \text{ kN/m}$	$F_{y,G+Q,k} = 27.66 \text{ kN/m}$
Determining maximum strains:	
$\dfrac{k_{s,k} \cdot L_{w,x}^2}{J_{x,110a}} = 0$ $\dfrac{F_{x,G+Q,k} / b_{Ers.}}{J_{x,110a}} = \dfrac{27.66 / 0.62}{1,520} = 0.0294$ From Figure 9.16: max $\varepsilon_{x,G+Q,k} = 4.0\%$	$\dfrac{k_{s,k} \cdot L_{w,y}^2}{J_{y,110a}} = 0$ $\dfrac{F_{y,G+Q,k} / b_{Ers.}}{J_{y,110a}} = \dfrac{27.66 / 0.62}{3,040} = 0.0147$ From Figure 9.16: max $\varepsilon_{y,G+Q,k} = 2.48\%$
Tensile forces from membrane action:	
$E_{M,x,G+Q,k} = \dfrac{4.0}{100} \cdot 1,520 = 60.80 \text{ kN/m}$	$E_{M,y,G+Q,k} = \dfrac{2.48}{100} \cdot 3,040 = 75.39 \text{ kN/m}$
Effects from spreading forces:	
–	$\Delta E_{y,G+Q,k} = 45.31 \text{ kN/m}$
Total effect in the geosynthetics:	
max $E_{x,G+Q,k} = 60.80 \text{ kN/m}$	max $E_{y,G+Q,k} = 75.39 + 45.31$ $= 120.70 \text{ kN/m}$

9.10.6 Design Values of Effects in the Geosynthetic Reinforcement

The design values of the effects are derived from the characteristic values of the effects resulting from permanent actions $E_{k,G}$ and resulting from permanent and variable actions $E_{k,G+Q}$, applying the partial safety factors for the STR limit state to DIN 1054, Table 2.

In general:

$$E_d = E_{k,G} \cdot \gamma_G + (E_{k,G+Q} - E_{k,G}) \cdot \gamma_Q.$$

The design values of the effects E_d specified in DIN 1054, Table 2 are summarised below as functions of the Action Combinations (AC) given in DIN 1054, which are allocated to Design Situations 1 to 4:

	Load case	x-axis	y-axis
Situation 1[*] (construction state)	2	$16.20 \cdot 1.20$ $+ (58.57 - 16.20) \cdot 1.30$ $E_{x,d} = $ **74.52 kN/m**	$22.16 \cdot 1.20$ $+ (77.61 - 22.16) \cdot 1.30$ $E_{y,d} = $ **98.68 kN/m**
Situation 2[*] (construction state)	2	$29.99 \cdot 1.20$ $+ (45.16 - 29.99) \cdot 1.30$ $E_{x,d} = $ **55.71 kN/m**	$52.69 \cdot 1.20$ $+ (90.91 - 52.69) \cdot 1.30$ $E_{y,d} = $ **112.91 kN/m**
Situation 3 (final state)	1	$29.03 \cdot 1.35$ $+ (52.74 - 29.03) \cdot 1.50$ $E_{x,d} = $ **74.76 kN/m**	$51.47 \cdot 1.35$ $+ (113.10 - 51.47) \cdot 1.50$ $E_{y,d} = $ **161.93 kN/m**
Special case (w/o subgrade)	3	$36.48 \cdot 1.1$ $+ (60.80 - 36.48) \cdot 1.1$ $E_{x,d} = $ **66.88 kN/m**	$59.07 \cdot 1.1$ $+ (120.70 - 59.07) \cdot 1.1$ $E_{x,d} = $ **132.77 kN/m**

[*]   It may be necessary to analyse additional construction conditions with different embankment heights

9.10.7 Design Values of Resistances

The general design values of the resistances are given by:

In general:

$$R_{B,d} = \frac{\eta_M}{\gamma_M} \cdot \frac{R_{B,k0.5\%}}{A_1 \cdot A_2 \cdot A_3 \cdot A_4 \cdot A_5} \ [\text{kN/m}]$$

where:

$\gamma_M = 1.40$ (LC 1) $\quad \gamma_M = 1.30$ (LC 2) $\quad \gamma_M = 1.20$ (LC 3),
$\eta_M = 1.10$,
$A_1 = 1.26$ $\quad$ t = 10 h, $\quad \varepsilon_{\text{Failure}} = 13\%$),
$A_1 = 1.34$ $\quad$ t = 500 h, $\quad \varepsilon_{\text{Failure}} = 13\%$),
$A_1 = 1.65$ $\quad$ t = 1,000,000 h, $\quad \varepsilon_{\text{Failure}} = 13\%$),
$A_2 = 1.10$ $\quad A_3 = 1.00$ $\quad A_4 = 1.00$ $\quad A_5 = 1.00$ (see Section 9.10.1).

	Load case	x-axis $R_{B,k0.5\%} = 200$ kN/m	y-axis $R_{B,k0.5\%} = 400$ kN/m
Situation 1[*] (construction state) (t = 10 h)	2	$R_{x,B,d} = \frac{1.1}{1.3} \cdot \frac{200}{1.26 \cdot 1.1 \cdot 1 \cdot 1 \cdot 1}$ $= 122.10$ kN/m	$R_{y,B,d} = \frac{1.1}{1.3} \cdot \frac{400}{1.26 \cdot 1.1 \cdot 1 \cdot 1 \cdot 1}$ $= 244.20$ kN/m
Situation 2[*] (construction state) (t = 500 h)	2	$R_{x,B,d} = \frac{1.1}{1.3} \cdot \frac{200}{1.34 \cdot 1.1 \cdot 1 \cdot 1 \cdot 1}$ $= 114.81$ kN/m	$R_{y,B,d} = \frac{1.1}{1.3} \cdot \frac{400}{1.34 \cdot 1.1 \cdot 1 \cdot 1 \cdot 1}$ $= 229.62$ kN/m
Situation 3 (final state) (t = 1,000,000 h)	1	$R_{x,B,d} = \frac{1.1}{1.4} \cdot \frac{200}{1.65 \cdot 1.1 \cdot 1 \cdot 1 \cdot 1}$ $= 86.58$ kN/m	$R_{y,B,d} = \frac{1.1}{1.4} \cdot \frac{400}{1.65 \cdot 1.1 \cdot 1 \cdot 1 \cdot 1}$ $= 173.16$ kN/m
Special case (w/o subgrade) (t = 1,000,000 h)	3	$R_{x,B,d} = \frac{1.1}{1.2} \cdot \frac{200}{1.65 \cdot 1.1 \cdot 1 \cdot 1 \cdot 1}$ $= 101.01$ kN/m	$R_{y,B,d} = \frac{1.1}{1.2} \cdot \frac{400}{1.65 \cdot 1.1 \cdot 1 \cdot 1 \cdot 1}$ $= 202.02$ kN/m

[*] It may be necessary to analyse additional construction conditions with different embankment heights

9.10.8   Analysing Bearing Capacity

The geosynthetic reinforcement is analysed in the STR limit state as described in Section 9.7.1.2.

In general:

$$R_{B,d} \geq \Delta E_d + E_{M,d} \, .$$

	Load case	x-axis $R_{B,k0.5\%} = 200$ kN/m	y-axis $R_{B,k0.5\%} = 400$ kN/m
Situation 1[*] (construction state)	2	122.10 kN/m > 74.52 kN/m	244.20 kN/m > 98.68 kN/m
Situation 2[*] (construction state)	2	114.81 kN/m > 55.71 kN/m	229.62 kN/m > 112.91 kN/m
Situation 3 (final state)	1	86.58 kN/m > 74.76 kN/m	173.16 kN/m > 161.93 kN/m
Special case (w/o subgrade)	3	101.01 kN/m > 66.88 kN/m	202.02 kN/m > 132.77 kN/m

[*]   It may be necessary to analyse additional construction conditions with different embankment heights

Adequate bearing capacity is thus demonstrated.

If adequate bearing capacity cannot be demonstrated the analysis is repeated using higher short-term strengths and the resulting axial stiffnesses until adequate bearing capacity is demonstrated.

Note: See Section 9.3 for analysis of anchorage and overlapping.

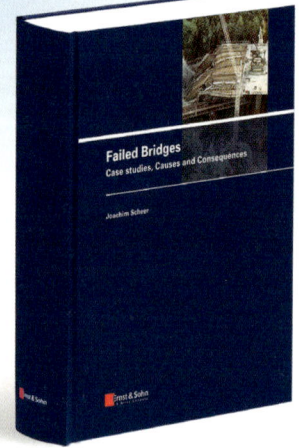

10 Foundation Systems using Geosynthetic-encased Columns

10.1 Definitions

In foundation systems using geosynthetic-encased columns, columns of granular material are installed on a load-bearing stratum to transmit static and variable loads in soft ground. A number of different installation methods are used. The columns are encased in geosynthetics, guaranteeing filter stability between the column fill and the surrounding ground. Thanks to the geosynthetic tube the column is supported radially together with the soft soil, and the tube is subjected to circumferential tensile forces. Figure 10.1 shows a schematic example of an embankment foundation.

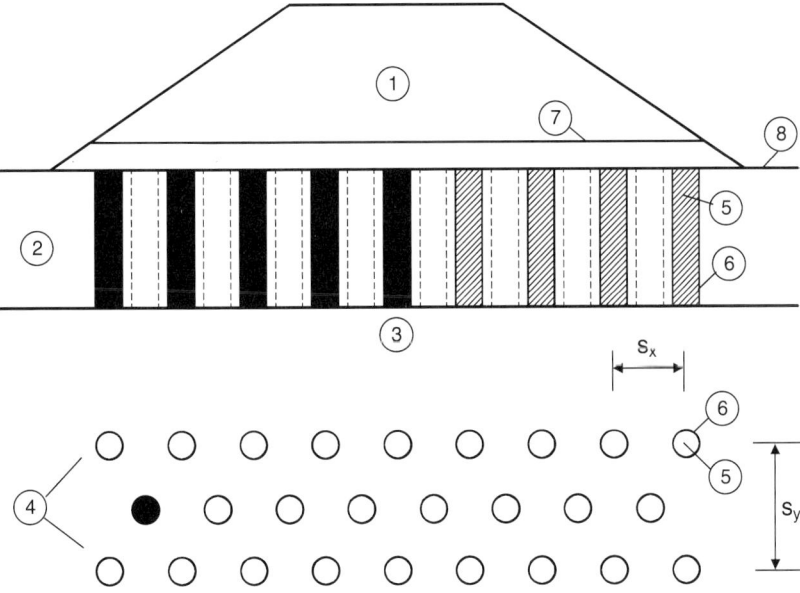

1 Embankment
2 Soft strata
3 Load-bearing ground
4 Column grid (e.g. triangular grid)
5 Column fill (granular)
6 Geosynthetic casing
7 Horizontal geosynthetic reinforcement
8 Column head plane

Figure 10.1 Schematic diagram of the 'geosynthetic-encased column' foundation system

Recommendations for Design and Analysis of Earth Structures using
Geosynthetic Reinforcements (EBGEO). German Geotechnical Society.
© 2011 Ernst & Sohn GmbH & Co. KG.
Published by Ernst & Sohn GmbH & Co. KG.

In addition to the live load, the load imposed on the foundation system is represented by the structure for which the foundation is intended (e.g. embankment fill).

The columns are arranged in a uniform column grid. Either rectangular or triangular grids are generally used. Triangular grids can be formed by either a square grid rotated through 45° or by other grid formats (e.g. a 60° grid, where three columns form respective equilateral triangles and all columns are equally spaced). The grid spacing is given by s_x and s_y (Figure 10.1) and the column area by A_S (Figure 10.2).

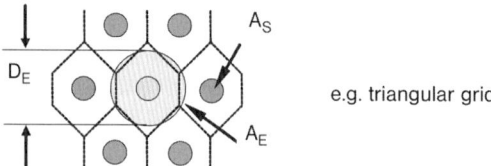

e.g. triangular grid

Figure 10.2 Definition of A_S, A_E and D_E

The influence area A_E of a column describes the area as that part of the overall area allocated to a column in the column head plane (Figure 10.2).

The unit cell is given by converting the influence area A_E into an equivalent area cylinder of diameter D_E (Figure 10.2).

The area ratio a_S describes the ratio of the column area A_S to the influence area A_E allocated to the individual column, where: $a_S = A_S / A_E$ (Figure 10.2).

The column fill consists of a high shear strength granular material (e.g. gravel, sand, crushed stone, crushed gravel).

The geosynthetic casing consists of geowovens, geogrids or geocomposites, where the final envelope is created using a number of techniques (e.g. circular weaving, stitching, bonding, welding, etc.). The envelope is adapted to the diameter of the columns (finished diameter), depending on the installation method and application, and envelops the entire length of the column in the soft stratum.

Activation widening represents the difference between the finished diameter of the geosynthetic casing D_{geo} and the installation diameter of the column D_{Column} (unloaded) (Figure 10.4). The column diameter is enlarged under load prior to the activation of circumferential tensile forces until it corresponds to the tube diameter D_{geo}.

The circumferential tensile force in the geosynthetic casing is mobilised when activation widening is achieved.

The load redistribution factor E describes the stress concentration above the column heads and represents the ratio of the load component Q_S transferred through the column to the total load Q_E in the entire area of influence of a column, where: $E = Q_S / Q_E$.

The stiffness ratio $k_{s,T}/k_s$ describes the ratio of the stiffness of the encased column (bearing element) $k_{s,T}$ to the stiffness of the surrounding soft stratum k_s. In terms of the resulting stiffness ratios, the horizontal reinforcement above the column heads is configured either for engineering purposes only or is additionally designed for membrane forces as described in Section 9.6.3. The regulations stipulated in Section 9.2.1 are used to determine the stiffness ratios, where the stress concentration above the column heads or the load redistribution factor E is adopted.

The horizontal geosynthetic reinforcement is located close to the column heads and is used for general stability, for transferring spreading forces or to facilitate load transfer into the geosynthetic-encased columns as described in Section 10.1, as well as to compensate for settlements.

10.2 Modus Operandi and Applications

10.2.1 Modus Operandi

Geosynthetic-encased columns:

– reduce absolute and differential settlements,
– accelerate settlement and excess pore water pressure dissipation, and
– increase stability in the construction and final states.

The stress concentration above the column heads invokes an additional, radial, horizontal stress directed outward from the columns, or increased earth pressure. A horizontal support of the same magnitude as the load on the columns or the stress concentration above the column heads is required in the soft stratum.

In uncased columns this support is entirely mobilised by the passive earth pressure in the soft strata as a result of the increase in column diameter (bulging). In very soft soils this leads to considerable deformations. Using the geosynthetic-encased column system the radial, horizontal column support is guaranteed by the geosynthetic casing in conjunction with the support provided by the surrounding soft stratum.

The increase in column diameter under load in the geosynthetic casing leads to strains (after achieving activation widening) and thus to circumferential tensile forces. The magnitude of the respective circumferential tensile force is determined by the geosynthetic material behaviour, among other things, and is proportional to further horizontal or radial column deformation.

The magnitude of the support provided by the surrounding soil is also linked to the increase in the column diameter, which is restricted to a large degree by the circumferential tensile forces in the geosynthetic. Due to the reduction in the required support provided by the soft stratum made possible by this, only a fraction of the passive earth pressure is activated as a support in the ground. The reduction in surcharge stresses above the soft stratum facilitates lower settlement values in the soft strata.

Because the stiffness ratio of the columns to the surrounding soft stratum is generally lower than when using the point bearings described in Section 10.1, for example, approximately uniform settlement can be assumed for the column and the surrounding soft stratum when designing the columns. Load transfer into the encased columns is achieved by way of the formation of stress arches in the cover fill. Flexible and self-regulating load-bearing behaviour is the result, because if the columns yield the load is redistributed to the soft strata, thereby increasing the ground resistance supporting the columns, leading to interactive load redistribution back to the columns.

The effectiveness of the system (for reducing settlement and increasing general stability, among other things) and the unloading effect in the soft strata are thus increased with:

– increasing area ratio,
– increasing axial stiffness of the tube,
– increasing shear strength of the column fill.

Generally, designing the horizontal reinforcement layers located above the column heads for membrane forces can be dispensed with (see Table 10.2). If activation widening is decreased, tube stiffness increased and column fill shear strength increased, the stiffness ratio increases and the load-bearing behaviour approaches that of the point bearing elements described in Section 9. This may make it necessary to configure and design horizontal reinforcement layers as described in Section 9.6.3 (Section 10.6.3).

Overall, only minor settlements occur once construction is complete. On the one hand, this is due to settlement reduction caused by stress concentration above the columns and the associated reduction in stresses above the soft stratum, and due to accelerated settlement by the action of the columns as vertical drains on the other. Generally, this leads to a large proportion of settlements being compensated for during the construction period.

10.2.2 Applications

Geosynthetic-encased columns may be employed as a foundation system for transferring static and variable loads in soft soils (soft strata). One special application and an advantage compared to non-encased columns is based on the support effect of the geosynthetic tube in very soft soils ($c_u < 15$ kN/m^2), e.g. in peat or in very soft silts/clays, such as mud and ooze.

The suitability of the method shall be examined on a case-by-case basis. It is not possible to define generally recognised limits because of the flexibility of this foundation system.

Some boundary conditions, representing the limits of the foundation system on an empirical basis, are summarised below:

- Governing circumferential tensile forces in the geosynthetic casing cannot be activated in very stiff soil strata, because the support provided by the surrounding ground can be much higher than the support required by the tube. This application is therefore only economical in soft soils. Special methods shall be employed in extremely soft soils (e.g. to stabilise the columns during the manufacturing process and thereafter). The usual application range can be given as follows, as a function of the constrained moduli of the soft strata, for a reference stress of 100 kN/m^2:
  $0.5 \text{ MN/m}^2 < E_{s,100 \text{ kN/m}^2} < 3.0 \text{ MN/m}^2$.
  In terms of the undrained shear strength c_u the following application limits are recommended:
  $3 \text{ kN/m}^2 < c_u < 30 \text{ kN/m}^2$.
  If special engineering measures are adopted it may be possible to manufacture and use geosynthetic-encased columns even in soft strata with $c_u < 3 \text{ kN/m}^2$ or in stiffer strata with $c_u > 30 \text{ kN/m}^2$.

- The geosynthetic-encased columns shall bottom in load-bearing ground to ensure that the foundation system functions as described. The stiffness of the load-bearing stratum should be at least a factor of 10 higher than the overlying soft stratum. The load-bearing ground can therefore be described by the following, simplified parameters:
  $E_{s,Ground} > 5.0 \text{ MN/m}^2$ (for a reference stress of 100 kN/m^2)
  to limit settlement of the load-bearing ground and
  $\varphi'_{s,k,Ground} > 30°$
  to prevent bearing capacity failure at the column feet.

- The column lengths l_{Column} and the maximum soft strata thickness are given by the machine boundary conditions.
  Common manufacturing lengths l_{Column} are:
  $3 \text{ m} < l_{Column} < 20 \text{ m}$.

- The column diameter D_{Column} depends on the manufacturing method, among other things. The circumferential tensile forces increase with increasing column diameter and constant area ratio. A minimum column diameter of 0.4 m is recommended to safely activate the governing circumferential tensile forces.
  Common column diameter limits are:
  $0.5 \text{ m} \leq D_{Column} \leq 1.5 \text{ m}$.
  Activation widening as a result of surcharge should not exceed approx. 3% of the diameter in order to restrict the deformations occurring before the tube is activated. Preliminary activation widening, i.e. activation widening occurring during system manufacturing and before loading, can generally be ignored.

- The magnitude of the axial stiffness of the tube is theoretically unlimited, but has a governing impact on the stiffness ratio of the columns to the surrounding soft strata, *cf.* Section 10.2.1. If the horizontal reinforcement is not designed for membrane forces as described in Section 9.6.3 (*cf.* Section 10.2.1),

the column stiffness shall be adapted to suit the stiffness of the surrounding soft strata. The axial stiffness of the tube has a governing impact.
The axial stiffness of the tube J is generally between 1,000 kN/m and 4,000 kN/m.

- The effectiveness of the column foundation is determined by the ratio of the stiffness of the column fill and the surrounding soft stratum. A column fill constrained modulus at least ten times larger than the constrained modulus of the soft strata is recommended ($E_{s,Column} > 10 \cdot E_{s,Soft\ stratum}$).

10.3 Manufacturing Methods

10.3.1 General Recommendations

Various geosynthetic-encased column manufacturing methods are introduced below as shown in Figure 10.3. They are known as excavation methods and displacement methods, according to the principle employed. The primary difference between the methods lies in the void occupied by the columns in the ground.

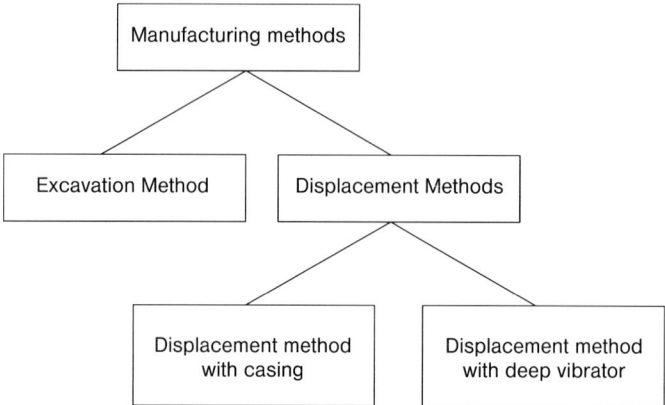

Figure 10.3 Manufacturing methods

The individual methods are described. Notes on method selection are then provided, including the risks involved in each method, in terms of the foundation quality and impacts on the surroundings.

10.3.2 Excavation Method

Using the excavation method an open casing (support casing) of diameter $D \approx 0.5$ to 1.5 m is generally driven to load-bearing ground with the aid of a leader-guided vibrator. The soft strata within the casing are then excavated.

Following excavation the tube, finished with the correct length as per requirements, is installed in the casing and the column filled. The casing is then pulled by the vibrator, simultaneously compacting the column fill by vibration.

10.3.3 Displacement Methods

Using displacement methods the ground is displaced during column manufacture. No soil is removed. The ensuing vertical and horizontal displacements shall be adopted for planning and in the design approaches. Special equipment and casings are required for use with displacement methods. They display advantages and disadvantages specific to each application.

Displacement method with casing

Using this method a casing is driven to load-bearing ground with the aid of a leader-guided vibrator. The ground at the base of the casing is laterally displaced by two flaps forming a cone. The tube is then installed analogous to the excavation method, the column filled and the casing pulled under vibration, after opening the bottom flaps.

Displacement method with deep vibrator

Using this method the geosynthetic-encased columns are manufactured with the aid of a deep vibrator. The geosynthetics are drawn in via the outer sleeve of the deep vibrator and then driven to load-bearing strata. The column sleeve is filled by continuously feeding material in while simultaneously compacting as the deep vibrator is extracted.

10.3.4 Method Selection

The excavation method is preferred in soils with high resistance to penetration or where vibration in adjacent buildings, transport infrastructure, etc, need to be minimised.

The advantage of the displacement method over the excavation method is based on the faster and more efficient column manufacture and the transfer of a prestress into the soft stratum. Additionally, no disposable soil is extracted.

An economical disadvantage of the displacement method is the increased geosynthetics required due to the smaller column diameter. Empirically, this is balanced by faster column manufacture. The excess porewater pressures, vibrations and deformations occurring in the soft strata shall be taken into consideration when adopting displacement methods. Overall, the displacement method makes increased demands on experience, and care in design and execution to ensure adequate bearing capacity and foundation quality.

When using the displacement method with a casing special care shall be taken to exactly fit the diameter of the geotextile tube to the diameter of the casing.

Due to the stress applied to the soft strata the columns may be pinched to less than the internal diameter of the displacement tube during execution and before applying a load when using this method. Correct incorporation of any activation widening caused by constriction requires special experience in execution and design, because this is adopted in the analysis model (*cf.* Figure 10.4) to facilitate design and settlement estimates.

If deep vibrators are used to manufacture the columns, activation widening can generally take place during infilling, so it may be ignored. In contrast to other methods greater circumferential tensile forces occur in the construction state,

Table 10.1 Characteristics of column manufacturing methods

	Excavation method	Displacement methods	
		with casing	with deep vibrator
Possible manufactured diameter	More than 1.5 m	Generally up to 0.8 m	Generally up to 0.6 m
Removal and disposal of soil material	Necessary	Unnecessary	Unnecessary
Time required for column manufacture	More	Less	Less
Manufacture with very high penetration resistances[1]	Possible	Generally not possible	Generally not possible
Vibrations and excess pore-water pressures as a result of column manufacture	Low	High[2]	High[2]
Column constriction during manufacture	No	Generally yes[2]	Generally no[2]
Horizontal and vertical displacement as a result of column manufacture	No	Yes[2]	Yes[2]
Prestressing of soft stratum during installation	No	Yes[2]	Yes[2]
Effects on geosynthetic casing during installation	Low	Low	Generally high
Examination of strata and column end depth	Possible visually	Via machine parameters	Via machine parameters

[1] For example, dense intermediate sand layers
[2] Depending on ground stiffness and grid spacing

because the tube is pre-strained during infilling. When designing the geosynthetic tube the anticipated effects (among others: preliminary strains and preliminary loads in the geosynthetics and corresponding A_2 values as described in Section 2) shall be taken into consideration, in addition to the load from the structure. If, after manufacturing the foundation system, it is uncertain whether constriction or activation widening has already occurred, measures shall be taken, for example measurements, to examine whether additional deformations are negligible as a result, or need to be taken into consideration.

The characteristic properties and the advantages and disadvantages of the manufacturing methods described here are summarised in Table 10.1.

It is recommended to perform feasibility testing of the selected method by manufacturing test columns prior to execution, if necessary with accompanying instrumented monitoring.

10.4   Design Recommendations and Engineering Notes

When designing a foundation using geosynthetic-encased columns it shall be noted that the bearing system must undergo deformation to activate the required circumferential tensile forces. It should therefore not be adopted for structures highly sensitive to settlement or only when certain boundary conditions are observed (e.g. long settlement periods, temporary cover fill).

The following recommendations and empirical data shall be observed:

– There should be a minimum, granular cover above the columns. The minimum cover should correspond approximately to the clear column spacing, but be at least approx. 1 m.
– A minimum column diameter of 0.4 m should be used.
– An area ratio $a_s = 10\%$ (cf. Section 10.1) should be adhered to.
– A horizontal geosynthetic reinforcement should be arranged either directly or generally up to 0.3 m above the columns heads to provide global stability, transfer spreading forces or support load transfer into the geosynthetic-encased columns, and to compensate for settlements (also see Section 10.6.3).
– If settlement requirements are specified settlement periods under load should be planned, because the foundation system settlements are delayed corresponding to consolidation (vertical drains).
– In high embankments the deformations required for geosynthetic casing activation are generally compensated for during embankment tipping.
– During the construction phase the system cover fill should at least correspond to the anticipated subsequent loads. The necessary settlement periods can be reduced further by applying a surcharge to the temporary fill.
– Examine whether the foundation deformations during and after load application require monitoring or inspection by a suitable programme of measurements on a case-by-case basis.

10.5 Building Materials

The materials required for geosynthetic-encased columns are the geosynthetics themselves and granular soil. In addition, the foundation system is completed by the cover fill (e.g. embankment) with horizontal geosynthetic reinforcement (see Figure 10.1). In terms of embankment fill the reinforced earth structure as described in Section 9.1 and the zone above this are differentiated. The following requirements apply to materials, inasmuch as no higher demands are made as a result of analysis and design, or for reasons specific to the application:

Geosynthetic casing (including seams, connections, etc.):
- woven, geogrid or geocomposite,
- higher or similar permeability in terms of the minimum demands on the column fill,
- axial design resistance of the geosynthetics (vertical following installation)
- $R_{Bd} \geq 20$ kN/m and short-term strength $R_{B,k0} \geq 60$ kN/m,
- radial design resistance of the geosynthetics (circumferential)
- $R_{Bd} \geq 30$ kN/m and short-term strength $R_{B,k0} \geq 80$ kN/m,
- radial stiffness $J \geq 700$ kN/m (circumferential).

Column fill:
- granular, coarse-grained soil to DIN 18196 as a function of the grain size distribution of the natural ground (if a filter criterion needs to be adhered to),
- effective friction angle $\varphi'_k \geq 30°$,
- permeability greater than $k_f = 10^{-5}$ m/s, but at least two orders of magnitude more permeable than the surrounding soft stratum,
- at least loose to medium-dense compaction after column manufacture.

Horizontal geosynthetic reinforcement (including seams, connections, etc.):
- design resistance $R_{Bd} \geq 30$ kN/m and short-term strength $R_{B,k0} \geq 80$ kN/m.

Reinforced earth structures:
- demands as in Section 9.3.

If other materials are used (e.g. recycled materials) they shall be shown to be suitable for the specific method and environment.

10.6 Notes on Analysis and Design

10.6.1 General Recommendations

Analysis and design of a foundation using geosynthetic-encased columns is possible using either an analytical method after [3] and [4] or a numerical method. They require in-depth knowledge by the user when taking the interaction between the column and the surrounding soft stratum into consideration, as well as knowledge of the specific procedure boundary conditions.

The design depends on local conditions, procedure boundary conditions and the specific loads as a function of the manufacturing method, etc. The basic ideas and design principles are described below. The notes on boundary conditions, design requirements and the necessary analyses shall be followed.

10.6.2   Actions and Resistances

Actions on the foundation system include permanent and variable loads as described in Section 1.2 and DIN 1054. Any special actions during manufacture (e.g. geosynthetic casing pre-strains, column fill compaction, etc., *cf.* Section 10.3.4) shall be taken into consideration. If no empirical data for similar conditions is available specimens are taken from test columns and the A_2 values determined.

Resistances include:

- shear strength, and column and embankment fill stiffness,
- shear strength, and soft strata and load-bearing ground stiffness,
- strength and axial stiffness of the column casing,
- strength and axial stiffness of the horizontal geosynthetic reinforcement.

10.6.3   Designing the Horizontal Geosynthetic Reinforcement

As explained in Section 10.2.1 approximately equal settlements can generally be assumed between the column and the surrounding soft stratum (Zone I in Table 10.2), and the arching in the cover fill is adequate to transfer loads to the encased columns. The horizontal reinforcement is installed as a structural element to guarantee global stability or to transfer spreading forces as described in Section 9.6.3.

However, stiffness ratios may occur between the columns and the surrounding ground as a function of the system parameters, which make it necessary to design the horizontal reinforcement for membrane forces as described in Section 9.6.3.

Large increases in settlement and changes in load-bearing behaviour are not anticipated due to the remaining, self-regulating load-bearing behaviour of the columns in Zone II (in the transition zone between Zones I and II in Table 10.2). However, larger settlements than those calculated may occur due to the lower stress concentration in the column heads possible in such cases. In individual cases the design of horizontal reinforcement layers may be necessary to ensure load transfer. Design may be based on Section 9.6.3, where the tensile forces in the horizontal reinforcement are calculated adopting the stiffness of the column or the stiffness ratio between the column and the surrounding ground.

If the stiffness ratios are higher still (Zone III in Table 10.2) the effectiveness of the foundation system is no longer guaranteed without designing the horizontal reinforcement for membrane forces. Design to Section 9.6.3 is therefore necessary in such cases.

Table 10.2 Necessity of designing a horizontal geosynthetic reinforcement for membrane forces to Section 9 as a function of stiffness ratios

Zone	Stiffness ratio	Design of horizontal geosynthetic reinforcement[1]
I	$k_{s,T}/k_s \leq 50$	Design unnecessary
II	$50 < k_{s,T}/k_s \leq 75$	Design recommended
III	$k_{s,T}/k_s > 75$	Design necessary

[1] The minimum reinforcement using design resistance R_{Bd} to Section 10.5 is required for all zones!

If it is necessary to design the horizontal reinforcement for membrane forces the grid spacing of triangular grids shall be changed to that of a rectangular grid or a square grid rotated through 45°, *cf.* Section 9.1.

10.6.4   Column Design

10.6.4.1   Analysis Model

The analysis model after [3] and [4] described below is based on the interaction between the columns and the surrounding soft strata, where approximately equal settlements are assumed in the column head plane (Figure 10.1). Some minimum demands are made on the analysis model regardless of the analysis method used (Figure 10.4):

– The interactive load-bearing behaviour, i.e. the time- and load-dependent stress concentration above the column heads, shall be recorded.
– The consolidation of the soft strata shall be recorded for analysis of the settlement history, adopting the column as a vertical drain.
– The stiffness of the soft stratum shall be adopted as a function of the prevalent effective stress (stress-dependent constrained modulus).
– If activation widening occurs the geosynthetic casing activation shall be correctly recorded as a function of the load and time.
– Where the displacement method is used the effect of displacement on the stress levels in the soft stratum shall be taken into consideration.
– In stratified ground all governing soil strata shall be adopted separately with their respective parameters (no summarising or averaging the parameters of individual soil strata).
– If compaction effects occur within the geosynthetic casing without a supporting, external column fill casing (e.g. when infilling using a deep vibrator), the load imposed on the geosynthetic casing is taken into consideration.
– The following parameters with a governing impact on the calculated circumferential tensile forces and settlements are introduced into the analyses:

- area ratio a_s,
- radius of column r_s and the geosynthetic casing r_{geo},
- time-dependent, radial stiffness of the geosynthetic casing J,
- thickness of the governing soft strata h_i,
- stress-dependent stiffnesses of the soft strata
  (e.g. constrained modulus $E_{s,i}$),
- shear parameters of the soft strata φ'_B, c'_B and the column fill φ'_S,
- unit weights of the soft strata γ_B, γ'_B and the column fill γ_S, γ'_S,
- at-rest earth pressure coefficients $K_{O,B}$ and primary stress state of the soft stratum (increased due to displacement effect if necessary),
- mean surcharge stress σ_0 in the column head plane of the unit cell,
- permeability of the ground $k_{f,B}$ and the columns $k_{f,S}$.

A schematic axis-symmetrical analysis model for a unit cell is shown in Figure 10.4.

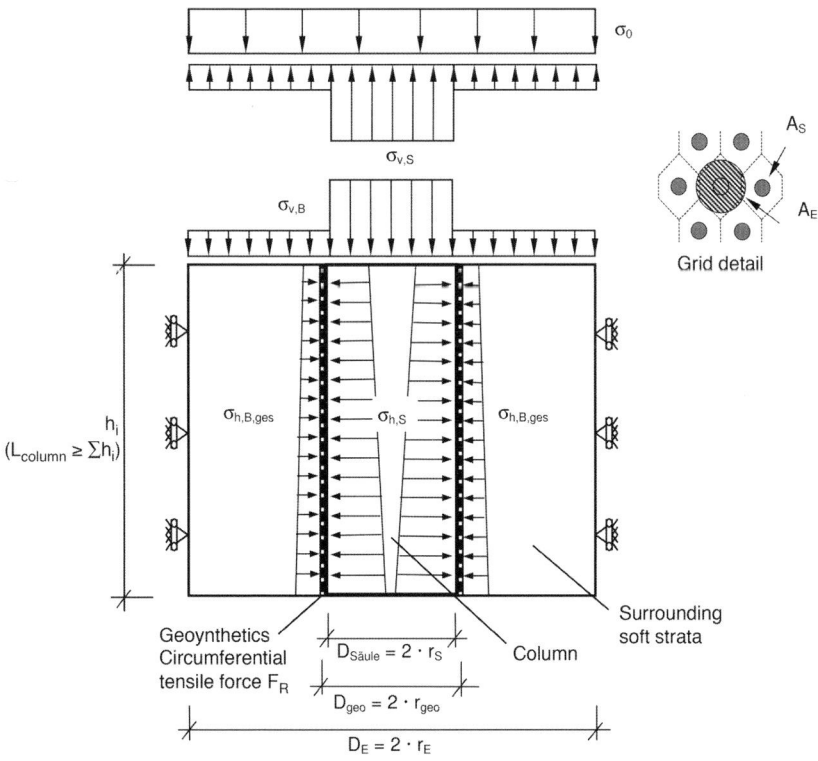

Figure 10.4 'Geosynthetic-encased column' load-bearing system and analysis model

213

10.6.4.2 Analysis Methods

The results of the analysis are the maximum circumferential strain and circumferential tensile force of the geosynthetic casing, and the primary settlements in the column head plane.

The derivation described below was taken from [3] and shows the analysis of a ground slab. Where possible, no single ground slab should be thicker than $h_i = 1$ m for the analysis. Depending on the geological structure of the soft stratum this provides an analysis model consisting of several ground slabs. The individual ground slabs are selected on the basis of the geological strata boundaries. The total settlement is given approximately by cumulating the settlements of the individual ground slabs. [3] gives more detailed information.

Assuming equilibrium between the loads in the column head plane σ_0 and the corresponding vertical loads above the column $\Delta\sigma_{v,S}$ and the soft stratum $\Delta\sigma_{v,B}$ we get:

$$\sigma_0 \cdot A_E = \sigma_{v,S} \cdot A_S + \sigma_{v,B} \cdot (A_E - A_S). \qquad \text{Eq. (10.1)}$$

The horizontal stresses are given by the normal stresses resulting from the surcharge and the soil dead weights, where $\sigma_{ü,S}$ and $\sigma_{ü,B}$ stand for the surcharge stresses in the column and in the soft stratum:

$$\sigma_{h,S} = \sigma_{v,S} \cdot K_{a,S} + \sigma_{ü,S} \cdot K_{a,S} \qquad \text{Eq. (10.2)}$$

$$\sigma_{h,B} = \sigma_{v,B} \cdot K_{0,B} + \sigma_{ü,B} \cdot K_{0,B}^* \qquad \text{Eq. (10.3)}$$

The geosynthetic casing (installation radius r_{geo}) is characterised by linear-elastic material behaviour with axial stiffness J:

$$F_R = J \cdot \Delta r_{geo} / r_{geo} \qquad \text{Eq. (10.4)}$$

The circumferential tensile stress can be converted to a radial horizontal stress $\Delta\sigma_{h,geo}$, which is allocated to the geosynthetics, using the *boiler equation*:

$$\sigma_{h,geo} = F_R / r_{geo} \qquad \text{Eq. (10.5)}$$

The individual horizontal stresses result in a differential stress $\Delta\sigma'$. This corresponds to the mobilisation of an additional earth pressure component in the soft stratum, until horizontal stress equilibrium is achieved.

$$\sigma_{h,Diff} = \sigma_{h,S} - (\sigma_{h,B} + \sigma_{h,geo}) \qquad \text{Eq. (10.6)}$$

The column strain is the result of the differential stress. The radial horizontal deformation Δr_S and the settlement of the soft stratum s_B with the constrained modulus $E_{S,B}$ are derived after [5] for a radially and longitudinally loaded hollow cylinder of height h_0 (v_B = Poisson's ratio of soft stratum):

214

$$\Delta r_S = \frac{\sigma_{h,Diff}}{E^*} \cdot \left(\frac{1}{a_S} - 1\right) \cdot r_S \qquad \text{Eq. (10.7)}$$

$$s_B = \left(\frac{\sigma_{v,B}}{E_{s,B}} - 2 \cdot \frac{1}{E^*} \cdot \frac{v_B}{1-v_B} \cdot \sigma_{h,Diff}\right) \cdot h\,0 \qquad \text{Eq. (10.8)}$$

where:

$$E^* = \left(\frac{1}{1-v_B} + \frac{1}{1+v_B} \cdot \frac{1}{a_S}\right) \cdot \frac{(1+v_B) \cdot (1-2v_B)}{(1-v_B)} \cdot E_{s,B} \qquad \text{Eq. (10.9)}$$

and

$$a_S = A_S / A_E . \qquad \text{Eq. (10.10)}$$

The following relationship exists between the settlement of the column s_S and the radial deformation at the column edge Δr_S for a constant volume of the column material as a function of the original radius r_0 (generally the installed column radius) or the original height h_0 (generally the installed height):

$$s_S = \left[1 - \frac{r_0^2}{(r_0 + \Delta r_S)^2}\right] \cdot h_0 \qquad \text{Eq. (10.11)}$$

The horizontal deformations shall be acceptable and the activation widening taken into consideration:

$$\Delta r_S = \Delta r_{geo} + (r_{geo} - r_S) \qquad \text{Eq. (10.12)}$$

In approximation, relative settlements between the column and the surrounding soft stratum do not occur:

$$s_S = s_B \qquad \text{Eq. (10.13)}$$

Based on the previous 11 equations the following equation can be derived for the horizontal deformation at the edge of the column:

$$\Delta r_S = \frac{K_{a,S} \cdot \left(\frac{1}{a_S} \cdot \sigma_0 - \frac{1-a_S}{a_S} \cdot \sigma_{v,B} + \sigma_{\ddot{u},S}\right) - K_{0,B} \cdot \sigma_{v,B} - K_{0,B}^* \cdot \sigma_{\ddot{u},B} + \frac{(r_{geo} - r_S) \cdot J}{r_{geo}^2}}{\frac{E^*}{(1/a_S - 1) \cdot r_S} + \frac{J}{r_{geo}^2}} \qquad \text{Eq. (10.14)}$$

Adopting this deformation the equation below for a ground slab now only includes the unknown variable $\sigma_{v,B}$. The conditional equation can be solved iteratively by estimating $\sigma_{v,B}$. Use of a software application is recommended due to the relatively time-consuming calculations involved.

$$\left\{ \frac{\sigma_{v,B}}{E_{s,B}} - \frac{2}{E^*} \cdot \frac{v_B}{1-v_B} \left[\begin{array}{c} K_{a,S} \cdot \left(\dfrac{1}{a_S} \cdot \sigma_0 - \dfrac{1-a_S}{a_S} \cdot \sigma_{v,B} + \sigma_{\ddot{u},S} \right) - \\[2mm] K_{0,B} \cdot \sigma_{v,B} - K_{0,B}^* \cdot \sigma_{\ddot{u},B} + \dfrac{(r_{geo} - r_S) \cdot J}{r_{geo}^2} - \dfrac{\Delta r_S \cdot J}{r_{geo}^2} \end{array} \right] \right\} \cdot h$$

$$= \left[1 - \frac{r_S^2}{(r_S + \Delta r_S)^2} \right] \cdot h \hspace{3cm} \text{Eq. (10.15)}$$

To incorporate activation widening it is first necessary to estimate an activation load and then to calculate the horizontal deformation at the edge of the column, adopting $J = 0$. By adopting $\Delta r_S = r_{geo} - r_S$ as the iteration condition, the activation load can be determined and the analysis completed with an activated geosynthetic casing for the load on the foundation system over and above this. The settlements before and after activation are then added.

Because the stiffness of the soft strata is generally highly dependent on the stress state, the constrained modulus of the soft stratum $E_{s,B}$ is introduced into the analysis as realistically as possible as a function of the governing mean effective stress p^* in the soft stratum. For analyses assuming activation widening it is necessary to determine the adoptable constrained modulus in the stress range between primary stress and activation loading and then between activation widening and final loading.

Generally, a simple power function after Ohde, with a reference constrained modulus $E_{s,B,ref}$, a reference stress p_{ref} and a stiffness exponent m (in normally consolidated, cohesive and organic soils ≈ 1) is used to indicate the stress dependence.

$$E_{s,B} = E_{s,B,ref} \cdot \left(\frac{p^* + c_B' \cdot \cot \varphi_B'}{p_{ref}} \right) m \hspace{3cm} \text{Eq. (10.16)}$$

The expression $c_B' \cdot \cot\varphi_B'$ takes the impact of cohesion into consideration and is used in analogy to the constitutive equations in numerical analyses. A non-loaded soil element (ground surface) would not possess stiffness without this term, in contrast to real conditions. If the loads change the effective stresses in the soft stratum before ($p1^*$) and after ($p2^*$) the load change are determined. The adopted stress is calculated using $p^* = (p2^* - p1^*)/\ln(p2^*/p1^*)$. The mean value $(p1^* + p2^*)/2$ is often sufficiently precise. It shall be noted that the horizontal stress in a one-dimensional compression test is defined by the at-rest earth pressure coefficient, while for a geosynthetic-encased sand column a horizontal stress increased by the differential stress $\sigma_{h,Diff}$ acts in the surrounding soil stratum. Ignoring this increase results in a conservative analysis. An approximate consideration of the effects of the size of the respective constrained modulus and the cohesion is described in [3].

An application can be developed for practical analysis using the methods introduced here. It simulates incremental load increases and thus calculates the complete load-settlement curve or stress-strain curve of the selected system. The horizontal deformation at the column edge during load increase is continuously determined and the geosynthetic casing activated just after achieving widening resulting from the surcharge. [3] gives more detailed information.

The analysis model after [3] and [4] can be simplified for circumferential tensile force design as described in Section 10.6.4.3 by ignoring the lateral support provided by the soft stratum $\sigma_{h,B}$. This gives the equations used in [6]. A greater circumferential tensile force is determined and design is conservative as a result. This simplification is not suitable for determining the settlements of the foundation system.

10.6.4.3 Analysing the Transfer of Circumferential Tensile Forces

Analysis of the transfer of circumferential tensile forces in the geosynthetic casing is performed for the STR limit state. The effects from permanent (dead weights) and variable (live loads) actions are taken into consideration.

For geosynthetic-encased columns note that non-linear behaviour occurs, where increases in the circumferential tensile forces are increasingly smaller for an increasing surcharge.

Because the characteristic effects are not proportional to the actions the principle of superposition, i.e. separate determination of the circumferential tensile forces for individual load components and the subsequent addition of effects or circumferential tensile forces, cannot be adopted. Instead, the effects are determined separately for the permanent and the total actions (permanent and variable) respectively, taking into consideration temporal and load-dependent non-linearity. The increase in the effect from variable actions only can be determined from the difference.

The calculated circumferential tensile forces are designated by E as effects for analysis in the STR limit state.

The design value for the effects E_d is given by multiplying by the partial safety factors for actions from permanent and variable loads:

$$E_d = E_{G,k} \cdot \gamma_G + (E_{G+Q,k} - E_{G,k}) \cdot \gamma_Q \qquad \text{Eq. (10.17)}$$

Analysis of the geosynthetic casing is performed for the STR limit state on the basis of the design tensile strength, which is regarded as the design value of a structural component resistance $R_{B,d}$.

$$E_d \leq R_{B,d} \qquad \text{Eq. (10.18)}$$

The design resistance of the geosynthetic casing $R_{B,d}$ is determined as described in Section 2.2, adopting the appropriate reduction factors A_i, and modified for the case in hand as follows:

$$R_{B,d} = \frac{R_{B,k,5\%} \cdot \eta_M}{\gamma_M}$$
Eq. (10.19)

where:

η_M calibration factor for modifying the STR limit case safety level
$\eta_M = 1.1$.

If the geosynthetic casing was produced with a seam or other joint, a strength reduction is adopted accordingly.

If large cyclic/dynamic actions on the foundation system are anticipated, they are adopted as described in Section 10.6.6.2.

10.6.5   Analysing Overall Stability

The stability of the overall system in the GEO limit state is analysed. If slip planes intersect the geosynthetic-encased columns and/or the horizontal geosynthetic reinforcement above the column heads, the resistances of these elements acting to increase stability may be adopted. However, if no special investigations have been carried out, it is recommended not to adopt the stability-enhancing action of the geosynthetic casing in the column axis (axial or vertical) for analysis of the overall stability in the GEO limit state.

The horizontal geosynthetic reinforcement is introduced into the analysis model by the design resistance as described in Section 1.2 (for a definition of the characteristic value and the design value of the pull-out resistance see Section 2.2.4.11).

The encased columns lead to an increase in shear resistance along the slip plane. The increase is caused by the stress concentration above the column heads and the high friction angle of the column fill.

The following is noted for determining the resistances to be adopted:

- The methods described in the literature (e.g. [1] and [2]) for uncased column foundations can be used for analysis of overall stability (general failure). They take into consideration the stress concentration above the column and thus the increase in shear strength by using mean parameters, i.e. equivalent values for unit weight, cohesion and friction angle for homogeneous ground. However, in addition to an analysis using mean shear parameters for the entire ground under consideration, the equivalent column fill parameters described in the literature [3] may be determined.
- As resistances the equivalent shear parameters are reduced using the appropriate partial safety factors.
- The time- and load-dependent stress concentration or load redistribution is given for the investigated states (initial states, construction states, final states) based on the demands on the column foundation analysis model as described in Section 10.6.4.1.

- In order to investigate the initial and construction states (e.g. c_u analysis) only the emplaced, granular column fill is adopted, taking the area ratio a_S into consideration for determining the increase in shear strength, inasmuch as no separate investigations on system consolidation are available at that time.
- With increasing effective stresses in the soft strata as a result of consolidation the increasing load redistribution to the columns can be recorded in the overall stability analysis model by an appropriate load- and time-dependent increase in the resistances.

10.6.6 Serviceability Limit State Analysis

10.6.6.1 Determining Settlement

The magnitude of subsequent settlement is critically important for practical design of the foundation method used and for success forecasts. Empirically, settlement occurring during the construction phase (e.g. in the course of filling) is compensated, while subsequent settlements may lead to delays in the construction phase or to damage to the building. Accordingly, building progress has a considerable impact on serviceability.

Settlement of foundations on geosynthetic-encased columns can be modelled on the basis of the axis-symmetrical analysis model as described in Section 10.6.4.1. In load-bearing ground settlement can generally be ignored.

Simplified, the load-settlement curve and the final settlement, and subsequently, via a separate analysis, the consolidation can be determined by adopting the geosynthetic-encased column as a vertical drain. The non-linearity of the load-settlement curve is taken into consideration. That is, the settlement is determined as a function of the respective stress when completely consolidated.

Secondary settlements (creep settlements) are anticipated in soft, cohesive soils and organic or organogenic soils. Creep settlements can be reduced by using geosynthetic-encased columns, in contrast to unimproved ground. There are currently no generally recognised analysis methods in terms of the creep behaviour of a foundation on geosynthetic-encased columns (a few notes are given in [3]). A rough estimate can generally be made by applying an approximate 50% reduction to the creep settlement values determined for unimproved ground conditions. If large creep settlements are anticipated instrumented monitoring and, in individual cases, application of the observational method to DIN 1054 are recommended.

10.6.6.2 Cyclic-dynamic Actions

Empiricism indicates that greater settlements are initiated compared to static loads if the foundation system is subjected to large cyclic-dynamic actions (live loads). No generally recognised analysis methods are known in this context. A temporary cover fill over the foundation system with a higher load than the anticipated subsequent live load is therefore recommended.

In systems subjected to governing cyclic-dynamic loads and those of Geotechnical Category GC 3 to DIN 1054, instrumented monitoring or, in individual cases, the observational method, should always be applied to verify deformation forecasts. In addition, the information and recommendations made in Section 12 shall be observed.

10.6.6.3 Overall Deformations

Generally, the deformation forecast based on the axis-symmetrical model of the unit cell (section of an assumed infinite field of columns) may be regarded as adequate to estimate the maximum anticipated settlement.

However, as soon as larger load differentials in the foundation zone, special demands on differential settlements or large horizontal loads are anticipated, additional deformation considerations and forecasts may be necessary.

No reliable analytical methods for forecasting overall deformations, including horizontal deformations, are currently known. A rough estimate of the anticipated horizontal deformations can be made on the basis of empirical data collected under similar conditions.

The overall deformations can be approximately determined using numerical methods. The installation process and the peculiarities of the manufacturing method used, as well as the requirements specified in Section 10.6.4.1 are taken into consideration. If a two-dimensional analysis model is used the columns may be replaced by slabs, retaining the area ratio a_S. The deformation behaviour of the foundation with geosynthetic-encased columns can then be approximately modelled with defined equivalent parameters for the column fill. Additional information can be taken from the literature [3].

Instrumented monitoring or, in individual cases, the observational method to DIN 1054 may be used to verify the forecast.

10.7   Inspection Criteria, Tolerances and Quality Assurance

The planning and execution of foundation systems with geosynthetic-encased columns requires appropriate inspection criteria and make special demands in terms of both acceptable tolerances and quality assurance.

Particular attention in terms of quality assurance shall be paid to control of reaching load-bearing ground. When using the excavation method this can generally by done by inspecting the excavated material. When using the displacement method it shall be ensured that the load-bearing ground is reached by recording suitable machine parameters. If necessary, the connection to the load-bearing ground shall be verified by specifying installation criteria. They are specified as obligatory installation instructions on site, taking the stratification of the respective ground into consideration.

A freely suspended casing with a top vibrator requires special monitoring to ensure correct installation.

A log is kept for each column. The following minimum data are recorded, checked and documented in the column log:

- column ID and consecutive manufacture number,
- manufacturing method,
- type and relevant data of the vibrator used,
- manufacturing time,
- column diameter,
- finished diameter of geosynthetic casing,
- type of geosynthetic casing (type and strength),
- column fill ID,
- end depth,
- height of column head.
- When using the excavation method:
  - length of support casing,
  - excavation data (e.g.: load-bearing ground reached and checked),
- When using the displacement method:
  - length of displacement casing or deep vibrator,
  - machine parameters to verify reaching load-bearing ground or a confirmation that the installation criteria were adhered to.

In individual cases, the following additional tolerances or demands must be met when manufacturing the foundations to guarantee the effectiveness of the method used.

The following variables are examined at the factory:

- finished diameter of geosynthetic casing:
  $\pm 1\%$ of D_{geo} but a maximum of ± 1 cm,
- installation diameter of column (support or displacement casing):
  $\pm 0.5\%$ of D_{Column} but a maximum of ± 0.5 cm.

The following variables are randomly inspected on site on at least three columns, or at least 3% of all columns:

- end depth or column toe embedment in load-bearing ground,
- column head integration in cover fill,
- position coordinates of column: max. 15 cm deviation from planned position (if it is anticipated that this value will be exceeded the deviations caused as a result of the process used are taken into consideration in design),
- column fill relative density.

In addition to the tolerances discussed above the demands on the column fill and the geosynthetic casing (including the minimum requirements as described in Section 10.5) are examined by appropriate suitability, self-monitoring and verification tests. The values discussed above are restricted further if special demands need to be met, e.g. deformation restrictions.

If the casing is subjected to special, process-specific effects during column manufacture the reduction factor A_2 for installation damage adopted during design is verified in the course of quality assurance (or self-monitoring/verification test) by testing appropriate specimens taken from test columns.

A quality assurance programme adapted to the specific application is compiled taking the points discussed above into consideration.

10.8 Bibliography

[1] Priebe, H. (1976): Abschätzung des Setzungsverhaltens eines durch Stopfverdichtung verbesserten Baugrundes. Die Bautechnik 5, pp. 160–162.

[2] Soyez, B. (1987): Bemessung von Stopfverdichtungen. Ingénieur des TPE Section des ouvrages en terre, Laboratoire Central des Ponts et Chaussées, Paris, translated into German by H. Priebe, Baumaschine + Bautechnik BMT, pp. 170–185.

[3] Raithel, M. (1999): Zum Trag- und Verformungsverhalten von geokunststoffummantelten Sandsäulen. Schriftenreihe Geotechnik, Heft 6, University of Kassel.

[4] Raithel, M., Kempfert H.-G. (1999): Bemessung von geokunststoffummantelten Sandsäulen. Die Bautechnik, 76, Heft 12.

[5] Ghionna, V., Jamiolkowski, M. (1981): Colonne di ghiaia, X Ciclo di conferenze dedicate ai problemi di meccanica dei terreni e ingegneria delle fondazioni metodi di miglioramento dei terreni. Politecnico di Torino Ingegneria, atti dell'istituto di scienza delle costruzioni, n° 507, Nov. 1981.

[6] Van Impe, W. (1986): Improving of the Bearing Capacity of Weak Hydraulic Fills by Means of Geotextile. Third International Conference on Geotextiles, Vienna.

10.9 Worked Example: Geotextile-encased Columns

10.9.1 Input data

Geotextile-encased columns:

10%-grid

$\rightarrow a_S = A_S / A_E = 0.1$

$r_s = 0.4$ m

Displacement methods

Geosynthetic casing:

J = 1,500 kN/m

$r_{geo} = 0.4$ m (no activation widening)

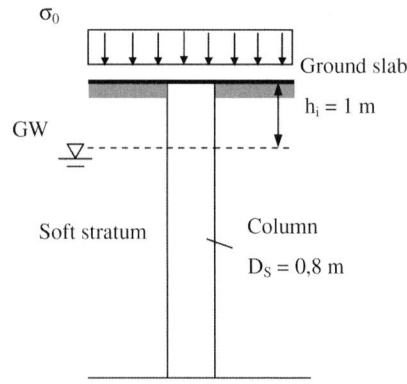

σ₀

Ground slab

$h_i = 1$ m

GW

Soft stratum

Column

$D_S = 0.8$ m

222

Soft stratum:

$h_{B,I} = 1.0$ m, i.e. the uppermost ground slab to a depth of 1 m below ground level is considered.

Groundwater level is below the ground slab.

$\varphi'_B = 15°$

$c'_B = 10$ kN/m^2

$\gamma_B / \gamma'_B = 15 / 5$ kN/m^3

$K_{0,B} = 1 - \sin \varphi'_B = 1 - \sin 15° = 0.741$

$K^*_{0,B} = 1.0$ displacement methods

$\nu_B = 0.4$

$E_{s,B} = E_{s,B,ref} \cdot \left(\dfrac{p^*}{p_{ref}} \right) m$

where:

$p_{ref} = 100.0$ kN/m^2

$E_{s,B,ref} = 750$ kN/m^2

$m = 1$

Column material:

$\psi'_S = 32.5°$

$\gamma_S / \gamma'_S = 20 / 10$ kN/m^3

$K_{a,S} = K_{agh} = 0.301$ for $\alpha = \beta = \delta_a = 0$

Load:

Actions in the column head plane:

$\sigma_0 = 100$ kN/m^2

10.9.2   Analysis

10.9.2.1   Determining Primary Stresses

In the centre of the ground slab:

$\sigma_{ü,B} = 0.5 \cdot 1.0 \cdot 15 = 7.5$ kN/m^2

$\sigma_{ü,S} = \gamma_S \cdot h = 0.5 \cdot 1.0 \cdot 19.0 = 9.5$ kN/m^2

10.9.2.2 Assuming the Load Redistribution Factor E

Here: initially determined after iteration is complete

$E = 0.66$

$$\sigma_{v,B} = \frac{\sigma_0 - E \cdot \sigma_0}{(1 - a_S)} = \frac{100 - 0.66 \cdot 100}{(1 - 0.1)} = 37.8 \text{ kN/m}^2$$

10.9.2.3 Determining the Stiffness Parameter

$$p^* = \frac{p_2 - p_1}{\ln (p_2 / p_1)}$$

where:

$p_1 = \sigma_{\ddot{u},B} = 7.5 \text{ kN/m}^2$

$p_2 = \sigma_{v,B} + \sigma_{\ddot{u},B} = 37.8 + 7.5 = 45.3 \text{ kN/m}^2$

$$p^* = \frac{37.8}{\ln (45.3 / 7.5)} = 21.0 \text{ kN/m}^2$$

$$p^* + c_B' \cdot \cot \varphi_B' = 21.0 + 10 \cdot \cot 15° = 58.3 \text{ kN/m}^2$$

$$E_{s,B} = E_{s,B,ref} \cdot \left(\frac{p^* + c_B' \cdot \cot \varphi_B'}{p_{ref}} \right) m = 750 \cdot \left(\frac{58.3}{100} \right) 1.0 = 437 \text{ kN/m}^2$$

$$E^* = \left(\frac{1}{1 - \nu_B} + \frac{1}{1 + \nu_B} \cdot \frac{1}{a_S} \right) \cdot \frac{(1 + \nu_B) \cdot (1 - 2 \nu_B)}{(1 - \nu_B)} \cdot E_{s,B}$$

$$= \left(\frac{1}{1 - 0.4} + \frac{1}{1 + 0.4} \cdot \frac{1}{0.1} \right) \cdot \frac{(1 + 0.4) \cdot (1 - 2 \cdot 0.4)}{(1 - 0.4)} \cdot 437 = 1,797 \text{ kN/m}^2$$

10.9.2.4 Deformation at the Column Edge

$$\Delta r_S = \frac{K_{a,S} \cdot \left(\frac{1}{a_S} \cdot \sigma_0 - \frac{1 - a_S}{a_S} \cdot \sigma_{v,B} + \sigma_{\ddot{u},S} \right) - K_{0,B} \cdot \sigma_{v,B} - K_{0,B}^* \cdot \sigma_{\ddot{u},B} + \frac{(r_{geo} - r_S) \cdot J}{r_{geo}^2}}{\frac{E^*}{(1 / a_S - 1) \cdot r_S} + \frac{J}{r_{geo}^2}}$$

$$= \frac{0.301 \cdot \left(\frac{1}{0.1} \cdot 100.0 - \frac{1 - 0.1}{0.1} \cdot 37.8 + 9.5 \right) - 0.741 \cdot 37.8 - 1.0 \cdot 7.5 + 0}{\frac{1,797}{(1 / 0.1 - 1) \cdot 0.4} + \frac{1,500}{0.4^2}}$$

$$= 0.0168 \text{ m}$$

224

10.9.2.5 Determining Settlement

$s_B = s_S$

$$\left\{ \frac{\sigma_{v,B}}{E_{s,B}} - \frac{2}{E^*} \cdot \frac{\nu_B}{1-\nu_B} \left[\begin{array}{l} K_{a,S} \cdot \left(\dfrac{1}{a_S} \cdot \sigma_0 - \dfrac{1-a_S}{a_S} \cdot \sigma_{v,B} + \sigma_{\ddot{u},S} \right) - \\[2ex] K_{0,B} \cdot \sigma_{v,B} - K_{0,B}^* \cdot \sigma_{\ddot{u},B} + \dfrac{(r_{geo} - r_S) \cdot J}{r_{geo}^2} - \dfrac{\Delta r_S \cdot J}{r_{geo}^2} \end{array} \right] \right\} \cdot h$$

$$= \left[1 - \frac{r_S^2}{(r_S + \Delta r_S)^2} \right] \cdot h$$

$$\left\{ \frac{37.8}{437} - \frac{2}{1,797} \cdot \frac{0.4}{1-0.4} \left[\begin{array}{l} 0.301 \cdot \left(\dfrac{1}{0.1} \cdot 100.0 - \dfrac{1-0.1}{0.1} \cdot 37.8 + 9.5 \right) - \\[2ex] 0.741 \cdot 37.8 - 1.0 \cdot 7.5 + 0 - \dfrac{0.0168 \cdot 1,500}{0.4^2} \end{array} \right] \right\} \cdot 1.0$$

$$= \left[1 - \frac{0.4^2}{(0.4 + 0.0168)^2} \right] \cdot 1.0$$

$0.080 \text{ m} \approx 0.079 \text{ m}$

(Uniform settlement given an analysis precision of 1 mm. If not met, repeat analysis with a different load redistribution factor until iteration is complete.)

Note: This settlement only corresponds to the settlement of the ground slab under consideration. The total system settlement is given by adding the respective settlements of the individual ground slabs.

10.9.2.6 Analysing Circumferential Tensile Force

$$F_r = J \cdot \frac{\Delta r_{geo}}{r_{geo}} = J \cdot \frac{\Delta r_S - (r_{geo} - r_S)}{r_{geo}} = 1,500 \cdot \frac{0.0168}{0.40} = 63 \text{ kN/m}$$

Note: The maximum circumferential tensile stress of all ground slabs is the governing design value!

11 Overbridging Systems in Areas Prone to Subsidence

11.1 General Recommendations

Collapses are represented by **crater-like subsidence** on the ground surface, generally appearing suddenly. They occur though the collapse of subterranean cavities, stoping upwards with time until they finally break through to the surface. Dolines, sinkholes, glory holes and collapsed shafts are differentiated according to the manner in which they were created.

Dolines are cavities created by natural dissolution and subrosion processes in soluble or erosion-sensitive rocks (see Figure 11.1).

Sinkholes and **glory holes** form by the upward stoping of inadequately stabilised subterranean mine works, such as headings, galleries and chambers. In contrast to glory holes, sinkholes are limited to a few metres in diameter.

Collapsed shafts form by a generally sudden collapse of backfill or other material in shafts or from shaft caps and the overlying fill material near the top of shafts.

Note: In addition to the crater-like subsidence described here, fissure-like collapses and movements are known (e.g. slope cracks or terraces). Bridging such subsidence with geosynthetics is not dealt with here.

Determining the general regional **sinkhole hazard, estimating the degree of hazard**, delineating the **threatened area** and gathering information on the **type, geometry and dimensions** of the potential collapse is generally performed by the geological surveys of the federal German states, mining agencies and local ground investigation institutes. They should have years of experience in this field and be in possession of reliable statistics. These data are indispensible input variables for designing and analysing bridging structures using geosynthetics. Depending on the properties of the ground, the geometry of the funnel above mining-induced cavities can be forecast, or at least estimated. See FENK (e.g. [1]) for details.

If **areas prone to subsidence hazards** are crossed by new build transport routes, **preventive stabilisation** is required. Because complete stabilisation of subterranean cavities is seldom possible, this generally consists of bridging structures leading the transport route across the entire length of the subsidence prone area.

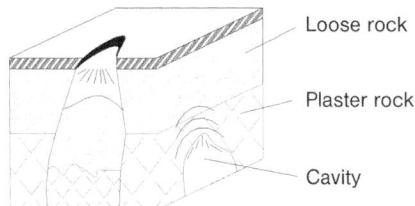

Loose rock

Plaster rock

Cavity

Figure 11.1 Doline evolution

Recommendations for Design and Analysis of Earth Structures using Geosynthetic Reinforcements (EBGEO). German Geotechnical Society.
© 2011 Ernst & Sohn GmbH & Co. KG.
Published by Ernst & Sohn GmbH & Co. KG.

227

The following methods may be considered:

- bridge-like reinforced concrete structures above or on the ground,
- reinforced concrete slabs below or within the transport route pavement,
- geosynthetic-reinforced earth structures below the transport route subgrade.

If **local, limited collapses** occur in an existing transport route the following stabilisation options are known and are often employed in combination:

- stabilising the subterranean cavity,
- backfilling the collapse crater,
- sealing the upper region of the collapse crater,
- bridging the collapse by means of a reinforced concrete structure (bridge, slab),
- bridging the collapse by means of a geosynthetic-reinforced earth structure.

All stabilisation options discussed above can be supplemented by **monitoring and warning systems**, making it possible to localise collapses at any time during the entire design working life of the structure.

11.2 Design

11.2.1 Principles and Definitions

Two stabilisation principles are differentiated in terms of the extent of stabilisation measures:

- The principle of **complete stabilisation** assumes that stability is guaranteed during the entire **design working life** t_b (t_b = lifetime = operating period) of the highway and no restrictions in highway use occur as a result of collapses. Accordingly, the allowable subsidence in the highway pavement is very low. Highway remediation following collapse damage is not planned. If the use of geosynthetics is planned the geosynthetics do not generally represent the only stabilisation element, but are usually used in combination with other stabilisation elements, such as stabilisation of the overlying soil strata, for example.
- The principle of **partial stabilisation** assumes that local subsidence is allowed in the highway pavement. However, it may not exceed a defined maximum subsidence limit $d_{s,max.}$ or the ratio $d_{s,max.}/D_S$ ($d_{s,max.}$ = maximum subsidence; D_S = diameter of subsidence depression) within a defined **load duration** t_d (t_d = minimum load duration) (designations: see Figure 11.2). The subsidence limit is specified according to the type of vehicle and the anticipated live loads. The load duration is selected on a project-specific basis such that the highway subsidence or the collapse is reliably detected during this period by the installed monitoring system or in the course of regular inspections of the structure. In highways engineering this is generally a few weeks. Planned traffic safety and collapse remediation measures are initiated immediately following subsidence detection. The partial stabilisation principle favours the use of geosynthetics.

Geotechnical Engineering Handbook

Editor: Ulrich Smoltczyk

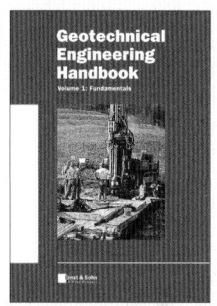

**Volume 1:
Fundamentals**
2002.
829 pages, 616 fig.
Hardcover.
€ 179,-*/ sFr 283,-
ISBN 978-3-433-01449-3

This is the English version of the Grundbau-Taschenbuch - a reference book for geotechnical engineering. The first of three volumes contains all information about the basics on the field of geotechnical engineering. The book is written by authors from Germany, Belgium, Sweden, the Czech Republic, Australia, Italy, U.K., and Switzerland.

**Volume 2:
Procedures**
2002.
679 pages, 558 fig.
Hardcover.
€ 179,-*/ sFr 283,-
ISBN 978-3-433-01450-9

Volume 2 of the Geotechnical Engineering Handbook covers the geotechnical procedures used in manufacturing anchors and piles as well as for improving or underpinning the foundations, securing existing constructions, controlling ground water, excavating rocks and earthworks. It also treats such specialist areas as the use of geotextiles and seeding.

**Volume 3:
Elements and
structure**
2002.
646 pages, 500 fig.
Hardcover.
€ 179,-*/ sFr 283,-
ISBN 978-3-433-01451-6

Special Set Price
(three volumes)
€ 499,-* / sFr 788,-
ISBN   978-3-433-01452-3

Volume 3 of the Geotechnical Engineering Handbook deals with foundations. It presents spread foundations starting with basic designs right up the necessary proofs. There is comprehensive coverage of the possibilities for stabilizing excavations, together with the relevant area of application, while another section is devoted to the useful application of trench walls. The entire book is an indispensable aid in the planning and execution of all types of foundations found in practice, whether for academics or practitioners.

Ernst & Sohn
Verlag für Architektur und
technische Wissenschaften GmbH & Co. KG

Für Bestellungen und Kundenservice:
Verlag Wiley-VCH
Boschstraße 12
69469 Weinheim
Telefon: +49(0) 6201 / 606-400
Telefax: +49(0) 6201 / 606-184
E-Mail: service@wiley-vch.de

A Wiley Company
www.ernst-und-sohn.de

* €-price is valid in Germany only.
  001423016...my Prices are subject to change without notice.

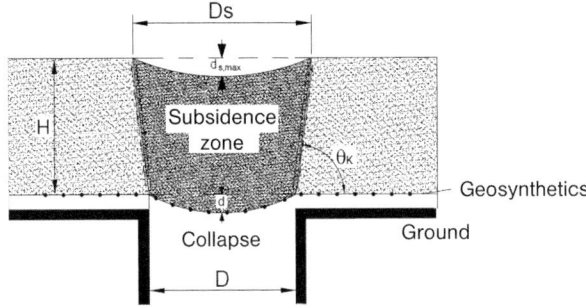

Figure 11.2 Designations

High tensile strength or low-creep geogrids, geowovens or composites are generally used for geosynthetic bridging structures. Their load-bearing synthetic elements (ribbons, ribs, fibres) are arranged orthogonally.

Geosynthetics are usually installed in parallel and overlapping layers to form a **geosynthetic layer**. If high tensile forces need to be transferred, **double layers** are also possible. The geosynthetic layer then consists of two sheets in direct contact, which also overlap in the transverse direction.

The **geosynthetic reinforcement** consists of at least one layer of geosynthetics (one-ply **installation**). For a **two- or more ply installation** the layers may be installed in **parallel** or **orthogonally**. The right side of Figure 11.3 shows an example of a one-ply geosynthetic reinforcement installed longitudinally only. The left side shows a **two-ply, installed longitudinally and transversely**.

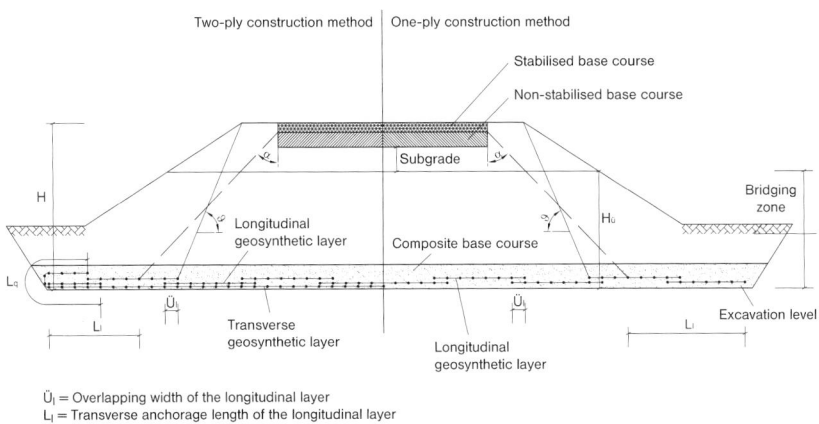

$Ü_l$ = Overlapping width of the longitudinal layer
L_l = Transverse anchorage length of the longitudinal layer
L_q = Anchorage length of the transverse layer
$H_ü$ = Height of bridging zone

Figure 11.3 Two examples of geosynthetic reinforcements

229

Isotropic and anisotropic conditions are also differentiated for geosynthetic reinforcements. An **isotropic geosynthetic reinforcement** is assumed for the following construction methods, for example:

- one-ply reinforcement, consisting of an isotropic geosynthetic ($J_L = J_Q$),
- two-ply reinforcement, consisting of two identical, anisotropic geosynthetics ($J_L \neq J_Q$), installed orthogonally.

An **anisotropic geosynthetic reinforcement** is assumed for the following construction methods, for example:

- one-ply reinforcement, consisting of an anisotropic geosynthetic ($J_L \neq J_Q$),
- two-ply reinforcement, consisting of two anisotropic geosynthetics ($J_L \neq J_Q$), installed in parallel.

An extremely anisotropic geosynthetic reinforcement is assumed if the axial stiffness in the machine direction (J_L) is at least ten times that in the cross machine direction (J_Q) and the limiting strain in the cross machine direction is at least twice the limiting strain in the machine direction.

The geosynthetic reinforcement is installed within an earth structure, known as the **composite base course**. The composite base course is installed in the natural **ground** after an excavation is created and following stabilisation of the **excavation level**. The **excavation level** is located at the required engineering depth below ground level (see Figure 11.3, for example).

The composite base course forms part of the **bridging zone**. The height of the bridging zone is dependent on the selected structural model and can surpass the height of the composite base course. The region above the bridging zone corresponds to the subgrade of the respective highway.

The geosynthetic reinforcement is adequately **anchored** in both the longitudinal and transverse highway directions. The geosynthetic reinforcement must therefore be installed longitudinally to outside of the subsidence prone area. In the transverse direction the geosynthetic reinforcement is installed to sufficiently protrude laterally such that the anchorage remains within the subsidence prone area, but outside of the area impacting on the highway (see Section 11.2.2.2). The anchorage is usually executed as a **planar anchorage** (see Figure 11.3, for example, right side); **wrap-around anchorages** are also common to lessen the lateral excavation width (see Figure 11.3, for example, left side). **Anchorage lengths** parallel (longitudinal) to the highway are designated by L_L. In the transverse direction the anchorage lengths of parallel geosynthetic layers are designated by L_l and those of transverse geosynthetic layers by L_q (see Figure 11.3, for example).

The longitudinal geosynthetic layers are installed with **overlaps** transverse to the highway and – for longer web lengths – also parallel to the highway. The transverse geosynthetic layers are generally only overlapped parallel to the highway. The **overlapping length** parallel to the highway is designated by $\ddot{U}_L$ and that transverse to the highway by $\ddot{U}_l$ for longitudinal geosynthetic layers.

The overlapping length parallel to the highway is designated by $\ddot{U}_Q$ for the transverse geosynthetic layers (see Figure 11.3, for example).

11.2.2 Design Notes

The design consists of all necessary diagrams, all stability analyses and an explanatory report with a quality assurance plan. The explanatory report describes the geotechnical and structural situation, and the demands on the structure and its execution. The load-bearing structural system is selected. The system and the load assumptions are described. A statement on the monitoring and warning systems is made.

11.2.2.1 Explanatory Report

The explanatory report includes the following points:

1. Results of the geotechnical, historical and structural investigations with information on:
   · the cause, geometry and size of the collapse being bridged,
   · the impacts effectively causing the failure (geogenic, anthropogenic),
   · the ground properties as they are relevant to the bridging structure.
   This information can generally be taken from the geotechnical report.

2. Information on:
   · the dimensions of the governing collapse in plan,
   · the geometry of the collapse in section,
   · the geotechnical properties of the support (ground) in the collapse boundary zone.
   A geometrical model of the collapse is developed from the data (Point 1) to facilitate designing the bridging structure. The geometry of the crater boundary is also given. Two examples of crater boundary forms are shown in Figure 11.4. The value of D_0 can generally be taken from the geotechnical report. The value of the design diameter D is specified by the designer. Further information can be found in [1], [2].

3. Information:
   · on the demands on the structure in terms of the required structural safety (failure probability),
   · on serviceability,
   · on the required operational life (complete or partial stabilisation, see Section 11.2.1).

4. Developing the Structural Model
   The bridging structure is designed to comply with the stability analysis results in conjunction with the selected structural model, based on the demands made (Point 3) and taking the support conditions on the ground (Point 2) into consideration (see Section 11.2.6).

231

Ground:
cohesive soil

Ground:
granular soil over cohesive soil

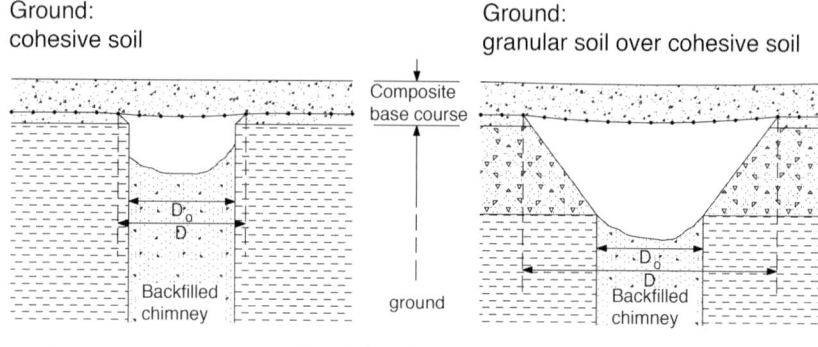

D = design diameter

Figure 11.4 Two examples of crater boundary forms

5. Monitoring the completed structure
Monitoring measures are required to assess the safety and serviceability of the bridging structure (see Section 11.4).
The logging, registration and evaluation of monitoring data within and below the bridging structure forms the basis for any additional measures required.

6. Planning stabilisation and remediation measures after the collapse has occurred
If collapses occur planned safety measures (e.g. road closures) and remediation measures (e.g. backfilling the detected collapse) are carried out (see Section 11.4), similar to the procedure used in the observational method (DIN 1054).

11.2.2.2 Determining the Width of the Stabilised Area

The excavation level, upon which the composite base course is installed, is located at the required engineering depth. If the highway is located on the embankment, the geosynthetic reinforcement is located in or below the embankment contact zone. The excavation depth for transport routes at ground level and in cuttings is given by the selected structural model.

The composite base course extends below the highway for an **installation width** B_{total}, which is given by the **action widths B** and the anchorage lengths of the geosynthetic reinforcement L_l on both sides of the highway (see Figure 11.5):

$$B_{total} = B + L_l + 2 .$$ Eq. (11.1)

The width B requiring stabilisation is given by:

$$B = B_V + 2 \cdot B_\alpha = B_V + 2 \cdot (H_D + H) \cdot \tan \alpha ,$$ Eq. (11.2)

where a load distribution angle of $45° < \alpha < 60°$ is selected for **lateral load distribution** B_α, depending on the soil type and with a starting point between the non-stabilised and the stabilised base course (highways engineering) or between

232

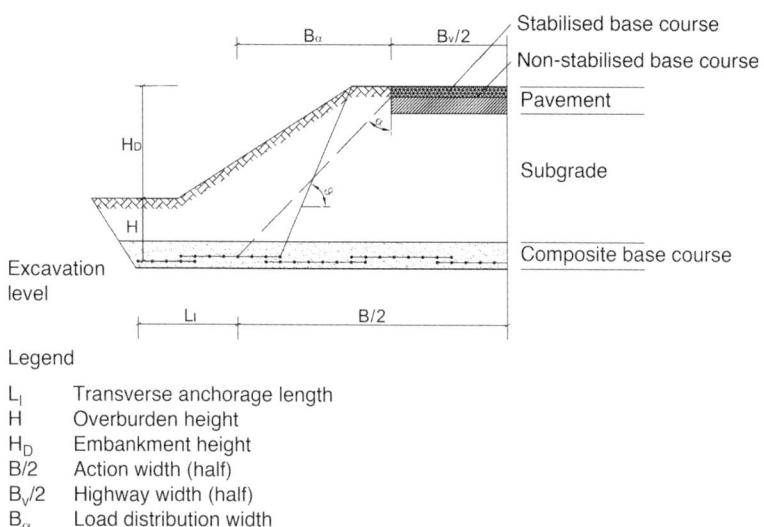

Legend

L_I	Transverse anchorage length
H	Overburden height
H_D	Embankment height
$B/2$	Action width (half)
$B_v/2$	Highway width (half)
B_α	Load distribution width
α	Load distribution angle
ϑ	$45 + \varphi/2$

Figure 11.5 Exemplary composite base course on a highway embankment

the non-stabilised base course and the track ballast (railways engineering). If on top of the embankment (see Figure 11.5), the end point of the lateral load distribution line remains outside of the imaginary slip plane, which intersects the upper slope edge at an angle of $\vartheta = 45° + \varphi/2$.

Figure 11.6 shows an example of subsidence stabilisation below an embankment in plan. The installation width is given by the width of the stabilised zone and the transverse anchorage lengths.

If the geosynthetic reinforcement consists of a longitudinally installed, extremely anisotropic geosynthetic, the installation width is increased structurally (see Table 11.4).

To ensure horizontal frictional connections in geosynthetic webs installed in parallel, sufficient transverse overlapping width or length shall be demonstrated, as well as longitudinally, if necessary (see Section 11.3.2.6). A minimum overlapping length of 50 cm is necessary perpendicular to the roll-out axis, depending on the structural model used. The required overlap length must generally be demonstrated in terms of the respective application for overlapping parallel to the roll-out axis and the route ($Ü_L$). The overlapping is staggered for geosynthetic webs installed in parallel.

Methods allowing compensation of unavoidable deformations of the bridging structure are especially suitable for use with pavement structures (e.g. bitumen pavement on highways, ballast bed on railways).

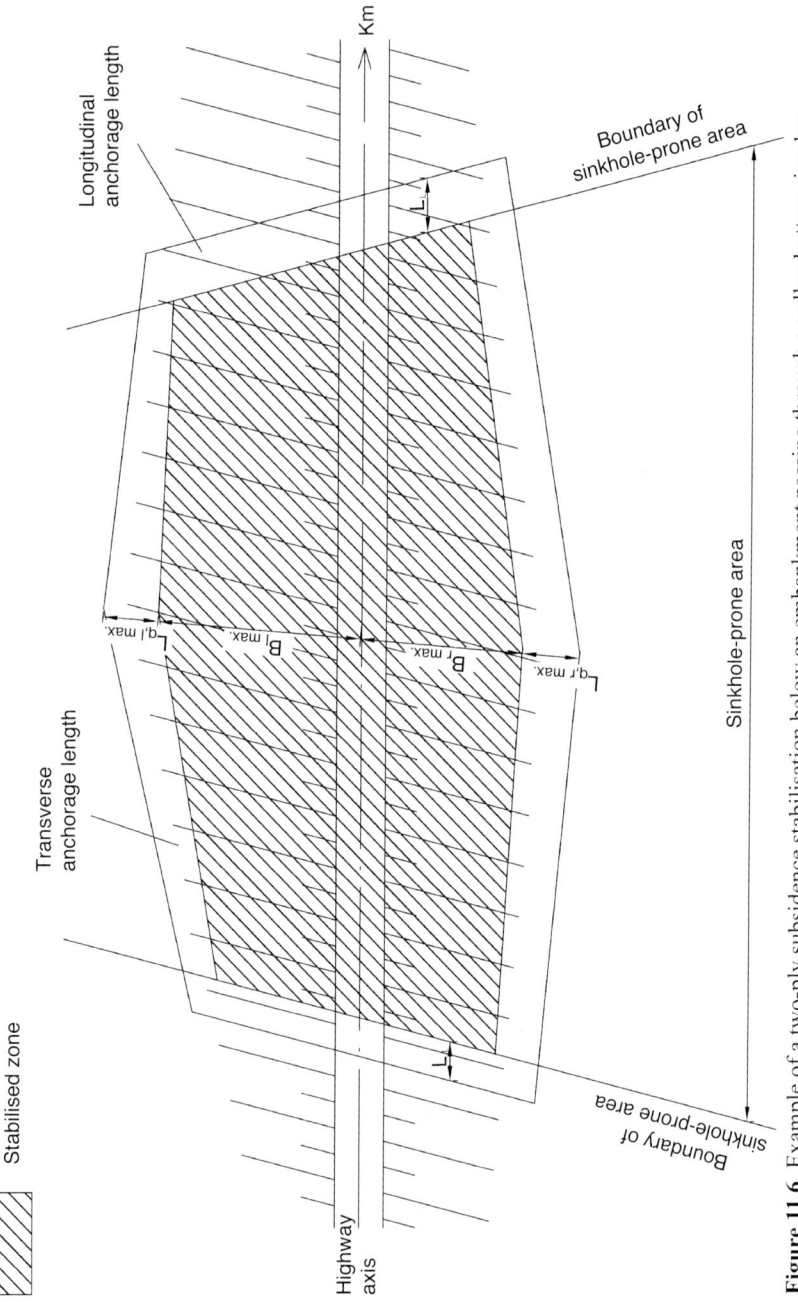

Figure 11.6 Example of a two-ply subsidence stabilisation below an embankment passing through a valley bottom, in plan

234

11.2.3 Ground and Materials

11.2.3.1 Excavation Level

The excavation level must display adequate bearing capacity to allow safe site traffic and allow the required compaction of the composite base course (see Section 11.2.3.2), as well as accepting the loads from the bridging structure (see Section 11.5).

11.2.3.2 Composite base course Materials

The materials used in the composite base course are fill soil and geosynthetics.

Materials described in Section 2.1.2 are used as fill soils, as well as soils and aggregates improved and stabilised with binders. These fill soils are compacted in accordance with the demands developed in the structural design phase. Minimum requirements are controlled by the relevant rail operator's regulations (e.g. ZTVT-StB, RiL 836).

11.2.3.3 Geosynthetic Reinforcement

It is assumed that the reinforcement is not subjected to tensile stresses before the collapse occurs. After the collapse occurs the reinforcement must prevent failure of the structure in conjunction with the fill soil and limit surface deformation to a defined value. Both requirements shall be met for a specified duration (load duration t_d for partial stabilisation, design working life t_b for complete stabilisation).

The information listed in Section 2.2 and 3 is required to design and select the geosynthetic products.

11.2.3.4 Bridging Zone Materials

The properties of the bridging zone materials depend on the structural model used (see Section 11.2.6). If the bridging zone needs to display tensile strength properties, it is often manufactured from binder-stabilised soils. If other soils are used they must demonstrate at least the shear properties of the composite base course.

11.2.3.5 Subgrade

The subgrade above the bridging zone is manufactured in accordance with the relevant operator's regulations.

11.2.4 Load Assumptions and Load Cases

The design approaches are defined relative to DIN 1054, in particular in terms of the Action Combinations (AC), Safety Classes (SC) and Load Cases (LC).

The characteristic values for live loads are defined as discussed in DIN Fachbericht 101 [13].

11.2.5 Allowable Deformations

The allowable deformations on the highway pavement depend on the operator's project-specific requirements and are specified case-by-case.

For example, the following criteria are adopted for highways:

- motorways, A-roads and similar routes, extra urban:
  $0.01 \leq d_s/D_s \leq 0.017$,

- A-roads and similarly trafficked urban routes and other extra urban highways:
  $0.017 \leq d_s/D_s \leq 0.025$,

- other urban highways and trafficked areas
  (e.g. car parks, access routes, escape and rescue routes):
  $0.025 \leq d_s/D_s \leq 0.07$

where:

d_s max. subsidence on the road surface,
D_s max. diameter of subsidence depression on the road surface.

On railway lines with ballast beds and maximum permitted speeds up to 200 km/h the following values can be assumed for preliminary drafting of partial stabilisation measures:

- $d_s/D_s \leq 0.002$ and
- $d_s \leq 1$ cm.

Otherwise, the necessary information for design and execution is provided in the course of the approval procedure.

The choice of supplementary monitoring and warning systems to allow routes to be closed in a timely manner, and their type and configuration, are specified to suit the structural model adopted and the project boundary conditions.

11.2.6 Structural Models

A structural system will develop within the bridging zone if a collapse occurs. The type and geometry of the structural system is determined by the following characteristics, among others:

- size and shape of the collapse,
- axial stiffness of the geosynthetic reinforcement,
- shear strength and stiffness of the soils/aggregates in the bridging zone,
- system stiffness of the bridging structure,
- structural system support conditions (stiffness and shear strength of the support).

Formation of the structural system above the collapse depends on the material properties of the geosynthetic reinforcement and the earth structure in the bridging zone, the configuration of the reinforcement layers and the ratio H/D (height of bridging zone and any earth structure above it to the design diameter).

Upward collapse stoping provokes deformation in the bridging zone and may lead to partial or complete failure. The physical models shown in Figure 11.7 were primarily observed.

Failure model	
Without lateral reaction	**With lateral reaction**
Granular soil Complete failure Full surcharge on membrane $H/D < 1$	Granular soil Complete failure Partial load transfer via lateral reaction Reduced surcharge on membrane $1 < H/D < 3$
Figure 1a)	Figure 1b)

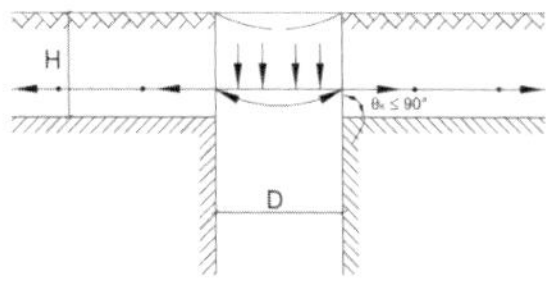

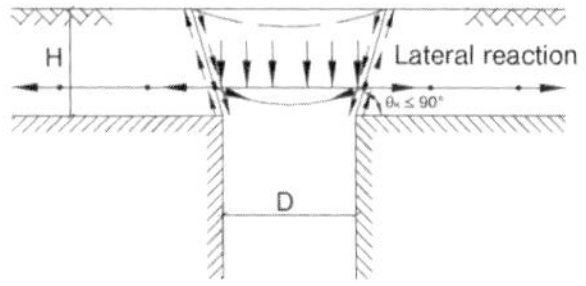

Arch model	
Temporarily limited surcharge	**Permanently limited surcharge**
Granular soil with high relative density Delayed upward stoping Temporarily limited surcharge on membrane $H/D > 3$	Stabilised soil Permanently limited upward stoping Permanently limited surcharge on membrane Any H/D
Figure 2a)	Figure 2b)

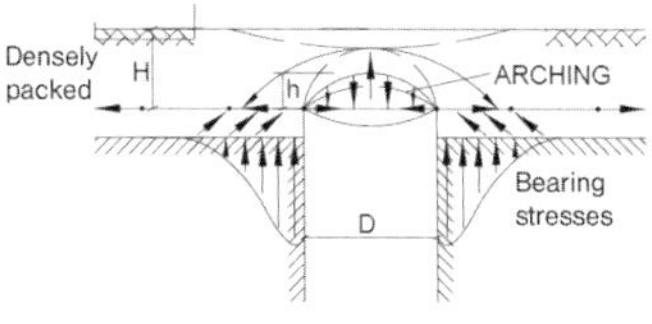

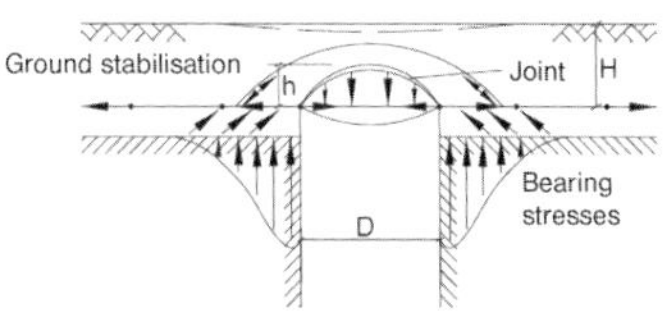

Figure 11.7 Structural models

Two formations are observed in soils with little or no cohesion and deep geosynthetic reinforcement (known as a membrane):

- If the soil material in the bridging zone has a low relative density large plastic zones occur above the collapse. In addition to the relative density, lateral delineation of the failure zone depends on the shear strength of the soil material; it is generally very steep. Models with and without adopted lateral reaction (friction in the failure plane) are differentiated (see Figure 11.7, 1a and 1b).
- If the relative density in granular soils is high (high relative compaction) an arch structure can form in the bridging zone given adequate thickness. The sheared material below the compression zone increasingly impacts on the geosynthetic reinforcement by means of its dead load and any surcharge (Figure 11.7, 2a). With time the arch can collapse, in particular when subjected to dynamic loads, such that the reinforcement is subjected to the full surcharge (in analogy to Figure 11.7, 1a).

An arch is also formed in structures comprising binder-stabilised soils with high tensile strength geosynthetics (Figure 11.7, 2b). The sheared material below the arch also increasingly subjects the geosynthetic reinforcement to its load. Tensile forces promoting the load-bearing behaviour of the apex of the arch are activated in geosynthetics with very high axial stiffness. The structural system then resembles an arch with a high tensile strength tie-beam. Eventually the geosynthetic reinforcement is subjected to a permanent, uniform surcharge.

Analytical methods have been developed for the physical models. Table 11.1 shows a summary of the physical characteristics employed by the best known analysis methods (also see Section 11.6).

Table 11.1 Physical characteristics of the best known analysis methods

Name of structural model	Geometry of failure body	Name of method, literature name	Method characteristics			Structural model shown in Figure 11.7
			With arching	With lateral reaction	With decompaction	
Failure model	Truncated cone	BS 8006	No	No	No	1a
		B.G.E.	Yes	Yes	Yes	1a/1b
	Cylinder	GIROUD	No	Possible	No	1a/1b
		R.A.F.A.E.L.	No	Yes	Yes	1b
Arch model	Spherical segment	A.S.T.	Yes	No	No	2b
		BGE	Yes	No	Yes	2a

11.3 Analyses

11.3.1 Analysis Principles

Analysis methods for bridging subsidence using geosynthetic reinforcements are the subject of continuous scientific development. Only current knowledge and a few selected approaches and analysis methods can therefore be described below. It is the responsibility of the analyst to find the most suitable method for the application at hand. It shall be selected such that as realistic a model as possible is achieved (see Section 11.2.6) and all analyses described below can be performed with the requisite accuracy and economy.

According to DIN 1054 analysis of the stability of geosynthetic-reinforced bridging systems must encompass analysis of the ultimate limit state (ULS) and the serviceability limit state (SLS).

Generally, only failure planes intersecting the reinforcement layers (analysis of reinforcement bearing capacity) or enveloping the reinforcement layers (analysis of reinforcement anchorage) need be investigated for analysis of the bearing capacity of geosynthetic-reinforced subsidence bridging in the STR limit state. Investigation of large-scale failure planes, for example running from deeper within the subsidence cavity diagonally upwards around the geosynthetic reinforcement and to the surface, is not generally necessary. This is because it is assumed that the excavation level used to support the composite base course is adequately stable. Moreover, the design diameter of the collapse is selected from the outset such that any crumbling or widening at the upper edge of the collapse crater are already incorporated. Finally, it is assumed that the anchorage lengths of the geosynthetic layers are adequately dimensioned to prevent pull-out of the geosynthetic anchorage and slippage into the funnel.

The serviceability analysis comprises analysis of the allowable surface deformations as a result of a deformation of the geosynthetic reinforcement immediately above the collapse. In addition, the total system deformation may also include ground settlement and intrinsic settlements in the composite base course, as well as the overlying layers.

In contrast to the usual geotechnical procedure, it has proven reliable in sinkhole bridging applications to specify an allowable characteristic deformation in the pavement plane and then to carry out the STR limit state analyses in terms of failure and strain. The serviceability analysis is thus confirmed. Figure 11.8 shows an analysis flow diagram.

Adopting the limit value for the maximum surface subsidence $d_{s,max.}$ and the adopted structural model (see Section 11.2.6) the maximum sag $d_{max.}$ of the selected geosynthetic reinforcement is determined and assessed in terms of a specified sinkhole diameter. The maximum geometric strain ε_{geom} of the geosynthetic elements in the longitudinal and transverse directions is determined from this. This represents a guide value for preselection of the geosynthetic reinforcement.

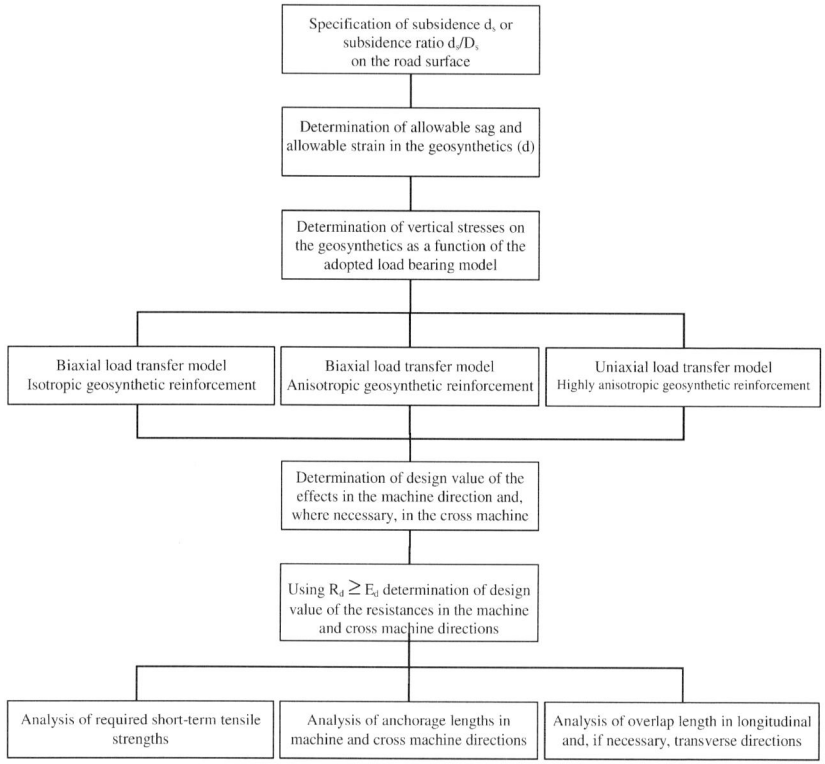

Figure 11.8 Analysis flow diagram

Adopting the axial stiffness of the selected reinforcement it is demonstrated that the geometric strain ε_{geom} is no greater than the allowable strain ε_B of the selected geosynthetic during the design period. The allowable strain ε_B is given by the following stress analysis as a function of the structural model and the load duration from the isochrones of the selected geosynthetic. The smaller of the two values ε_{geom} and ε_B is the governing design strain ε_d.

Based on the normal stresses acting on the geosynthetic reinforcement due to the dead weight of the ground and live loads, the design values of the normal stresses are first calculated by multiplying by the partial safety factors γ for actions to DIN 1054, Table 2 using Equation (11.3):

$$\sigma_{v,d} = \sigma_{v,G,k} \cdot \gamma_G + \sigma_{v,Q,k} \cdot \gamma_Q \hspace{2cm} \text{Eq. (11.3)}$$

Depending on the load transfer model and the type of geosynthetic reinforcement three cases are differentiated when determining the tensile effects E_d (see Table 11.2):

Table 11.2 Types of analysis method

Load transfer model	Biaxial	Biaxial	Uniaxial
Reinforcement	Isotropic	Anisotropic	Extremely anisotropic
Schematic principle			
Analysis methods	BS 8006 [3] Giroud et al. [4] B.G.E. [5] A.S.T. [6]	B.G.E. [5]	Giroud et al. [4] R.A.F.A.E.L. [8] BS 8006 [3]

- biaxial load transfer model, isotropic geosynthetic reinforcement,
- biaxial load transfer model, anisotropic geosynthetic reinforcement,
- uniaxial load transfer model, extremely anisotropic geosynthetic reinforcement.

It is possible to determine the characteristic tensile stress E_k with the aid of several analytical design methods for the 'biaxial load transfer model' and the 'isotropic geosynthetic reinforcement'. It is assumed that the stress in the reinforcement is equal in the machine and cross machine directions. Only the B.G.E. analysis method is described in detail in these recommendations (see Section 11.3.2.1). The A.S.T. method is briefly described as a special method (see Section 11.3.2.3).

Differing tensile stresses result for the 'biaxial load transfer model' and the 'anisotropic geosynthetic reinforcement' as a function of the axial stiffnesses in the machine and cross machine directions. Only the B.G.E. method is available as analytical design method for this case (see Section 11.3.2.2).

The case of a 'uniaxial load transfer model' and an 'extremely anisotropic geosynthetic reinforcement' is assumed if the reinforcement is rolled out parallel to the highway and the axial stiffness in the machine direction is at least ten times that in the cross machine direction, and the limiting strain in the cross machine direction is at least twice the limiting strain in the machine direction. Load is then primarily transferred in the machine direction and only subordinately in the cross machine direction. Several analytical design methods are also available for this case. Analysis is carried out in the machine direction only. For reasons of structural preservation only a minimum tensile strength is assumed in the cross machine direction. Only the R.A.F.A.E.L. method is described in more detail in these recommendations (see Section 11.3.2.2).

Note: The tensile forces can also be determined with the aid of numerical analysis methods (e.g. finite element method). They are not the subject of these Recommendations; only analytical methods are dealt with here.

After determining the various tensile stress design values E_d the bearing capacity is analysed using the general limit state equation (Eq. (11.4))

$$E_d \le R_d, \qquad \text{Eq. (11.4)}$$

where R_d represents the respective design value of the various resistances.

In extreme cases the following analyses may be required, depending on the load transfer model and the geosynthetic reinforcement:

– analysis of the bearing capacity of the geosynthetic reinforcement parallel to the highway,
– analysis of the bearing capacity of the geosynthetic reinforcement transverse to the highway,
– analysis of the bearing capacity of the geosynthetic reinforcement in the overlap zone transverse to the highway,
– analysis of the bearing capacity of the geosynthetic reinforcement in the overlap zone parallel to the highway,
– analysis of the anchorage of the geosynthetic reinforcement parallel to the highway,
– analysis of the anchorage of the geosynthetic reinforcement transverse to the highway.

The design values of the friction resistances (see Section 11.3.2.6) govern analysis of the bearing capacity in the overlap zones and the design values of the pull-out resistances (see Section 11.3.2.5) govern analysis of the anchorages.

11.3.2 Design

11.3.2.1 Determining the Tensile Stress Design Values using the B.G.E. Method

The method of geosynthetic reinforcement design for overbridging systems in areas prone to subsidence (B.G.E. method (*Bemessung von Geokunststoffbewehrungen zur Erdeinbruchüberbrückung*), based on [4]) is described for determining the design value of the tensile stress in biaxial load transfer models. It may be applied to both isotropic and anisotropic geosynthetic reinforcements. The analysis flow diagram is shown in Figure 11.9.

a) Specifying the subsidence contours and determining the maximum depression on the road surface $d_{s,max}$

In failure models a funnel-shaped subsidence depression extending to the road surface is adopted over the edge of the subsidence. If no precise investigations were carried out, a value of $\theta = 85°$ is adopted for analysis. The diameter of the subsidence depression D_S at the road surface is given by Eq. (11.5):

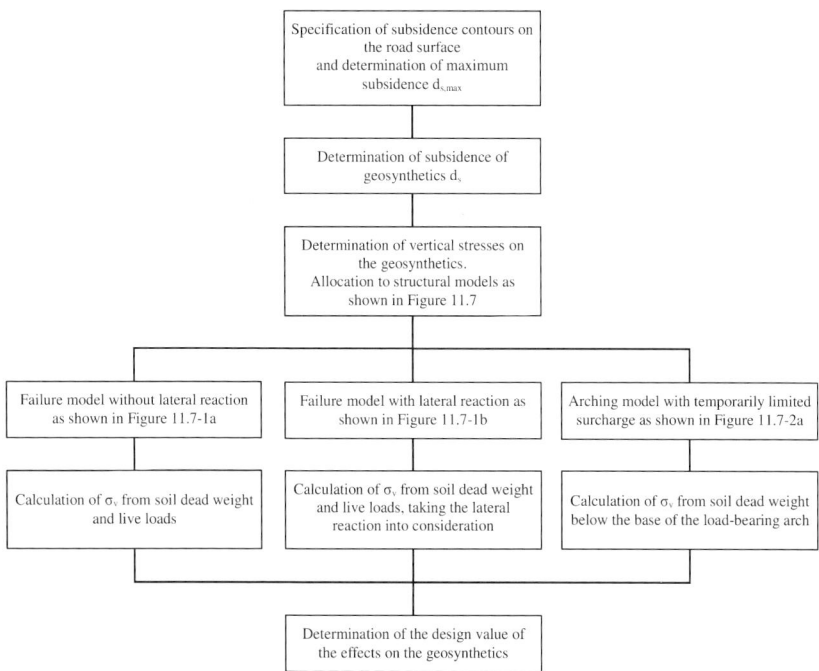

Figure 11.9 Flow diagram for determining actions using the B.G.E. method

$$D_S - D + \frac{2 \cdot H}{\tan \theta}. \qquad \text{Eq. (11.5)}$$

$d_{s,max.}$ is given by the ratio d_s / D_s, which is generally specified for the analysis (see Section 11.2.5).

When using the arching model approach the subsidence d_s at the road surface is set to zero for the STR limit state analysis. In this case an additional deformation analysis is required for the serviceability limit state at the road surface. This can be provided by numerical analysis, for example.

b) Determining the allowable subsidence at the geosynthetic reinforcement

Subsidence is determined for the failure models with the aid of Eq. (11.5), which was adopted from the R.A.F.A.E.L. method [8] (see Section 11.3.2.2):

$$d_{max} = d_{s.max} + 2 \cdot H \cdot (C_e - 1) \qquad \text{Eq. (11.6)}$$

The decompaction factor C_e can be adopted for granular soils, taking a minimum relative compaction $D_{pr} \geq 98\%$, with $C_e = 1.03$ for round-grained material and $C_e = 1.05$ for crushed aggregates.

243

If these values are not adopted the decompaction factor is derived from large-scale tests.

The following equation can be used for the temporary arching model:

$$d_{max} = h \cdot (C_e - 1) \qquad \text{Eq. (11.7)}$$

where:

h   distance between the reinforcement plane and the base of the load-bearing arch as shown in Figure 11.7, 2a),

C_e decompaction factor.

Assuming parabolic subsidence the analysis can be performed with the aid of the following relationship:

$$d_B = \sqrt{\frac{3}{8} \cdot \varepsilon_B \cdot D^2} \qquad \text{Eq. (11.8)}$$

where:

d_B maximum allowable sag,

ε_B allowable strain on the geosynthetic.

Analysis is successful if $d_B \leq d_{max}$.

Note: The maximum allowable sag is selected such that the allowable strain of the geosynthetic is not exceeded.

c) Determining the normal stresses

The three models shown in Figure 11.7 are utilised to determine the normal stresses. If no more detailed information is available these models are differentiated by the following ratios:

- $H/D < 1$ failure model without lateral reaction,
- $1 \leq H/D \leq 3$ failure model with lateral reaction,
- $H/D > 3$ temporary arching model.

These minimum values may be waived if sufficient empirical data is available for the chosen system of sinkhole diameter, coverage ratio, geosynthetic reinforcement and base course. The data may derived from practical experience or from large-scale tests.

Failure model without lateral reaction

The characteristic values of the normal stresses are calculated separately for the soil dead weight and live load components.

$$\sigma_{v,G,k} = \gamma_K \cdot H \qquad \text{Eq. (11.9)}$$

$$\sigma_{v,Q,k} = q_k \qquad \text{Eq. (11.10)}$$

where:

γ_k characteristic unit weight of the overlying soil strata,
q_k characteristic value of the live loads.

Failure model with lateral reaction

In this case the normal stress acting on the geosynthetics is reduced on the basis of the TERZAGHI method [9] and [8]. The characteristic values of the normal stresses are calculated separately for the soil dead weight and live load components.

$$\sigma_{v,G,k} = \frac{\dfrac{D}{2} \cdot \left(\gamma_k - \dfrac{4 \cdot c_k}{D} \right)}{2 \cdot K_{ak} \cdot \tan \varphi_k} \cdot \left[1 - e^{-K_{ak} \cdot \tan \varphi_k \cdot \left(\frac{4 \cdot H}{D} \right)} \right]$$

$$= q_k \cdot e^{-K_{ak} \cdot \tan \varphi_k \cdot \left(\frac{4 \cdot H}{D} \right)}$$ Eq. (11.11)

where:

γ_k characteristic unit weight of the soil strata above the geosynthetics,
c_k characteristic value for cohesion,
D collapse diameter,
K_{ak} characteristic active earth pressure coefficient ($\delta_a = 0$),
q_k characteristic value of the live load,
φ_k characteristic value of the friction angle.

If cohesion is adopted it shall be ensured that it can be permanently activated with the assumed magnitude. The value may not be greater than:

$$c_k = \frac{\gamma_k \cdot D}{4} .$$ Eq. (11.12)

Temporary arching model

If it is certain that the load-bearing arch remains stable for the design period and the decompacted zone does not *wander upwards*, e.g. as a result of dynamic or hydrological impacts, σ_v may be calculated from the soil dead weight below the arch. The height of the governing soil region below the load-bearing arch under a uniform live load of $q_k = 33.3$ kN/m^2 can be read off from diagrams [9].

A minimum of $h / D = 1.0$ is recommended. This recommendation may be waived if smaller ratios are demonstrated in large-scale tests under representative test conditions.

$$\sigma_{v,G,k} = \gamma_K \cdot h$$ Eq. (11.13)

Determining the design value of the effects in the geosynthetic reinforcement

A geosynthetic reinforcement with known axial stiffness in the machine and cross machine directions is selected. The axial stiffness ratio ω can then be determined.

$$\omega = \frac{J_{cmd}}{J_{md}} \qquad \text{Eq. (11.14)}$$

where:

J_{cmd} transverse axial stiffness for an assumed strain and load duration,
J_{md} axial stiffness in the machine direction for an assumed strain and load duration,
md geosynthetic machine direction,
cmd geosynthetic cross machine direction.

For an isotropic geosynthetic reinforcement $\omega = 1.0$.

Inserting ω gives the load components factors for the machine and cross machine directions as follows:

$$X_{md} = \frac{1}{1 + \omega} \qquad \text{Eq. (11.15)}$$

$$X_{cmd} = 1 - X_{md} \qquad \text{Eq. (11.16)}$$

The design values of the horizontal tensile forces are given by:

$$H_{md,d} = \frac{X_{md} \cdot (\gamma_G \cdot \sigma_{v,G,k} + \gamma_Q \cdot \sigma_{v,Q,k}) \cdot D^2}{8 \cdot d_{max}} \qquad \text{Eq. (11.17)}$$

$$H_{cmd,d} = \frac{X_{cmd} \cdot (\gamma_G \cdot \sigma_{v,G,k} + \gamma_Q \cdot \sigma_{v,Q,k}) \cdot D^2}{8 \cdot d_{max}} \qquad \text{Eq. (11.18)}$$

The design values of the actions are determined using the following equations:

$$E_{md,d} = F_{md,d} = \frac{H_{md,d}}{\cos \alpha_{md}} \qquad \text{Eq. (11.19)}$$

$$E_{cmd,d} = F_{cmd,d} = \frac{H_{cmd,d}}{\cos \alpha_{cmd}} \qquad \text{Eq. (11.20)}$$

where:

α angle at the geosynthetic boundary, dependent on type of geosynthetic, sinkhole radius, geometry of sinkhole boundary and sag in centre.

Determining angle α

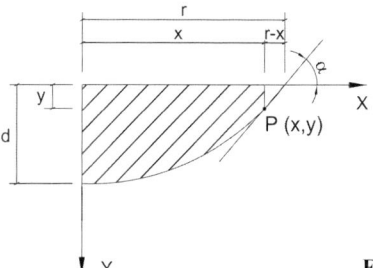

Figure 11.10 Sag in geosynthetic reinforcement

The angle α at the collapse boundary is determined for a value pair x, y, where the distance to the sinkhole boundary $(r - x) = 0.1 \cdot r$ is:

$$\alpha = \arctan\left(\frac{y}{0.1 \cdot r}\right) \qquad \text{Eq. (11.21)}$$

where:

r   radius of sinkhole depression $r = \dfrac{D}{2}$.

It can generally be assumed that the geosynthetic sag is parabolic. The y ordinate for calculating the angle α can be determined as follows:

$$y = a \cdot x^2 + d \qquad \text{Eq. (11.22)}$$

where:

d   geosynthetic reinforcement sag,

a   parabola coefficient: $a = \dfrac{-d}{r^2}$.

In anisotropic geosynthetic reinforcements with an axial stiffness ratio $J_{md}/J_{cmd} \geq 2$ the sinkhole depression is more accurately described by an elliptical function.

$$y = \frac{d}{r} \cdot \sqrt{r^2 - x^2} \qquad \text{Eq. (11.23)}$$

where:

d   geosynthetic reinforcement sag,

r   radius of sinkhole depression.

11.3.2.2 Determining the Design Value of the Tensile Stress Based on the R.A.F.A.E.L. Method [8]

When analysing 'extremely anisotropic' geosynthetic reinforcements it is assumed that the tensile forces are only transferred in one direction. This applies to reinforcements accepting forces predominantly in one direction due to their material-specific configuration. This condition is regarded as met if:

- the axial stiffness ratio $J_{md}/J_{cmd} \geq 10$
and
- the ratio of the short-term strength limit strains $\varepsilon_{md}/\varepsilon_{cmd} \leq 0.5$

The design process does not incorporate cross machine tensile forces analyses. This method can be particularly recommended for geosynthetic reinforcements where the results of practical tests are available.

The designations can be taken from Figure 11.11. Analysis assumes that a cylindrical failure body forms above the geosynthetic reinforcement.

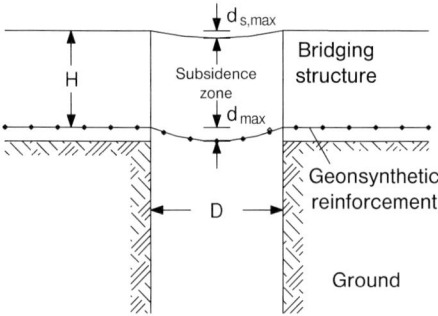

Figure 11.11 R.A.F.A.E.L. analysis model [8]

The allowable sag of the geosynthetic reinforcement is calculated with the aid of the allowable subsidence at the road surface using Eq. (11.24):

$$d_{max} = d_{s.max} + 2 \cdot H \cdot (C_e - 1) \qquad \text{Eq. (11.24)}$$

where:

C_e decompaction factor
      ($C_e = 1.03$ for round-grained material; $C_e = 1.05$ for crushed aggregates; larger values may be adopted if they were determined in large-scale testing.) The magnitude of the proposed decompaction factors is the result of experience gained in tunnel engineering (see [10]).

H cover depth, see Figure 11.2.

The allowable geometric strain of the geosynthetic can be calculated from the allowable sag $d_{max.}$ using the following equation:

$$\varepsilon_{geom} = \frac{8}{3} \cdot \left(\frac{d_{max}}{D} \right)^2 \qquad \text{Eq. (11.25)}$$

The normal stresses acting on the geosynthetic are calculated analogous to Section 11.3.2.1 with the aid of Eq. (11.11).

The design value of the actions is calculated using Eq. (11.26). ε_d is the smaller of the two values ε_{geom} and ε_B (cf. Section 11.3.1).

248

$$E_d = F_d = (\gamma_G \cdot \sigma_{v,G,k} + \gamma_Q \cdot \sigma_{v,Q,k}) \cdot \frac{D}{2} \cdot \sqrt{1 + \frac{1}{6 \cdot \varepsilon_d}}$$
Eq. (11.26)

11.3.2.3 Special Methods

The method for construction and design of the reinforced stabilised base course (*Armierten Stabilisierten Tragschicht* (A.S.T. method)) above collapses [5], [11] can be adopted for both isotropic and anisotropic geosynthetic reinforcements. An anisotropic reinforcement shall be installed orthogonally and in two-ply. The structural system is shown schematically in Figure 11.7, 2b.

A prerequisite for applying this analysis method is the production of a base course above the geosynthetic reinforcement, stabilised by binders. The reinforcement itself is embedded in a composite base course of coarse-grained soil or aggregates.

The thickness of the binder-stabilised base course is designed such that an arch forms if a collapse occurs. The compressive strength of the base course shall be high enough that the compressive stresses in the apex and the abutments are safely accepted.

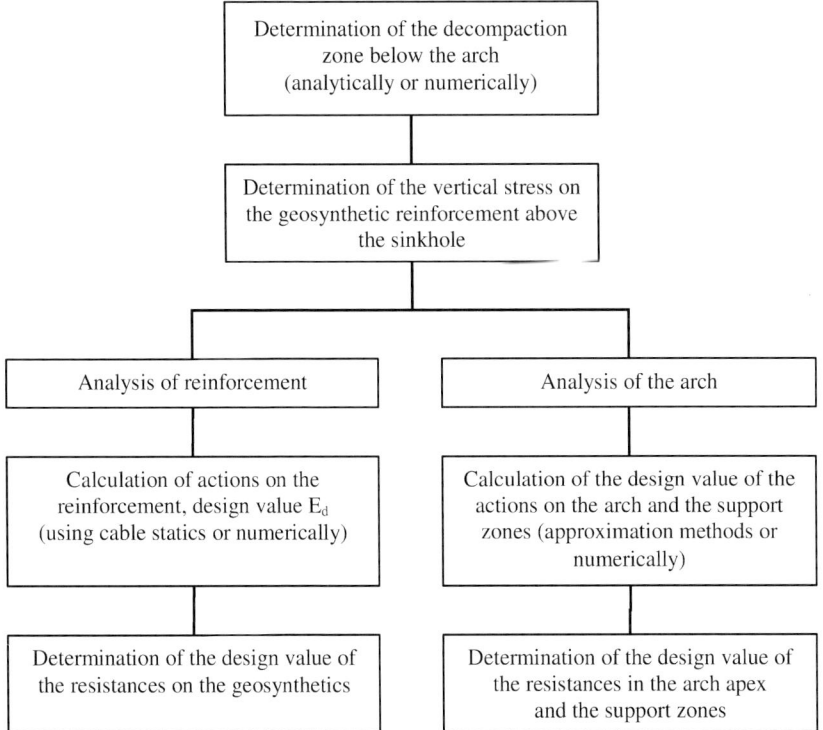

Figure 11.12 Flow diagram of A.S.T. method

249

Due to the stiffness of the surface of the stabilised base course the deformation above the collapse can be practically ignored.

Figure 11.12 shows a flow diagram of the analysis.

11.3.2.4 Determining the Required Short-term Tensile Strength

The design values of the effects were calculated in Sections 11.3.2.1 to 11.3.2.3. Now the working design resistance of the geosynthetic is analysed.

This consists of two different, individual analyses:

- strength-related determination of the required short-term tensile strength (see Section 3).
  This analysis is performed separately for the machine and cross machine directions for anisotropic geosynthetic reinforcements.
- strain-related determination of the required short-term tensile strength with the allowable long-term utilisation factor (see Section 3).
  The largest short-term tensile strength $F_{B,k0}$ determined at the given times is the governing strength.

The maximum value from the two individual analyses is adopted for design.

11.3.2.5 Analysing Anchorage Lengths

The geosynthetic reinforcement parallel to the highway must therefore be anchored outside of the subsidence prone area. Load distribution in the cross machine direction depends on the ground surface and, depending on the design method used, the design diameter of the potential collapse. The anchorage length can only begin at the edge of this imaginary collapse (see Section 11.2.2.2).

The design tensile forces E_d determined using the methods described in Sections 11.3.2.1 to 11.3.2.3 are secured against pulling out in the fill soil. If only little room is available for anchorage the geosynthetics may be wrapped around or installed in an anchor trench.

The required anchorage length for the respective anchored direction of movement is given by rearranging the equation for pull-out resistance as discussed in Section 2.2.4.11 with n = 2:

$$L_A \geq \frac{E_d \cdot \gamma_B}{\sigma_{v,k} \cdot f_{sg,k} \cdot 2} \qquad \text{Eq. (11.27)}$$

In the case of biaxial, anisotropic geosynthetics the sinkhole diameter D need not be included for anchorage in the cross machine direction if engineering measures are taken to ensure that no load transfer occurs in that direction. The minimum requirements of column 3 in Table 11.3 shall be adhered to accordingly.

The minimum requirements for the overlap lengths of extremely anisotropic reinforcements as given in Table 11.4 shall be noted for cross machine anchoring.

Table 11.3 Minimum requirements for anchorage lengths

Load transfer model	Biaxial	Biaxial	Uniaxial
Reinforcement	Isotropic	Anisotropic	Extremely anisotropic
	1	2	3
Schematic principle			
Anchorage length	$L_L \geq L_{A,md}$ $L_1 \geq L_{A,cmd} + D$	$L_L \geq L_{A,md}$ $L_1 \geq L_{A,cmd}(+D)$	$L_L \geq L_{A,md}$ $L_1 \begin{cases} \geq 0.5\text{ m} \\ \geq 0.1 \cdot B \\ \geq 2.0 \cdot \varepsilon_d \cdot D \end{cases}$ (The larger value is the governing value.)

11.3.2.6 Analysing Overlap Lengths

The overlap lengths in the machine and cross machine directions (see Section 11.2.1) are designed such that the tensile force in the respective section be transferred with adequate safety. The governing tensile force E_d is calculated using the methods described in 11.3.2.1 to 11.3.2.3.

Analysis of the overlap lengths is carried out analogous to analysis of the anchorage length (see Section 11.3.2.5). The number of adoptable friction planes n is found to be n = 1 for complete failure of the composite soil layers above the geosynthetic and n = 2 for pull-out from the anchorage zone. The required overlap length in the machine direction ($\ddot{U}_L$) and the cross machine direction ($\ddot{U}_1$) is given by Equation (11.28):

$$\ddot{U}_L(\ddot{U}_1) \geq \frac{E_d \cdot \gamma_B}{\sigma_{v,k} \cdot f_{sg,k} \cdot n} \qquad \text{Eq. (11.28)}$$

In the overlap zone the adopted characteristic value of the friction coefficient $f_{sg,k}$ is the smaller value of the friction angle between the geosynthetic reinforcement and the fill soil or between geosynthetic and geosynthetic. The friction bond is enhanced if fill soil with a higher friction angle is used in the reinforcement plane.

Table 11.4 Minimum requirements for overlaps

Load transfer model	Biaxial	Biaxial	Uniaxial
Reinforcement	Isotropic	Anisotropic	Extremely anisotropic
Schematic principle			
Overlap length	$\ddot{U}_L \geq \ddot{U}_{A,md} + D$ $\ddot{U}_l \geq \ddot{U}_{A,cmd}$	$\ddot{U}_L \geq \ddot{U}_{A,md} + D$ $\ddot{U}_l \geq \ddot{U}_{A,cmd}$	$\ddot{U}_L \geq \ddot{U}_{A,md} + D$ $\ddot{U}_l \begin{cases} \geq 0.5\,\text{m} \\ \geq 0.1 \cdot B \\ \geq 2.0 \cdot \varepsilon_d \cdot D \end{cases}$ (The larger value is the governing value.)

If an unfavourable friction coefficient is assumed for the geosynthetic/geosynthetic contact plane, the friction bond can be improved by introducing a thin, intermediate aggregate layer.

For geosynthetics installed in parallel the design diameter D is included in the overlap length parallel to the highway ($\ddot{U}_L$), calculated using Equation (11.28). In calculating the overlap length in the cross machine direction ($\ddot{U}_l$) enlargement by D is not necessary, because it is assumed that if a collapse occurs the load is redistributed longitudinally around the breach.

The following minimum overlap length requirements thus follow as a function of the selected design method (Table 11.4):

Anisotropic, orthogonally installed geosynthetic reinforcements guaranteeing biaxial load transfer are dealt with as shown in column 3 of Table 11.4.

11.3.3 Safety Theory Analysis

A bridging system reinforced with geosynthetics displays pronounced ductility. For short periods, and if larger deformations are allowed, the system can therefore bridge a collapse with a greater diameter than the design diameter (D) specified in the design. The tensile strength of the reinforcement is reached in the limit strain state, which is far removed from the allowable strain in the serviceability limit state. Thus, the bridging system displays residual safety and bearing capacity reserves which can be utilised in the event of larger than anticipated collapses occurring. In the design, the residual safety can be determined in an additional analyse and be taken into consideration in a hazard scenario.

It is recommended to first determine the reinforcement for the previously specified design state, i.e. for the design diameter (D) and the bridging duration (t_d). Both the serviceability and the STR limit states (complete stabilisation: Load Case 1; partial stabilisation: Load Case 2) are designed for. Analysis of the residual safeties can be useful in certain cases. This is the case, for example, if the anticipated sinkhole diameter does not appear to be sufficiently reliable statistically. The maximum bridgeable sinkhole diameter for LC 3 is determined for a specified bridging duration t_d (e.g. 1 week, 1 day). In contrast to the allowable design subsidence the maximum subsidence in the pavement plane in the serviceability limit state can be specified in cooperation with the client (rail, road, etc.).

Because the required reinforcement anchorage and overlap lengths increase with increasing reinforcement utilisation factor, the required reinforcement anchorage and overlap lengths are also determined during the safety analysis. If the anchorage and overlap lengths are designed for the maximum bridging collapse diameter, the residual safeties determined may be utilised, if need be. If an anchorage or overlap is designed for the collapse design diameter only, the residual safeties are very low, because the bearing capacity of the anchorages becomes the governing factor.

11.4   Applying the Observational Method

Using the observational method the numerical analyses are supplemented by visual and/or instrumented inspections, which are carried on through both the construction phase and the design working life. This allows subterranean collapses and critical situations at the road surface to be detected in time.

The basis for the observational method is given by DIN 1054. The monitoring systems employed may also comprise warning systems.

The monitoring and warning systems are generally used in the following cases:

– Design using the partial stabilisation principle:
  depending on the duration of the load on the partial stabilisation visual observation of the road surface may be sufficient, if the observation intervals guarantee that any deformations of the road surface are detected in time.

– Design taking arching into consideration:
  in this case the deformations within or below the plane of the geosynthetics are monitored by instruments.

– If the safety reserves are utilised as planned:
  instrumented observation in the plane of the geosynthetics is also necessary here.

The following methods may be used for inspection or in conjunction with a warning system:

– Monitoring systems on the road surface:
  · visual inspection of the road surface including subsidence and cracking by regular foot and mounted inspections, if possible by the same inspector,
  · pavement subsidence measurements with the aid of transportable flatness measuring equipment,
  · geodetic measuring system with the aid of zone markings or elevation measuring points arranged in a grid using automatic-display electronic monitoring systems (e.g. monitoring grids or strain gauges embedded in asphalt),

– Monitoring systems within or below the plane of the geosynthetics:
  · signalling nonwovens,
  · horizontal extensometer,
  · hydrostatic deformation monitoring devices,
  · geosynthetics with integrated deformation monitoring devices.

11.5   Notes on Execution

When manufacturing bridging structures the following points shall be noted in addition to the information in Section 11.2.3:

– The excavation level shall be adequately load bearing before applying the composite base course. It shall also be flat and include a gradient.
– The lowest geosynthetic reinforcement layer is installed on an at least 10 cm thick and adequately compacted layer of fill soil.
– The geosynthetics are installed flat and without folds.
– To achieve high installation efficiency and to minimise overlapping, web dimensions are optimised in terms of site boundary conditions by compiling an installation plan.
– The structural demands on the geosynthetics, soils, relative compaction, layer thicknesses, anchorage and overlap lengths in the longitudinal and transverse directions, and monitoring systems are specified in an installation plan. They are shown in cross section and in standard profiles and additionally noted in a quality assurance plan and controlled on site.
– Any deviations from the execution plans are logged.
– Particular care shall be taken that anisotropic geosynthetic reinforcement webs are correctly installed (e.g. direction of the principal tension in longitudinal or transverse direction, upper or lower layer in longitudinal or transverse direction).
– Parallel geosynthetic webs are preferably installed staggered longitudinally.
– Following installation geosynthetics shall be quickly covered or surcharged with cover fill to protect them from mechanical damage and weathering.

Uncovered geosynthetics may not be directly traversed. The geosynthetics may only be traversed by heavy equipment once an at least 25 cm thick, penetration-proof fill soil layer has been emplaced. Lesser thicknesses may be employed if this was taken into consideration when specifying the A_2 value.

– Where pipelines cross, penetration of the geosynthetic reinforcement shall be avoided wherever possible. If they are unavoidable, they shall be adopted in the structural analysis and included in the installation plan.

– The fill soil is installed by end-tipping, graded and compacted. The fill soil may not be tipped directly onto the geosynthetic layers. The degree of compaction of the composite base course is specified in the structural analysis. A relative compaction $D_{pr} = 100\%$ shall be aimed for. The relative compaction achieved shall be demonstrated.

– Adequate overlap length in the roll-out direction $\ddot{U}_L$ of geosynthetic webs installed parallel to the highway shall generally be demonstrated.

11.6   Bibliography

[1]   Fenk, J., Ast, W. (2004): Geotechnische Einschätzung bruchgefährdeten Baugrunds. In: Geotechnik, 27, No. 1, pp. 59–65.

[2]   Heckner, J., Herold, U., Strobel, G., Schönberg, G. (1998): Zum Baugrund und Subrosionsgeschehen in Sachsen-Anhalt. Mitteilungen des Geologischen Landesamtes Sachsen-Anhalt No. 4, pp. 101–121.

[3]   BS 8006 (1995), BSI – British Standard Institution: Code of practice for strengthened/reinforced earths and other fills. London.

[4]   Giroud, J. P., Bonaparte, R., Beech, J. F. (1990): Design of Soil Layer – Geosynthetic Systems overlying Voids. In. Geotextiles and Geomembranes, Vol. 9 (1990), Issue 1, pp. 11–50.

[5]   Schwerdt, S., Meyer, N., Paul, A. (2004): Die Bemessung von Geokunststoffbewehrungen zur Überbrückung von Erdeinbrüchen (B.G.E.-Verfahren). Bauingenieur 79, H. 9.

[6]   Ast, W., Hubal, H., Schollmeier, P. (2001): Bewehrter Erdkörper mit Erdfall-Warnanlage für den Eisenbahnknoten Gröbers. Sonderveröffentlichung der Eisenbahntechnischen Rundschau, Edition ETR Ingenieurbauwerke, 2001.

[7]   Alexiew, D., Ast, W., Elsing, A., Sobolewski, J. (2003): Erdfallüberbrückungssystem Eisenbahnknoten Gröbers – zur Bemessungsplanung und Bauausführung. 8. Informations- und Vortragstagung über 'Kunststoffe in der Geotechnik', Munich. February 2003, Geotechnik, Sonderheft 2003, pp. 235–248.

[8]   Blivet, et. al. (2002): Design method for geosynthetics as reinforcement for embankment subjected to localized subsidence. In: Delmas, Gourc, Girard (ed.): 7th ICG, 2002 Swets & Zeitlinger.

[9]   Therzaghi, K. (1943): Theoretical Soil Mechanics. John Wiley and Sons, Inc., New York, 1943.

[10] Schwerdt, S. (2004): Untersuchungen zur Ableitung eines Bemessungsverfahrens für die Überbrückung von Erdeinbrüchen unter Verwendung von Geokunststoffbewehrungen. Diss. TU Clausthal.

[11] Maidl, B. (1995): Handbuch des Tunnel- und Stollenbaus, Band II: Grundlagen und Zusatzleistungen für Planung und Ausführung. Verlag Glückauf Essen, 2nd edition, 1995.

[12] Alexiew, D., Elsing, A., Ast, W. (2002): FEM-analysis and dimensioning of a sinkhole overbridging system for high speed trains at Gröbers in Germany. In: Delmas, Gourc, Girard (ed.): 7th ICG, 2002 Swets & Zeitlinger.

[13] DIN Fachbericht 101: Actions on Bridges. 2003.

11.7   Worked Analysis Example 1

Preventive stabilisation of a trafficked area against collapse using a one-ply geosynthetic reinforcement.

Design using the *B.G.E.* method:

– biaxial load transfer,
– anisotropic geosynthetic reinforcement.

11.7.1   Specifications

Height of surcharge above reinforcement: $H = 2.0$ m

Diameter of circular collapse in the reinforcement plane as per specifications: $D = 1.0$ m

Wet unit weight of fill material above reinforcement (including pavement): $\gamma_k = 22.0$ kN/m^3

Friction angle of the soil emplaced in the bridging zone: $\varphi_k = 35°$

Cohesion of the soil emplaced in the bridging zone: $c_k = 0.0$ kN/m^2

Live load VHGV 60: $q_k = 33.3$ kN/m^2

Design working life: $t_b = 60$ years

Load duration: $t_d = 1$ week

Deformation criterion on the road surface as agreed with operator: $d_s \le \dfrac{1}{60} \cdot D_s \le 0.017 \cdot D_s$

Load case as agreed with operator: LC 2

Maximum allowable reinforcement strain for load duration $t_d = 1$ week: $\varepsilon_{md,max.} = 6\%$

11.7.2 Allowable Reinforcement Sag and Strain

Adopted sinkhole angle:

$$\theta_k = 80°$$

Diameter of subsidence depression on road surface:

$$D_s = D + \frac{2 \cdot H}{\tan(\theta_k)} = 1.71 \, \text{m}$$

Maximum allowable subsidence at the road surface:

$$d_{s,max} = 0.017 \cdot D_s = 0.029 \, \text{m}$$

Decompaction factor for the emplaced soil on the bridging zone (estimated):

$$C_e = 1.05$$

Reinforcement sag at the maximum allowable subsidence at the road surface:

$$d_{max} = d_{s,max} + 2 \cdot H \cdot (C_e - 1)$$
$$= 0.23 \, \text{m}$$

Allowable geometric strain in the reinforcement resulting from the maximum allowable subsidence at the road surface:

$$\varepsilon_{md,geom} = \left(\frac{8}{3}\right) \cdot \left(\frac{d_{max}}{D}\right)^2$$
$$= 0.141 = 14\%$$

Maximum allowable reinforcement strain for a load duration of 1 week:

$$\varepsilon_{md,zul} = \min(\varepsilon_{md,geom}, \varepsilon_{md,max})$$
$$= 0.06$$

Reinforcement sag at the maximum allowable strain:

$$d_{max} = D \cdot \sqrt{\frac{3}{8} \cdot \varepsilon_{md,zul}}$$
$$= 0.15 \, \text{m}$$

11.7.3 Preselecting the Geosynthetics

	Machine direction	Cross machine direction
Short-term strength	$F_{md,B,k0} = 200 \, \text{kN/m}$	$F_{cmd,B,k0} = 50 \, \text{kN/m}$
Strain for short-term strength	$\varepsilon_{md,k0} = 10.0\%$	$\varepsilon_{cmd,k0} = 10.0\%$
Axial stiffness	$J_{md} = \dfrac{F_{md,B,k0}}{\varepsilon_{md,k0}} = 2{,}000 \, \text{kN/m}$	$J_{cmd} = \dfrac{F_{cmd,B,k0}}{\varepsilon_{cmd,k0}} = 500 \, \text{kN/m}$

11.7.4 Determining the Actions

11.7.4.1 Normal Stresses

Partial safety factor for Load Case 2
to DIN 1054 (STR): $\qquad \gamma_G = 1.20, \gamma_Q = 1.30$

Characteristic earth pressure angle: $\qquad \delta_k = 0$

Ground angle: $\qquad \beta = 0°$

Characteristic earth pressure coefficient: $\qquad K_{agh,k} = K_{a,k} = \dfrac{1 - \sin(\varphi_k)}{1 + \sin(\varphi_k)}$

$$= 0.271$$

Because H/D = 2.0/1.0 = 2.0 the acting normal stresses are determined using the 'failure model with lateral reaction' $\left(1 \le \dfrac{H}{D} \le 3 \right)$:

$$\sigma_{v,G,k} = \frac{D \cdot \left(\gamma_k - \dfrac{4 \cdot c_k}{D} \right) \cdot \left\{ 1 - \left[\exp\left(-K_{a,k} \cdot \tan(\varphi_k) \cdot \dfrac{4 \cdot H}{D} \right) \right] \right\}}{4 \cdot K_{a,k} \cdot \tan(\varphi_k)} = 22.63 \text{ kN/m}^2$$

$$\sigma_{v,Q,k} = q_k \cdot \exp\left(-K_{a,k} \cdot \tan(\varphi_k) \cdot \frac{4 \cdot H}{D} \right) = 7.30 \text{ kN/m}^2$$

11.7.4.2 Load Component Factors

$$\omega_{vorh.} = \frac{J_{cmd}}{J_{md}} = 0.250$$

$$X_{md} = \frac{1}{1 + \omega_{vorh.}} = 0.80 \qquad X_{cmd} = 1 - X_{md} = 0.20$$

11.7.4.3 Design Values of Horizontal Tensile Forces

Adopting $d_{max.} = 0.15$ m:

$$H_{md,d} = \frac{X_{md} \cdot (\gamma_G \cdot \sigma_{v,G,k} + \gamma_Q \cdot \sigma_{v,Q,k}) \cdot D^2}{8 \cdot d_{max}} = 24.43 \text{ kN/m}$$

$$H_{cmd,d} = \frac{X_{cmd} \cdot (\gamma_G \cdot \sigma_{v,G,k} + \gamma_Q \cdot \sigma_{v,Q,k}) \cdot D^2}{8 \cdot d_{max}} = 6.11 \text{ kN/m}$$

11.7.4.4 Design Values of Actions

The design values of the actions are a function of the angle α of the geosynthetics in the collapse boundary zone. This depends on the type of geosynthetic, the sinkhole radius r and the sag in the centre d_{max}.

Adopting $r = 0.5$ m and $d_{max} = 0.15$ m, and assuming that the sinkhole depression is parabolic in both the machine and cross machine directions, the angle α at the edge of the geosynthetic reinforcement is:

$$\alpha = a\ tan\left[\frac{\dfrac{-d_{max}}{r^2}\cdot(r-0.1\cdot r)^2 + d_{max}}{0.1\cdot r}\right] = 32.66°$$

The actions in the machine and cross machine directions are then:

$$E_{md,d} = \frac{H_{md,d}}{\cos\alpha} = 29.02\ \text{kN/m}$$

$$E_{cmd,d} = \frac{H_{cmd,d}}{\cos\alpha} = 7.26\ \text{kN/m}\,.$$

11.7.5 Determining the Design Values of the Resistances in the Machine and Cross Machine Directions

11.7.5.1 Adopted Reinforcement

Geogrid A 200/50, one-ply installed parallel to the highway:

$F_{md, B, k0} = 200$ kN/m $\qquad F_{cmd, B, k0} = 50$ kN/m

11.7.5.2 Design Value of the Tensile Strength, Criterion 1: Reinforcement Creep Failure

Coefficient for creep, loading duration up to 1 month, data provided by geosynthetics manufacturer:	$A_1 = 1.5$
Coefficient for reinforcement damage during transportation, installation and compaction, data provided by geosynthetics manufacturer:	$A_2 = 1.05$
Coefficient for connections (no joins or connections):	$A_3 = 1.00$
Coefficient for environmental impacts (pH 2 to 12), using data provided by the geosynthetics manufacturer for a 60 year design working life:	$A_4 = 1.00$
Coefficient for dynamic actions (no actions):	$A_5 = 1.00$
Partial safety factor for flexible reinforcement elements, LC 2:	$\gamma_B = 1.30$

$$R_{md,B,d} = \frac{F_{md,B,k0}}{A_1 \cdot A_2 \cdot A_3 \cdot A_4 \cdot A_5 \cdot \gamma_B} = 97.68 \text{ kN/m}$$

$$R_{cmd,B,d} = \frac{F_{cmd,B,k0}}{A_1 \cdot A_2 \cdot A_3 \cdot A_4 \cdot A_5 \cdot \gamma_B} = 24.42 \text{ kN/m}$$

11.7.5.3 Design Value of the Tensile Strength, Criterion 2: Reinforcement Creep Strain

Allowable reinforcement utilisation factor for $\varepsilon_{max} = 6.0\%$
(for a loading duration of 1 month), based on isochrones: $\qquad \beta = 0.30$

$$R_{md,D,d} = \frac{F_{md,B,k0} \cdot \beta}{A_2 \cdot A_3 \cdot A_4 \cdot A_5 \cdot \gamma_B} = 43.96 \text{ kN/m}$$

$$R_{cmd,D,d} = \frac{F_{cmd,B,k0} \cdot \beta}{A_2 \cdot A_3 \cdot A_4 \cdot A_5 \cdot \gamma_B} = 10.99 \text{ kN/m}$$

11.7.5.4 Governing Design Value of the Tensile Strength of the Reinforcement

$R_d = \min (R_{B,d}; R_{D,d})$

$R_{md, d} = 43.96 \text{ kN/m}$ $\qquad$ $R_{cmd, d} = 10.99 \text{ kN/m}$

11.7.6 Analysing Adequate Tensile Strength

$R_{md,d} = 43.96 \text{ kN/m} > E_{md,d} = 29.02 \text{ kN/m}$

$R_{cmd,d} = 10.99 \text{ kN/m} > E_{cmd,d} = 7.26 \text{ kN/m}$

11.7.7 Analysing Anchorages

11.7.7.1 Specifications

Anchored forces, STR load case, LC 2: $\qquad$ $E_{md,d} = 19.05 \text{ kN/m}$
$E_{cmd,d} = 4.76 \text{ kN/m}$

Geosynthetic/soil composite coefficient: $\qquad$ $\alpha = 0.9$

Partial safety factor for the pull-out resistance
to DIN 1054, STR, LC 2: $\qquad$ $\gamma_B = 1.30$

11.7.7.2 Required Reinforcement Anchorage Lengths in Machine Direction Outside of the Sinkhole-prone Area

$$L_{L,req} = \frac{E_{md,d} \cdot \gamma_B}{2 \cdot \gamma_k \cdot \alpha \cdot H \cdot \tan(\varphi_k)} = 0.68 \text{ m} \qquad \text{Adopted: } L_{L,work} = 0.7 \text{ m}$$

11.7.7.3 Required Reinforcement Anchorage Lengths in Cross Machine Direction Outside of the Sinkhole-prone Area

Without doline diameter:

$$L_{Q,req,w/o} = \frac{E_{cmd,d} \cdot \gamma_B}{2 \cdot \gamma_k \cdot \alpha \cdot H \cdot \tan(\varphi_k)} = 0.17 \text{ m} \qquad \text{Adopted: } L_{Q,work,w/o} = 0.5 \text{ m}$$

11.7.7.4 Required Reinforcement Anchorage Lengths in Cross Machine Direction Inside of the Sinkhole-prone Area

If the collapse hazard also exists outside of the embankment area, the doline diameter is included:

$$L_{Q,req} = L_{Q,req,w/o} + D = 1.17 \text{ m} \qquad \text{Adopted: } L_{Q,work} = 1.2 \text{ m}$$

11.7.8 Overlap Analysis

11.7.8.1 Required Overlap Length in Machine Direction

$$\ddot{U}_{L,req} = L_{L,req} + D = 1.68 \text{ m} \qquad \text{Adopted: } \ddot{U}_{L,work} = 1.5 \text{ m}$$

11.7.8.2 Required Overlap Width in Cross Machine Direction

$$\ddot{U}_{l,req} = \frac{E_{cmd,d} \cdot \gamma_B}{\gamma_k \cdot \alpha \cdot H \cdot \tan(\varphi_k)} = 0.34 \text{ m} \qquad \text{Adopted: } \ddot{U}_{l,work} = 0.5 \text{ m}$$

11.8 Worked Analysis Example 2

Preventive stabilisation of a road against collapse using a one-ply geosynthetic reinforcement, Design using the *R.A.F.A.E.L.* method:

– uniaxial load transfer,
– extremely anisotropic geosynthetic reinforcement.

11.8.1 Specifications

Height of surcharge above reinforcement:	$H = 2.5$ m
Diameter of circular collapse in the reinforcement plane as per specifications:	$D = 3.0$ m
Wet unit weight of fill material above reinforcement (including pavement):	$\gamma_k = 22.0 \text{ kN/m}^3$
Friction angle of the soil emplaced in the bridging zone:	$\varphi_k = 35°$
Cohesion of the soil emplaced in the bridging zone:	$c_k = 0 \text{ kN/m}^2$

Live load VHGV 60:	$q_k = 33.3$ kN/m^2
Design working life:	$t_b = 60$ years
Load duration:	$t_d = 1$ month
Deformation criterion on the road surface as agreed with operator:	$d_s \leq 0.02\ D_s$
Load case as agreed with operator:	LC 2

11.8.2 Allowable Reinforcement Sag and Strain

Allowable pavement depression ($d_s/D \leq 0.02$): $\quad d_{s,max.} = 0.02 \cdot D = 0.06$ m

Decompaction factor for the emplaced soil on the bridging zone (estimated): $\quad C_e = 1.05$

Reinforcement sag at the allowable subsidence at the road surface $d_s/D \leq 0.02$:

$$d_{max} = d_{s,max} + 2 \cdot H \cdot (C_e - 1)$$
$$= 0.31 \text{ m}$$

Allowable geometric strain in the reinforcement, resulting from the allowable subsidence $d_s/D \leq 0.02$, ($\varepsilon_{max.}$ shall be adhered to for a 1 month loading duration):

$$\varepsilon_{max} = \frac{8}{3} \cdot \frac{d_{max}^2}{D^2} = 0.0285$$

Allowable strain in the reinforcement for the loading duration $t_d = 1$ month: $\quad \varepsilon_d = \varepsilon_{max}$

11.8.3 Determining the Actions

11.8.3.1 Normal Stresses

Partial safety factors for Load Case 2 to DIN 1054 (STR): $\quad \gamma_G = 1.20, \quad \gamma_Q = 1.30$

Characteristic earth pressure angle: $\quad \delta_k = 0°$

Ground angle: $\quad \beta = 0°$

Characteristic earth pressure coefficient:

$$K_{agh,k} = K_{ak} = \tan\left(45° - \frac{\varphi_k}{2}\right)^2 = 0.271$$

Vertical soil pressure resulting from the weight of the emplaced soil:

$$\sigma_{vgk} = \frac{\frac{D}{2} \cdot \left(\gamma_k - 4 \cdot \frac{c_k}{D}\right)}{2 \cdot K_{ak} \cdot \tan(\varphi_k)} \cdot \left(1 - e^{-K_{ak} \cdot \tan(\varphi_k) \cdot \frac{4 \cdot H}{D}}\right) = 40.76 \text{ kN/m}^2$$

Vertical soil pressure from live load:

$$\sigma_{vqk} = q_k \cdot e^{-K_{ak} \cdot \tan(\varphi_k) \cdot \frac{4 \cdot H}{D}} = 17.69 \text{ kN/m}^2$$

Design value of the normal stresses:

$$\sigma_{vd} = \sigma_{vgk} \cdot \gamma_G + \sigma_{vqk} \cdot \gamma_G = 71.91 \text{ kN/m}^2$$

11.8.3.2 Design Value of the Actions on the Geosynthetic Reinforcement

$$E_d = \sigma_{vd} \cdot 0.5 \cdot D \cdot \sqrt{1 + \frac{1}{6 \cdot \varepsilon_d}} = 282.27 \text{ kN/m}$$

11.8.4 Determining the Design Values of the Resistances in the Machine Direction

11.8.4.1 Adopted Reinforcement

Geogrid X 1000/100, one-ply, installed longitudinally

	Machine direction	Cross machine direction
Short-term strength	$F_{md,B,k0} = 1,000 \text{ kN/m}$	$F_{cmd,B,k0} = 100 \text{ kN/m}$
Strain for nominal strength	$\varepsilon_{md,k0} = 6.0\%$	$\varepsilon c_{md,k0} = 12.0\%$
Axial Stiffness	$J_{md} = \dfrac{F_{md,B,k0}}{\varepsilon_{md,k0}} = 16,700 \text{ kN/m}$	$J_{cmd} = \dfrac{F_{cmd,B,k0}}{\varepsilon_{cmd,k0}} = 833 \text{ kN/m}$

The web width is required to determine the overlap width in the cross machine direction: $B = 5$ m

11.8.4.2 Analysing Extreme Anisotropy

The two conditions in Section 11.3.2.2

$$\frac{J_{md}}{J_{cmd}} = 20 > 10 \quad \text{and} \quad \frac{\varepsilon_{cmd}}{\varepsilon_{md}} = \frac{12}{6} = 1$$

are adhered to, extreme anisotropy is thus demonstrated.

11.8.4.3 Design Value of the Tensile Strength, Criterion 1: Reinforcement Creep Failure

Coefficient for creep, loading duration up to 1 month: $A_1 = 1.35$

Coefficient for reinforcement damage during transportation, installation and compaction: $A_2 = 1.05$

Coefficient for connections (no joins and connections): $A_3 = 1.00$

Coefficient for environmental impacts (pH 2 to 12), for a 60 year design working life: $A_4 = 1.00$

Coefficient for dynamic actions (no dynamic actions): $A_5 = 1.00$

Partial safety factor for flexible reinforcement elements, LC 2: $\gamma_M = 1.30$

$$R_{d,B} = \frac{F_{B,k0}}{A_1 \cdot A_2 \cdot A_3 \cdot A_4 \cdot A_5 \cdot \gamma_M} = 542.67 \text{ kN/m}$$

11.8.4.4 Design Value of the Tensile Strength, Criterion 2: Reinforcement Creep Strain

Allowable reinforcement utilisation factor for $\varepsilon_{max.} = 2.85\%$ (for a loading duration of 1 month) based on isochrones, $\beta_d = 0.40$:

$$R_{d,D} = \frac{F_{B,k0} \cdot \beta_d}{A_2 \cdot A_3 \cdot A_4 \cdot A_5 \cdot \gamma_M} = 293.04 \text{ kN/m}$$

11.8.4.5 Governing Design Value of the Tensile Strength of the Reinforcement

$$R_d = \min(R_{d,B}, R_{d,D}) = 293.04 \text{ kN/m}$$

11.8.4.5 Analysing Adequate Tensile Strength

$$R_d = 293.04 \text{ kN/m} > E_d = 282.27 \text{ kN/m}$$

11.8.6 Analysing Anchorages

Anchored force, Load Case LC 2: $E_d = 282.27 \text{ kN/m}$

Partial safety factor for the pull-out resistance to DIN 1054: $\gamma_B = 1.30$

Geogrid/soil composite coefficient: $\alpha = 0.9$

11.8.6.1 Required Reinforcement Anchorage Lengths in Machine Direction Outside of the Sinkhole-prone Area

$$L_{L,req} = \frac{E_d \cdot \gamma_B}{2 \cdot \gamma_k \cdot \alpha \cdot H \cdot \tan(\varphi_k)} = 5.29 \text{ m} \qquad \text{Adopted: } L_{L,work} = 5.50 \text{ m}$$

11.8.6.2 Required Reinforcement Anchorage Length in Cross Machine Direction

The reinforcement is installed below the entire width of the embankment with 0.5 m overlap at the edges in the cross machine direction.

11.8.7 Overlap Analysis

11.8.7.1 Overlap Length in Machine Direction

$$\ddot{U}_L = L_{L,req} + D = 8.29 \text{ m} \qquad \text{Adopted: } \ddot{U}_L = 8.5 \text{ m}$$

11.8.7.2 Overlap Length in Cross Machine Direction

The following values from Table 11.4 result for the cross machine overlaps in the adopted uniaxial, extremely anisotropic bridging:

$$\ddot{U}_{l,1} = 0.5 \text{ m} \qquad \ddot{U}_{l,2} = 0.1 \cdot B = 0.5 \text{ m} \qquad \ddot{U}_{l,3} = 2 \cdot \varepsilon_d \cdot D = 0.17 \text{ m}$$

The governing value is the greater of these:

$$\text{max. } \ddot{U}_l = \text{max. } (\ddot{U}_{l,1}, \ddot{U}_{l,2}, \ddot{U}_{l,3}) \quad \text{max. } \ddot{U}_l = 0.5 \text{ m} \quad \text{Adopted: } \ddot{U}_l = 0.5 \text{ m}$$

12 Dynamic Actions on Geosynthetic-reinforced Systems

12.1 General Recommendations

Reliable current information on the behaviour of geosynthetic-reinforced structures under dynamic actions is presented. On the one hand, it is known that these structures display high load-bearing reserves, for example under seismic loads. On the other hand, where dynamic actions are adopted, the structures are currently dimensioned primarily on the basis of empirical data and/or experience in specific applications. The variety of impacts, including that of dynamic actions on the load-bearing behaviour of the structure, make the establishment of universal design approaches difficult. Below, current knowledge is explained together with approaches for adopting dynamic actions/stresses. The Recommendations are limited to considerations of the governing design approaches for practice-oriented applications.

12.2 Dynamic Actions

All actions variable with time are regarded as dynamic actions as described in Section 6.1.4 of DIN 1054. These actions are:

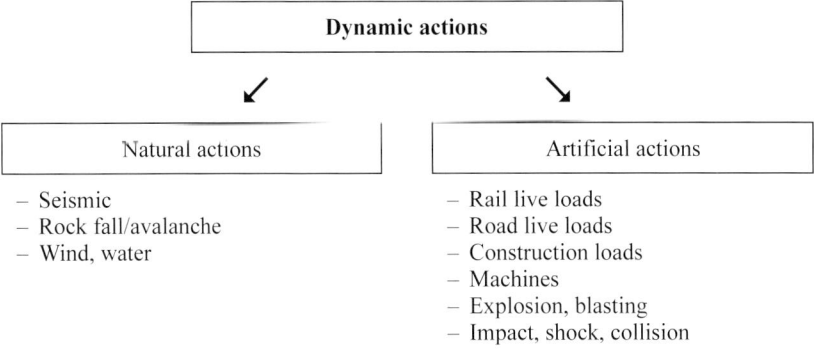

The actions in terms of DIN 1054 can be differentiated into dynamic, cyclic and shock-like actions.

- Dynamic actions:
  Refers to high-frequency actions. Inertial forces are not negligible, they can critically influence system behaviour.

- Cyclic actions:
  Refers to low-frequency actions where the inertial forces can generally be ignored (frequencies ≤ 1 to 2 Hz).

- Shock-like actions:
  Refers to actions acting over a short period only. The time may be in the range of milliseconds up to several seconds; they upper bound is not fixed. Inertial forces may also act.

Additional distinguishing criteria include load-time history characteristics, effective spatial direction of the actions, source and frequency of occurrence:

- Load-time history (Figure 12.1):
  harmonic, periodic, transient, pulsing,

- Effective direction:
  direction of action relative to geosynthetic orientation,

- Source:
  seismic, explosion, rail traffic, road traffic, compaction processes, vehicle impact, rock fall, machines,

- Frequency of action:
  The frequency describes the probable number of occurrences of the dynamic actions relative to the lifetime of the structure.

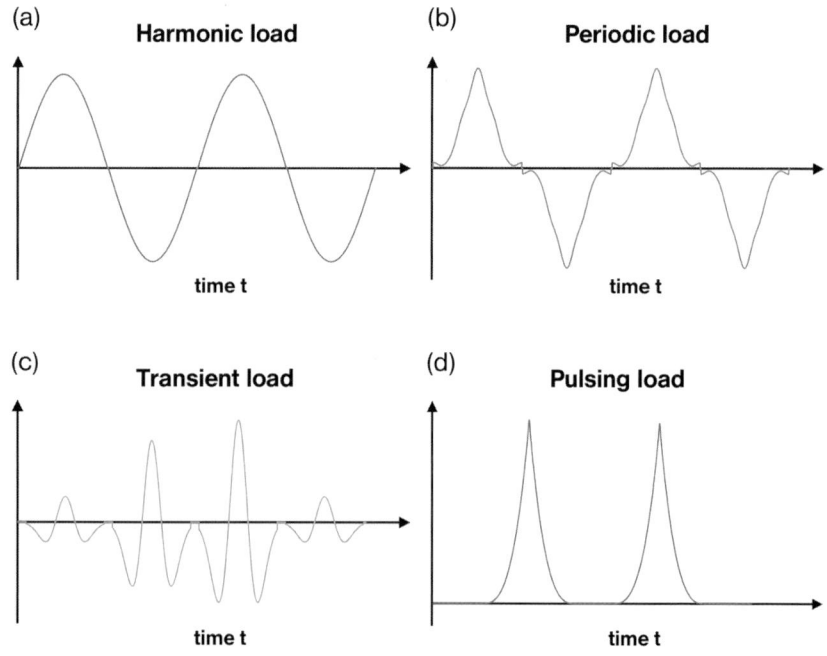

Figure 12.1 Dynamic actions with time [11]

12.3   Dynamic Effects

Dynamic effects refers to only those action effect components occurring as a result of dynamic actions as described in Section 12.2. They are determined as described in Section 12.7, depending on the limit state involved.

Note: The dynamic actions lead to dynamic effects in the geosynthetics. They are governed by geometric and material damping, porewater pressures and resonance phenomena. In saturated soils in particular, parts of the system or the overall system can be excited to produce vibrations at natural frequencies (resonance phenomena). It may be necessary to take mass moment of inertia effects from dynamic actions into consideration when determining dynamic effects.

12.4   Resistances

The design-relevant resistances of the composite geosynthetic-fill soil system can be impacted both individually and as composites:

– Geosynthetics:
  impacts on the structural resistance by fatigue and/or damage as a result of dynamic actions.

– Fill soil:
  impacts on the structural resistance (shear strength) by compaction, decompaction, changes in porewater pressures, grain destruction, grain redistribution, changes in the void ratio, deformation.

– Geosynthetic/fill soil composite system:
  impacts on the composite action between geosynthetic and fill soil (e.g. due to decompaction, unloading phases and grain destruction, grain redistribution, etc.).

12.5   Dynamic Design Cases

Dynamic actions can be taken into consideration in a number of ways when dimensioning geosynthetic-reinforced structures/structural elements, depending on the frequency and magnitude of the actions. To facilitate uniform design procedures three Dynamic Design Cases are differentiated. They allow assessment of whether design for static actions only is adequate in the case considered, whether dynamic actions need to be adopted using quasi-static methods, or whether special deliberations in terms of dynamic actions/effects and resistances are required. The frequency of occurrence of the dynamic action in the course of the assumed lifetime of the structure (load cycle) and the ratio of static action to dynamic action can be approximately defined as the distinguishing criterion.

Frequency:

n number of load cycles [–]   n < 10   (rare)
                                 $10 \leq n < 10^6$ (often)
                                 $n \geq 10^6$ (constant)

Relationship:

$$\zeta = \text{max. } F_{dyn,k} \, / \, \text{max. } F_{stat,k} \quad [-]$$ Eq. (12.1)

$$\text{max. } F_{dyn,k} = 2 \cdot F_{Ampl,k} + \text{max. } F_{stat,k}$$ Eq. (12.2)

where:

ζ	ratio,
max. $F_{stat,k}$	characteristic value of the maximum static action,
max. $F_{dyn,k}$	characteristic value of the maximum dynamic action,
$F_{Ampl,k}$	characteristic value of the dynamic load amplitude.

The definition of the dynamic load components is shown for a cyclic action in the example in Figure 12.2 (using the example of normal stress).

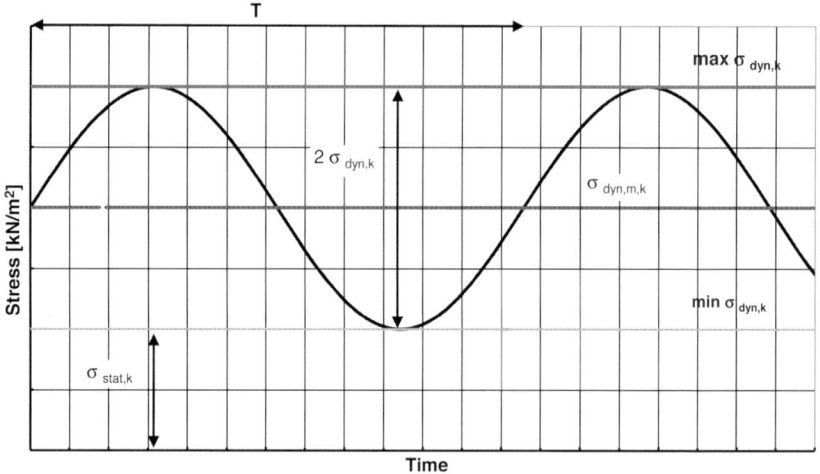

Figure 12.2 Definitions

Annotation in Figure 12.2:

max. $\sigma_{dyn,k}$	maximum dynamic stress,
min. $\sigma_{dyn,k}$	minimum dynamic stress,
$\sigma_{dyn,m,k}$	mean dynamic stress,
$\sigma_{dyn,k}$	dynamic stress amplitude,
$\sigma_{stat,k}$	maximum static stress,
$f = 1 / T$	frequency.

If an analysis of the static actions also takes the dynamic actions adequately into consideration in the terms of DIN 1054, no additional analysis steps are necessary (Dynamic Design Case 1).

If dynamic actions occur with a magnitude and frequency similar to that of the anticipated effects, the *quasi-static equivalent load* method is suggested as a general approximation method (Dynamic Design Case 2).

Where large dynamic actions and/or a large action frequency are involved, or when adopting large dynamic ground effects (e.g. resonance, liquefaction), geo-synthetic-reinforced structures are designed taking the actual load-time histories in the time or, alternatively, the frequency domain into consideration (Dynamic Design Case 3). Suitable analysis procedures and methods are required for this, as well as special software applications. They are generally highly complex and demand expertise in both analysis and in determining the soil mechanics parameters.

The individual Dynamic Design Cases are discussed below.

Dynamic Design Case 1:

No special deliberations necessary. All dynamic effects are sufficiently accounted for when analysing with static actions.

Used for:

- simple cases,
- only one governing load component orthogonal to the geosynthetic layer, planar applications,
- Applications where no dynamic ground effects are anticipated

Dynamic Design Case 2:

Dynamic actions shall be taken into consideration. They may be determined using quasi-static equivalent loads and approximation methods or, more accurately, using the recommendations for Dynamic Design Case 3.

Used for:

- as for Dynamic Design Case 1
  and
- reinforcement in zones subjected to dynamic actions.

Dynamic Design Case 3:

Dynamic actions impact the structure. The structures are designed using suitable procedures/methods in the time or frequency domain, including the necessary laboratory testing for the actual object, or suitable field testing performed to determine the necessary parameters.

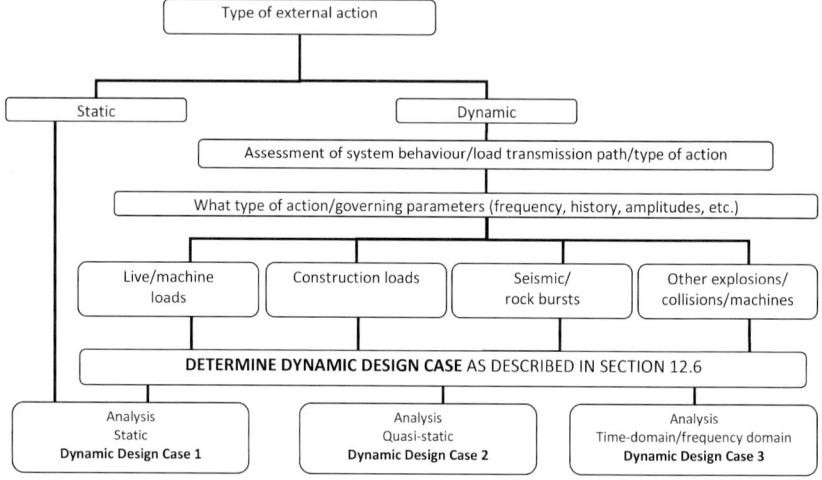

Figure 12.3 Design procedure

Used for:

– as for Dynamic Design Case 2
  and
– several governing load components, not only orthogonal to the geosynthetic layer,
– planar/spatial applications,
– applications for which statements on acceleration, vibration velocity and displacement are required and dynamic ground effects are anticipated.

The design procedure is shown schematically in the following organisation diagram (Figure 12.3). The individual design stages are described in the following sections.

12.6   Dynamic Actions

12.6.1   Dynamic Actions – Live Loads

12.6.1.1   Adopting Live Loads

One principal application is that of dynamic live loads. In assessing whether Dynamic Design Case 1, 2 or 3 is present, the orientation of the geosynthetic in relation to the point where the load is introduced is of decisive importance. The distance of the geosynthetic layer from the point of load introduction is defined as the effective depth z. The magnitude of the dynamic action depends on the effective depth. Therefore, to determine the Dynamic Design Case the dynamic actions (here: normal stresses) and the frequency of occurrence are determined.

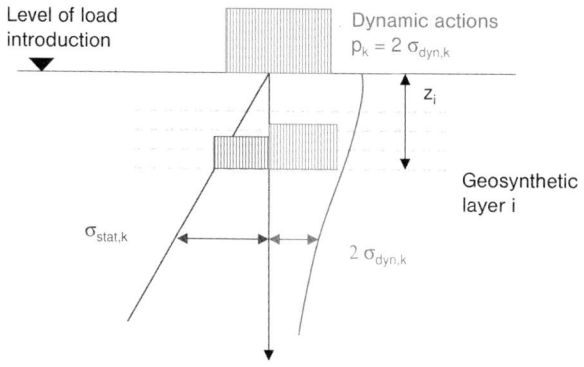

Figure 12.4 Dynamic and static normal stresses

The relationship between static and dynamic actions is defined below as described for normal stresses in Section 12.5:

$$\zeta = 2 \cdot \sigma_{dyn,k} / \sigma_{stat,k} \qquad \text{Eq. (12.3)}$$

Only permanent load components are adopted as effective static load components $\sigma_{stat,k}$ in the analysis. The dynamic normal stress component can be determined in detail after [16]. This approach for estimating the effective depth and the magnitude of dynamic normal stresses is recommended, taking the results and measurements in [7], [9] and [14] into consideration. The cone models that the approach is based on can be applied to determine the anticipated dynamic ground actions and may be used to practically determine dynamic surcharges in any geosynthetic layer. Impedance functions for any load transfer surface can be derived from the dynamic ground properties with the aid of an equivalent truncated cone. The accuracy of the results meets the demands for engineering applications [18]. More recent measurements made in large-scale tests confirm this [23]. The following diagram shows the depth-dependency of dynamic live loads for typical traffic-related frequencies and circular foundations. The stresses in the direction of the load decrease slightly faster with depth for higher constrained moduli, such that the simplified diagram shown (Figure 12.5) may be used [18]. A number of dynamic parameters are drawn for different stress amplitudes and frequencies in Section 12.11.

In addition, predominantly dynamic horizontal stresses and shear stresses at any depth can be roughly estimated using the diagrams in Section 12.11.

Dynamic normal stress amplitude:

$$\sigma_{dyn,k(z)} = F_{skal(z)} \cdot \max. \, \sigma_{dyn,k(z=0)} \qquad \text{Eq. (12.4)}$$

where:

$F_{skal\,(z)}$ amplitude factor as shown in Figure 12.5,

max. $\sigma_{dyn,k\,(z=0)}$ maximum dynamic action in the load transfer plane.

273

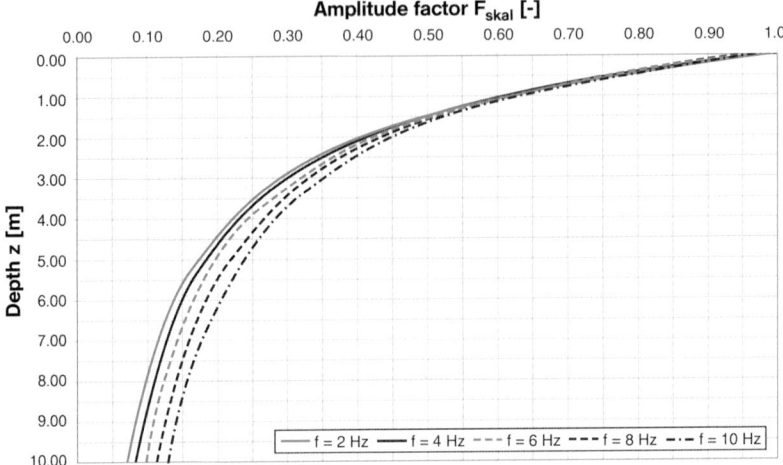

Figure 12.5 Depth effect with frequency variation for a rectangular influence area (3 m × 1 m) and $\sigma = 52$ kN/m^2 for $E_{s,k} = 100$ MN/m^2

The dynamic horizontal stress amplitude may be estimated with the aid of the following equation. The dynamic horizontal stress amplitude is approximately given by:

$$\sigma_{dyn,h,k} = K_0 \cdot \sigma_{dyn,k} \cdot \qquad\qquad \text{Eq. (12.5)}$$

12.6.1.2 Allocation to Dynamic Design Cases – Live Loads

Allocation to Dynamic Design Cases is done on the basis of the location of the geosynthetic using Table 12.1. The depth below the load transfer plane and the shear strain adopted for that depth are critical. The limit values are governed by the frequency, the geometry of the action surface and the load. The distinguishing criterion for the Dynamic Design Cases is the adopted shear strain amplitude. Differentiation is based on the principles in [25]. Table 12.1 shows examples of the subdivisions. Section 12.11 gives the shear strain amplitude diagrams for a variety of load amplitudes (intermediate values may be interpolated).

Table 12.1 Minimum spacing/limit criteria for allocating to the Dynamic Design Cases

Load type:	Live load	Construction load/compaction
Dynamic design case 1	$\gamma < 5 \cdot 10^{-5}$	$h_{min} > 0.30$, because not a permanent load
Dynamic design case 2	$5 \cdot 10^{-5} < \gamma < 1.3 \cdot 10^{-4}$	–
Dynamic design case 3	$\gamma > 1.3 \cdot 10^{-4}$	–

where:

γ shear strain as a result of dynamic load amplitude,

h_{min} distance between load transfer location and first geosynthetic layer.

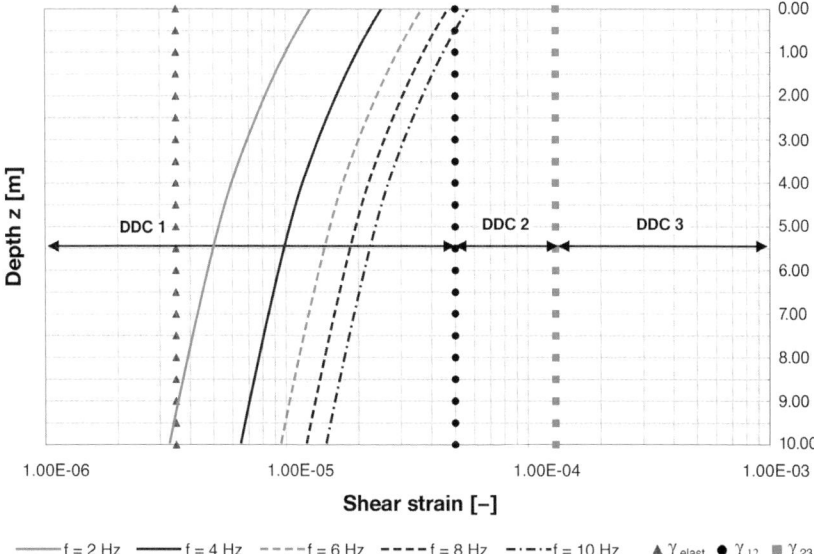

Figure 12.6 Example Dynamic Design Cases (DDC) for a rectangular influence area (3 m × 1 m) and σ = 52 kN/m² for $E_{s,k}$ = 100 MN/m²

12.6.1.3 Design Recommendations

Recommendations for analysis methods and approaches are given below. Both quasi-static equivalent methods and time/frequency domain analysis methods are discussed.

Dynamic Design Case 1:

In Dynamic Design Case 1 the actions are adopted by means of equivalent loads as discussed in DIN Fachbericht 101 (DIN TR101), DIN 1054 and RiL 836. These static equivalent loads take all dynamic impacts adequately into consideration.

Dynamic Design Case 2:

The actions from dynamic loads can be modelled in a Dynamic Design Case 2 analysis using a quasi-static approach as shown in Table 12.2. In contrast to Dynamic Design Case 1 the loads are introduced into the analysis using a load increase factor κ in order to incorporate the additional dynamic load components.

275

Table 12.2 Determining the external dynamic actions for live loads

Load type:	Road traffic	Rail traffic
Size of load max. $\sigma_{k,dyn\,(z=0)}$	$\kappa \cdot$ equivalent distributed load (to DIN TR 101)	$\kappa \cdot$ distributed load (to RiL 836)
Load increase factor κ	$1.0^{**)}$	$1.0^{*)}$
Frequency range for load transfer	0 to 10 Hz	0 to 10 Hz
Depth effect	Section 12.11	Section 12.11

Note: *) *For preliminary design $\kappa = 1.0$ may be adopted, see RiL 836 and [6], [23] for details.*
 **) *Current knowledge indicates that $\kappa = 1.0$ may be adopted for the equivalent distributed load.*
An equivalent distributed load is adopted conservatively for the load. It may be necessary to adopt wheel loads in combination with load increase factors to investigate shallow geosynthetics layers.

The impedance functions and shear strain amplitudes are analysed separately as a function of the footing geometry after [15] for machine footings.

Dynamic Design Case 3:

Dynamic design case 3 is adopted if high dynamic loads occur and the geosynthetic is very close to the load transfer plane. The analysis and generation of load-time functions for dynamic actions from live loads for adopting in time or frequency domain analyses are described in [14]. This allows almost any dynamic action to be derived and the appropriate analyses to be performed. Standardised load-time functions derived from measurements may also be used for special cases. Mass moment of inertia effects and softening phenomena may be taken into consideration. Alternatively, analysis using numerical methods in the time or frequency domain is possible [16].

12.6.2 Dynamic Actions – Explosions, Impact, Avalanches

Impact or explosion loads, although classified as 'rare', are allocated to Dynamic Design Case 3, because of the very high dynamic actions anticipated. On the whole, the dynamic actions from explosions or impacts on geosynthetic-reinforced structures have not been investigated. Approaches for rock fall ramparts based on testing are described in [17]. Approaches for modelling pulse-like loads on geosynthetic-reinforced systems as a result of explosions or rock fall/impact can be found in [24].

12.6.3   Dynamic Actions – Seismic Loads

Seismic loads and their impacts are taken into consideration if structures are built in seismically active zones. Refer to DIN 4149 for adopting seismic loads in Germany. Seismically active zones are marked in the earthquake zone map. If geosynthetic-reinforced structures are erected outside of earthquake zone 0, dimensioning surcharges are adopted. Quasi-static equivalent methods (Dynamic Design Case 2) are adequate for dimensioning. In special cases detailed investigations and an analysis in the time/frequency domain are necessary (Dynamic Design Case 3). Practical experience demonstrates that geosynthetic-reinforced structures under seismic loads behave very favourably and display high load-bearing reserves due to their flexibility and high friction angle [22]. The following procedures are recommended for adopting seismic loads:

Dynamic Design Case 1:

Not relevant.

Dynamic Design Case 2:

The surcharges resulting from seismic actions are suitably adopted for dimensioning. The impacts on the passive and active earth pressure and the slope or global stability in particular are investigated. LC 3 is generally analysed to DIN 1054 with increased earth pressures or surcharges as a function of the seismic zone. It is not necessary to adopt the composite friction coefficient, or fatigue and continuous loads (frequency: n < 10 (rare)). Statements on anticipated deformations are not possible using this method. DIN 4149, 12.2 for the basis for analysis. The ultimate limit state analyses are performed for LC 3 with the following earth pressure coefficients:

$$K_e = K + a_g + \gamma_1 \qquad \qquad \text{Eq. (12.6)}$$

where:

K	earth pressure coefficient,
K_e	seismic earth pressure coefficient,
a_g	maximum ground acceleration to DIN 4149:2005, Table 2,
γ_1	significance coefficient to DIN 4149:2005, Table 3,

Alternative approaches:

Active and passive earth pressure may be adopted in accordance with the Mononobe-Okabe method (M-O method) [12], [13], [18]. Determination of the active and passive dynamic earth pressure components, and the point of application of the load resultant for analysis are described in [12] and [13]. Particular attention is paid to water pressure, inasmuch as it impacts the structure's zone of influence. Corresponding recommendations can be found in [12].

Dynamic Design Case 3:

If statements on the anticipated deformations and/or water pressures require consideration or the structures are in Geotechnical Category 3, analyses in the time or frequency domain are necessary to allow mass moment of inertia effects to be suitably adopted. The same applies if liquefaction or resonance effects are anticipated. Analysis notes and recommendations can be found in [11], [12], [13] and [22]. Pilot and laboratory scale tests improve analysis accuracy [18]. Analysis can be carried out using suitable numerical program systems.

12.7 Determining the Dynamic Effects on the Geosynthetics

12.7.1 Dynamic Design Case 1

No separate analysis, design for static actions is adequate.

12.7.2 Dynamic Design Case 2

The dynamic effect in the geosynthetics layers is determined according to the limit state under consideration. The dynamic effect component is determined as the differential effect between the dynamic actions and the static actions only. Two computations are required. Where:

Computation 1: Effects without dynamic actions $B_{stat,k}$,

Computation 2: Effects with dynamic actions $B_{total,k}$.

The dynamic effect component is given by the difference and is approximately determined as follows for each layer of reinforcement:

$$B_{dyn,k(i)} = B_{total,k(i)} - B_{stat,k(i)}. \hspace{2cm} \text{Eq. (12.7)}$$

This analysis is performed conservatively using the composite friction angle $f_{s,k}$. The effects are then factorised according to limit state and load case and converted to design effects.

12.7.3 Dynamic Design Case 3

The surcharges are analysed separately with the aid of special design procedures and software applications (e.g. [14]). The results are examined using plausibility tests and comparative analyses.

12.8 Determining the Resistances for Dynamic Actions

12.8.1 Pull-out Resistance of Reinforcement

12.8.1.1 Dynamic Design Case 1

Separate analysis of the influence of dynamic actions on the pull-out resistance is not required. Analysis is performed as described in Section 3.3.3.

12.8.1.2 Dynamic Design Case 2

The pull-out resistance is directly influenced by dynamic actions [3], [4]. This is taken into consideration in analysis necessary. If the influence of the pull-out resistance is not determined in tests as described in Section 12.9, the following approach may be adopted for dimensioning:

Table 12.3 Range for λ_{Dyn}

λ_{dyn}	Range	
1.0	$\zeta = 0.0$	Static case
$1 - \zeta$	$0.0 < \zeta \le 0.80$	
0.20	$0.80 < \zeta$	

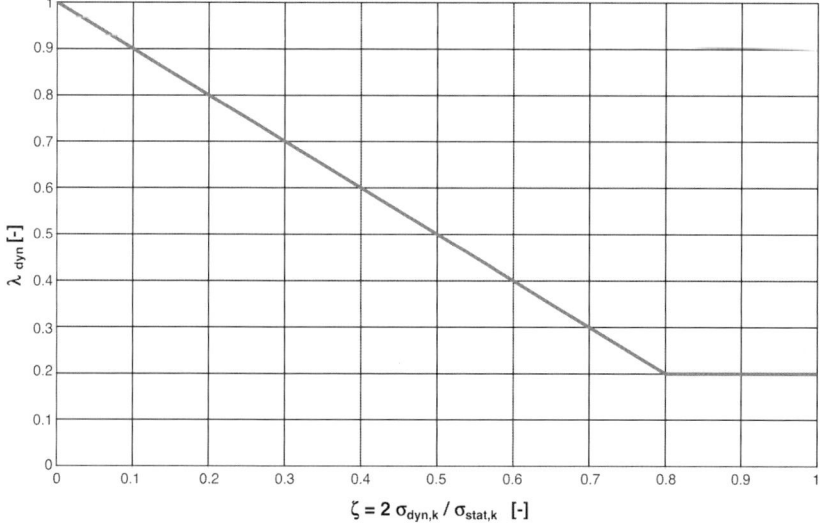

$$\zeta = 2\,\sigma_{dyn,k} / \sigma_{stat,k} \ [-]$$

Figure 12.7 Determining λ_{dyn}

279

The approximate interface friction coefficient required to determine the pull-out resistance is determined as follows:

$$f_{s,k} = \lambda_{dyn} \cdot \lambda_{stat} \cdot \tan \varphi'_k .$$ Eq. (12.8)

This relationship can be used to conservatively estimate the change in the friction coefficient. It takes into consideration a reduction in the friction bond as a result of dynamic actions. The pull-out resistance is determined as described in Section 3.3.3 using the above relationship and based on the total load ($\sigma_{stat} + \sigma_{dyn}$).

12.8.1.3 Dynamic Design Case 3

Object-related tests as described in Section 12.9 are used to determine the pull-out resistance in Dynamic Design Case 3.

12.8.2 Structural Resistance of Reinforcement

12.8.2.1 Dynamic Design Case 1

Separate analysis of the dynamic actions on the structural resistance of the reinforcement is not required ($A_5 = 1.00$).

12.8.2.2 Dynamic Design Case 2

The structural resistance of the reinforcement is calculated using the A_5 coefficient to include the dynamic actions. The A_5 coefficient takes account of fatigue phenomena and damage as a result of repeated actions, and ist determined in laboratory tests as described in Section 12.9. If no more precise data is available A_5 assumes values of $A_5 = 1.0$ to 1.5 as shown in Figure 12.8.

Table 12.4 Range for A_5

A_5	Range	
1.0	$\zeta = 0.0$	Static case
1.0 to 1.5	$0.0 < \zeta \le 1.00$	Linear interpolation
1.50	$1.00 < \zeta$	

Note: Tests have been carried out on a number of products to allow fatigue to be taken into consideration. Damage in the soil composite has barely been investigated.

12.8.2.3 Dynamic Design Case 3

Separate analyses of the fatigue and damage behaviour of the geosynthetics are required as described in Section 12.9. The determination of structural resistance may only be specified on the basis of tests.

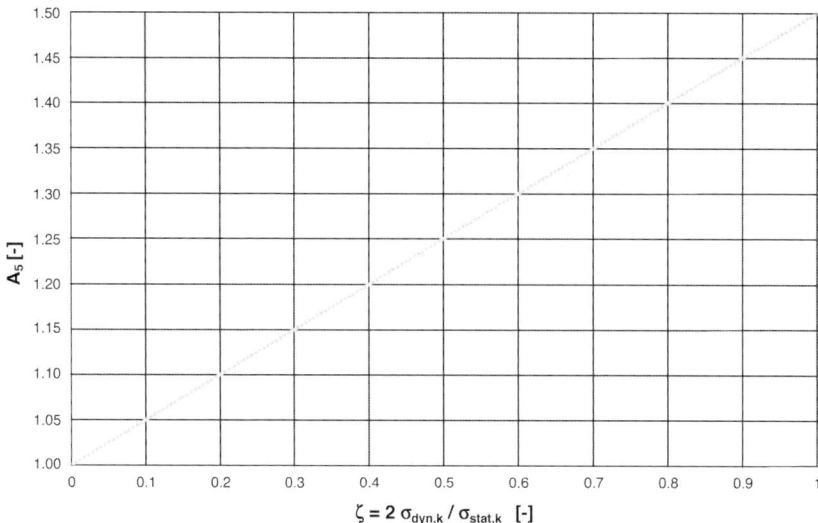

$$\zeta = 2\, \sigma_{dyn,k} / \sigma_{stat,k}\ [\text{-}]$$

Figure 12.8 Determining A_5

12.9　Demands on Building Materials under Dynamic Loads

The following recommendations apply to the areas covered by Dynamic Design Cases 2 and 3 only.

12.9.1　Fill Soil

In addition to the demands of soil mechanics on the fill soil compliant with EBGEO, Section 2.1.2, the following demands shall be met for these structures.

12.9.1.1　Grain Sizes

Percentage grain diameter less than 0.063 mm: 　$< 7.0\%$ wt. (installed)

Percentage grain diameter greater than 100 mm: 　$< 25.0\%$ wt.

Max. grain: 　　　　　　　　　　　　　　max. $d < 150$ mm

Grading curve: 　　　　　　　　　　　　as well graded as possible
　　　　　　　　　　　　　　　　　　　(close to the Fuller parabola)

12.9.1.2　Grain Shape, Grain Strength

Grain failure and edge wear can occur given very high dynamic load components and low grain strength. This leads to grain redistribution, impacts on the friction bond and deformation. The following recommendations relating to the fill soil grain properties aim to counteract these effects:

281

Grain shape: Squat (tested to DIN 52114 (DIN 4226)),

Grain strength: Impact destruction test/destruction test,
limit value as defined in
(TL-Min StB/TL Mineralstoffe DB AG).

12.9.1.3 Fill Soil Friction Coefficient

Use of soil with the required relative compaction and an installed friction angle of $\varphi'_k \geq 30.0°$ is recommended.

12.9.1.4 Relative Compaction

In contrast to *ZTV E-StB* and RiL 836 a relative compaction in the geosynthetic-reinforced earth structure of

$D_{Pr} \geq 100\%$

is demanded.

12.9.2 Geosynthetics

The basic demands on the reinforcement are defined in Section 2.2. In addition, the following material properties for Dynamic Design Case 3 shall be analysed using field and/or laboratory testing:

– fatigue behaviour,
– damage when installed,
– composite behaviour.

12.9.2.1 Determining Fatigue Behaviour

The fatigue behaviour of geosynthetics subjected to dynamic effects can be investigated and described after [1]. The reduction factor A_5, which takes the influence of fatigue into consideration, is adopted as described in Section 12.8.2 and is determined in a pulsating load test. The test is performed in accordance with the notes below. If necessary preliminary clarifying tests are performed to specify the final test boundary conditions:

– The load cycles selected for testing shall allow statements on the frequency of occurrence of the actions for the entire design working life of the structure.
– A sinusoidal pulsating load is applied. The testing frequency is specified taking material warming into consideration; $\vartheta_{Probe} =$ approx.. 20 °C ± 5 °C (if necessary specimens shall be cooled for testing).
– Any frequency-dependence in terms of fatigue behaviour shall be examined.

Measurements from [7], [20] and [28], displaying a dynamic effect component on the geosynthetic of up to 50% of the effect from the static base load, can be used to determine the dynamic effects on geosynthetics.

This leads to the following extreme load spectra:

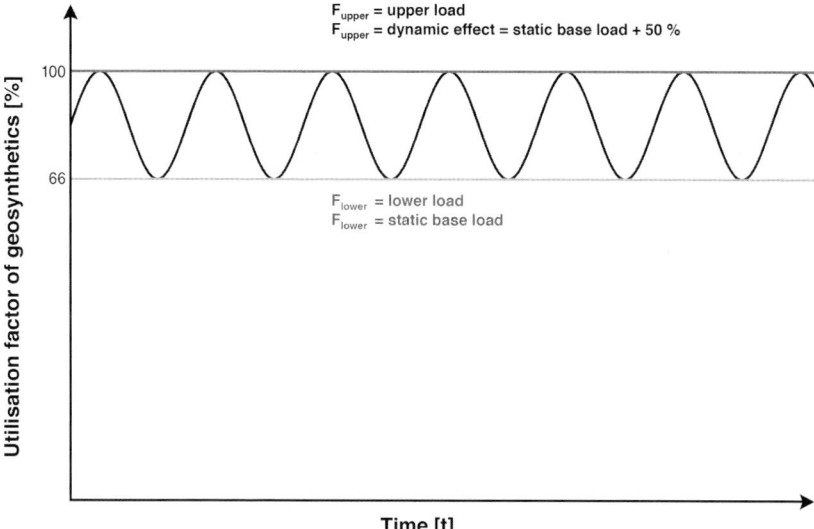

Figure 12.9 Load spectra for testing

Currently, the load spectra recommended for testing takes the maximum antici-pated dynamic effect into consideration. When determining the utilisation factor the creep rupture strength of the geosynthetics for the test duration is taken into consideration. That is, a reduction coefficient A_1 is taken into consideration for 11 days for the example combination of 10^7 load cycles applied at a frequency of 10 Hz, because this corresponds to the duration of the example test. Investigations by [20] have shown that it may be assumed that a load somewhere between the dynamic mean load and the upper load components in the described testing pro-cedure corresponds to the utilisation factor in the static creep tests. The reference value used is the mean value of the short-term strength $R_{B,k0}$. When determining the creep rupture strength using A_1 it can therefore be assumed that an upper load F_{upper} corresponding to the utilisation factor is adequate and no further impacts on creep rupture strength from dynamic effects need be taken into consideration.

Note: To determine A_5, $R_{B,k0}$ can also be determined from five individual measure-ments of a single geogrid strap or a 5 cm wide geotextile strip, analogous to DIN EN ISO 10319.

$R_{B,k,dyn}$ is the force corresponding to a 100% load on the geosynthetics for the test duration and thus the upper load F_{upper} during testing.

Testing programme:

Three specimens are loaded at frequencies up to 10 Hz (cooled, if necessary) for 10^7 load cycles with load spectra as shown in Figure 12.9. The lower the frequency in the test, the greater the load on the geosynthetics due to its relaxation behaviour. Therefore a frequency below the actual effect is selected. A test frequency of 10 Hz is generally regarded as the maximum, because this value may already approach the range of the live load excitation frequency. Seismic loads are not governing for these tests due to the small number of load intervals.

The creep rupture strength R_f following cyclic effects is determined in tensile tests similar to those for the reference values. A_5 is determined from the quotient of the reference value $R_{B,k0}$ and R_f:

$$A_5 = \frac{R_{B,k0}}{R_f} .$$
Eq. (12.9)

Note: Given sufficient experience three tests up to 10^5 cycles and one test up to 10^7 cycles may serve as analyses.

The number of cycles determined for the defined operating period can be adopted as test stresses for temporary construction projects.

A detailed description of the test can be found in [26].

12.9.2.2 Determining Damage

The additional damage caused by the dynamic action (not installation damage) is examined on an object-related basis taking the anticipated load cycle number (frequency of actions) into consideration. Where necessary it is determined in field or laboratory tests.

Note: This damage is not assessed using the A_5 coefficient investigations described above. In one case excavation revealed no additional damage to the geosynthetics under dynamic actions [27].

12.9.2.3 Determining the Geosynthetic/Fill Soil Composite Coefficient

However, recent investigations described in [3], [4] and [15] show that dynamic actions influence the composite coefficient. Current research indicates that the interface friction coefficient falls with increasing load amplitude. According to [4] and [16] this can be determined in the laboratory for the respective fill soil-geosynthetic combination if the approaches described in Section 12.8.1 are not adopted or accurate determination is required. Investigations of the friction bond are carried out based on DIN EN ISO 12957-1 or DIN EN 13738/DIN 60009 taking the dynamic load component into consideration.

The following points shall be observed when performing tests:

- The selected load cycle numbers shall allow a statement on the operational life of the structure.
- A sinusoidal pulsating load is applied.
- Any frequency-dependence in terms of the composite friction coefficient shall be examined.

Pull-out test:

- The cyclic (or sinusoidal) load component is transferred as a horizontal component into the geosynthetic layer.
- The acceptable pull-out force is a function of the maximum load level max. F_{dyn}, the magnitude of the sinusoidal component and the number of cycles, and depends on the use and effect conditions.
- Every sinusoidal load cycle generally causes deformation in the geosynthetic. The deformation resulting from the cyclic pull-out test may not be greater than the deformation from a comparable static pull-out test (failure value) at the required load cycle number.
- The tests may be evaluated to DIN EN 13738/DIN 60009.

Direct shear tests:

- The interface friction coefficient between the geosynthetics and the fill soil is determined based on DIN EN ISO 12957-1.
- A vertical sinusoidal pulsating load is applied. The frequency used depends on the application.

12.10   Bibliography

[1] Müller-Rochholz, J. (1999): Dynamisches Verhalten von HDPE Bewehrungsgittern. FS-KGeo 1999, Munich.

[2] Nimmesgern, M., Busch, D. (1991): The Effect of Repeated Loading on Geosynthetic Reinforcement Anchorage Resistance. Proc. of the Conference Geosynthetics 1991, Atlanta.

[3] Herold, A., Mannsbart, G. (1997): Dynamisches Kreisringschergerät zur Bestimmung des Grenzflächenscherverhaltens zwischen Geokunststoff und Lockergestein. FS-KGeo 1997, Munich.

[4] Herold, A. (1999): Geokunststoffe unter dynamischer Belastung II. FS-KGeo 1999, Munich.

[5] Martinek, M. (1976): Bodendruckmessungen bei den Schnellfahrversuchen zwischen Gütersloh und Neubeckum, ZEV-Glas.

[6] Göbel, C., Lieberenz, K., Richter, F. (1996): Der Eisenbahnunterbau. DB-Fachbuch, Band 8/20, Eisenbahnfachverlag, Mainz.

[7] Lieberenz, K., Weisemann, U. (2002): Geosynthetics in dynamically stressed earth structures of railway lines. 7th ICG Nizza, 2002.

[8] Gotschol, A. (2002): Veränderlich elastisches und plastisches Verhalten nichtbindiger Böden und Schotter unter zyklisch-dynamischer Beanspruchung. Schriftreihe Geotechnik Universität GH Kassel, Heft 12.

[9] Stöcker, T. (2002): Zur Modellierung von granularen Materialien bei nichtruhenden Lasteinwirkungen. Schriftreihe Geotechnik, Universität GH Kassel, Heft 13.

[10] Bauen in Bergschadengebieten. Editor: Ledwon, J. A., published by Ernst & Sohn, Berlin, 1983.

[11] Baudynamik praxisgerecht, Band 1: Berechnungsgrundlagen. Editor: Flesch, R., published in Bauverlag Wiesbaden and Berlin, 1993.

[12] Bodendynamik, Grundlagen, Kennziffern, Probleme. Editor: Studer, J. A., Koller, M. G., published by Springer, 2nd edition, 1997.

[13] Geotechnical Engineering Handbook, Part 1: Fundamentals. Editor: Smoltczyk, U., published by Ernst & Sohn, Berlin, 2002.

[14] Herold, A., Müller-Borotau, F. (2003): Die Ermittlung von dynamischen Spannungen/ Lasten innerhalb von kunststoffbewehrten Stützkonstruktionen. FS-KGeo 2003, Munich.

[15] Nernheim, A., Meyer, N. (2003): Verbundverhalten von Geokunststoff und Boden unter nichtruhender Belastung – Vorstellung eines Pull-Out-Gerätes. FS-KGeo 2003, Munich, pp. 69–73.

[16] Herold, A., Tamaskovic, N. (2004): Bestimmung von dynamischen Spannungen in Kunststoff-bewehrte-Erde-Konstruktionen unter Zuhilfenahme von Kegelmodellen. Bautechnik, 09/2004.

[17] Blovsky, S. (2002): Bewehrungsmöglichkeiten mit Geokunststoffen. Dissertation TU Vienna 04/2002.

[18] Klapperich, H. (1983): Untersuchungen zum dynamischen Erddruck. TU Berlin, Heft 14, Berlin.

[19] Popp, K., Schielen, W. (2003): System Dynamics and Long-Term Behaviour of Railway Vehicles, Track and Subgrade. Springer Verlag.

[20] Herold, A., Pachomow, D., Murray, H., Boones, B. (2006): Messung von statischen und dynamischen Dehnungen in KBE-kunststoffbewehrte-Erde-Konstruktionen mit faseroptischen Sensoren – Praxisbeispiele/Ergebnisse/numerische Simulationen und Rückrechnungen. Messen in der Geotechnik seminar, 23/24 February 2006 at TU Braunschweig.

[21] Harting, U. (1993): Kriechverhalten von Geogittern unter dynamischer Beanspruchung. Diploma thesis at the Münster University of Applied Sciences.

[22] Herold, A. (2006): Bauen in Erdbebengebieten mit Geokunststoffen. TU Freiberg lecture text, Fachsektion Geotechnik, dated 2006.

[23] Göbel, C., Lieberenz, K. (2004): Handbuch Erdbauwerke der Bahnen. 1st edition, Verlagsdruckerei Kessler, Boblingen.

[24] Peila, D., Oggeri, C., Casttiglia, C., Recalcati, P., Rimoldi, P., Magnus, M., Oehrl, M. (2005): Schutzwälle mit Geogitterbewehrung gegen Steinschlag – Erprobung und FEM Berechnung. In: Müller-Rochholz: Geokunststoffe im Erd- und Straßenbau, Werner Verlag, 1st edition, 2005.

[25] Hu, Y., Gartung, E., Prühs, H., Müllner, B. (2003): Bewertung der dynamischen Stabilität von Erdbauwerken unter Eisenbahnverkehr. Geotechnik 26, 2003, pp. 42–56.

[26] Retzlaff, J. (2007): Verhalten von Geokunststoffbewehrungen unter zyklischer Beanspruchung. Publication of Institut für Geotechnik at TU Freiberg, 2007-3.

[27] Lieberenz, K., Nimmesgern, M. (2001): Geogitterbewehrter Bahndamm – Ausgrabung nach 10-jähriger dynamischer Belastung. FS-KGeo 2001, Munich, pp. 173–176.

286

[28] Auersch, L., Rücker, W. (2005): Dynamic Loads of Railway Traffic. Proceedings of Geosynthetics 2004, ECI-Conference, Pillnitz/Germany, Glückauf Essen.

[29] Vucetic, M. (1994): Cyclic threshold shear strains in soils. Journal of Geotechnical Engineering, Vol. 120, No. 12, pages 2008–2228.

[30] Zanzinger, H., Hangen, H., Alexiew, D. (2007): Ermüdungsverhalten von Geogittern unter dynamischer Belastung. FS-KGEO 2007, Munich.

12.11 Diagrams

Input data:

Frequenz: $f = 2$ to 10 Hz   Surface area: Rectangle where:

                                       $B \cdot L$: $3.0 \cdot 1.0$ m

Load amplitude: $2\,\sigma_{dyn} = 52.0$ kN/m^2 Constrained modulus: $E_{s,k} = 50.0$ MN/m^2

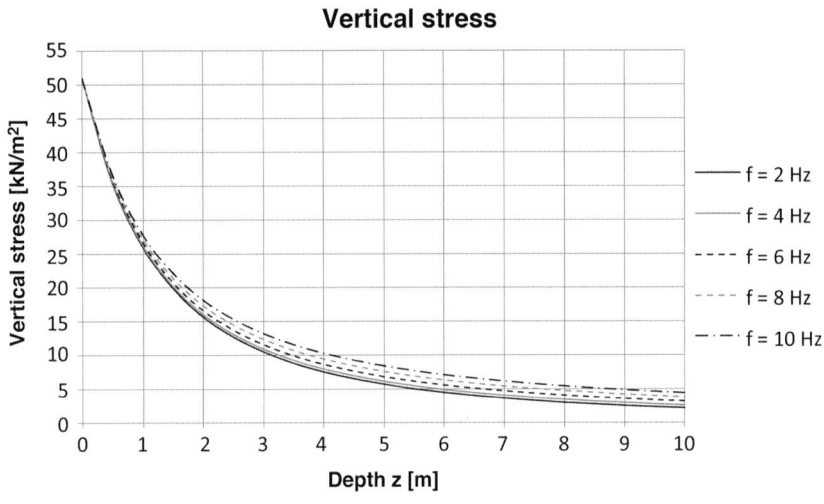

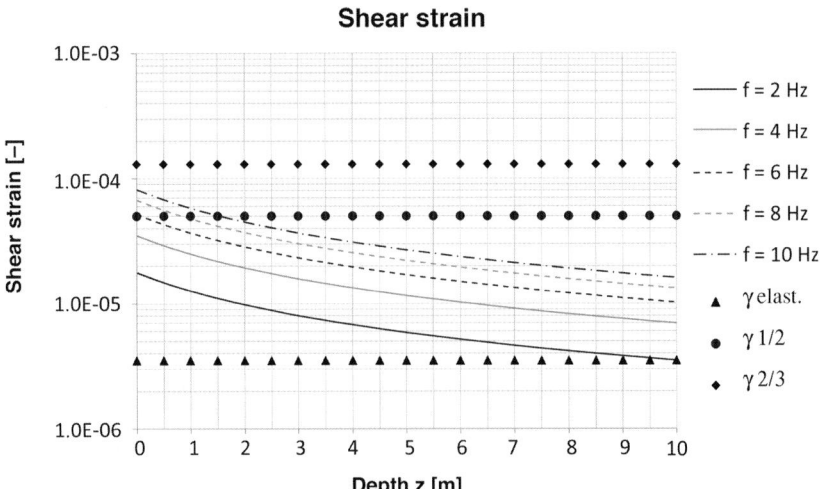

Input data:

Frequenz: $f = 2$ to 10 Hz   Surface area: Rectangle where:

                                                                              $B \cdot L: 3.0 \cdot 1.0$ m

Load amplitude: $2\,\sigma_{dyn} = 52.0$ kN/m^2 Constrained modulus: $E_{s,k} = 75.0$ MN/m^2

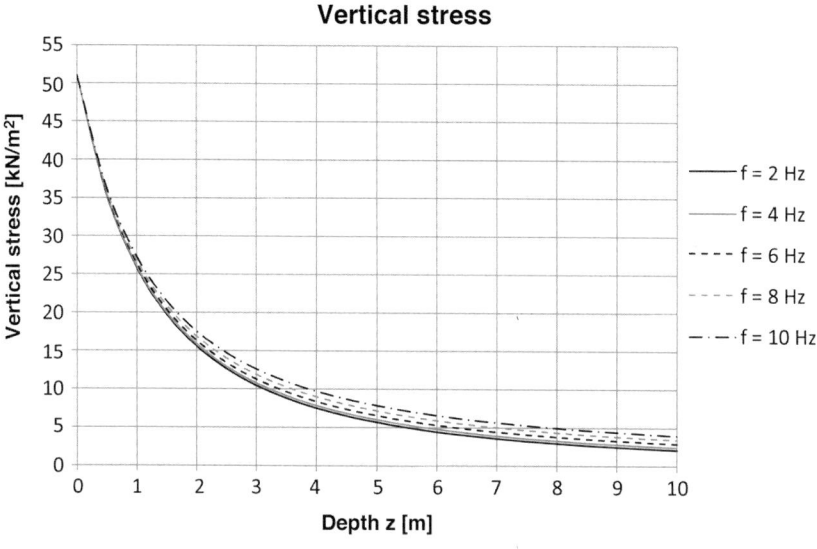

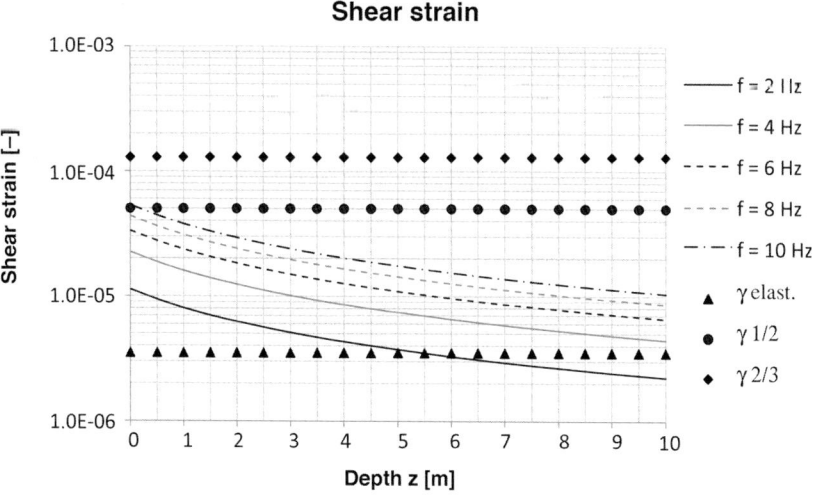

Input data:

Frequenz: $f = 2$ to 10 Hz   Surface area: Rectangle where:

                                               $B \cdot L : 3.0 \cdot 1.0$ m

Load amplitude: $2\,\sigma_{dyn} = 52.0$ kN/m^2 Constrained modulus: $E_{s,k} = 150.0$ MN/m^2

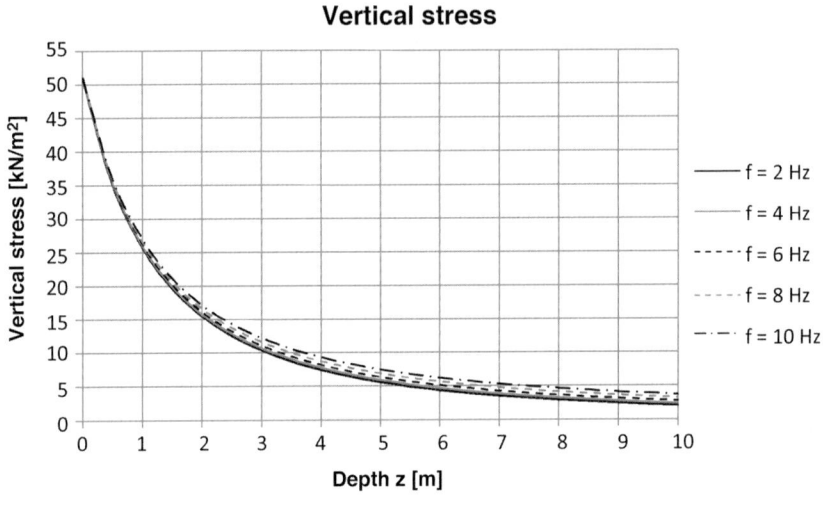

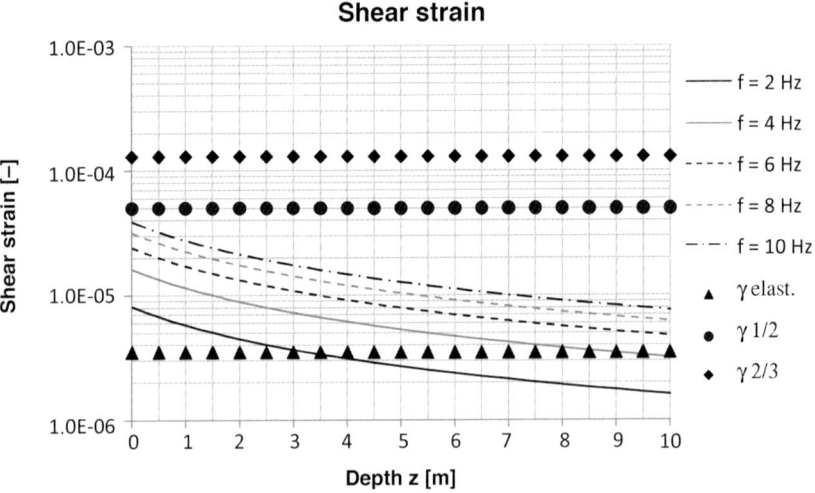

Input data:

Frequenz: $f = 2$ to 10 Hz   Surface area: Rectangle where:

                                                     $B \cdot L: 3.0 \cdot 1.0$ m

Load amplitude: $2\,\sigma_{dyn} = 41.0$ kN/m^2 Constrained modulus: $E_{s,k} = 50.0$ MN/m^2

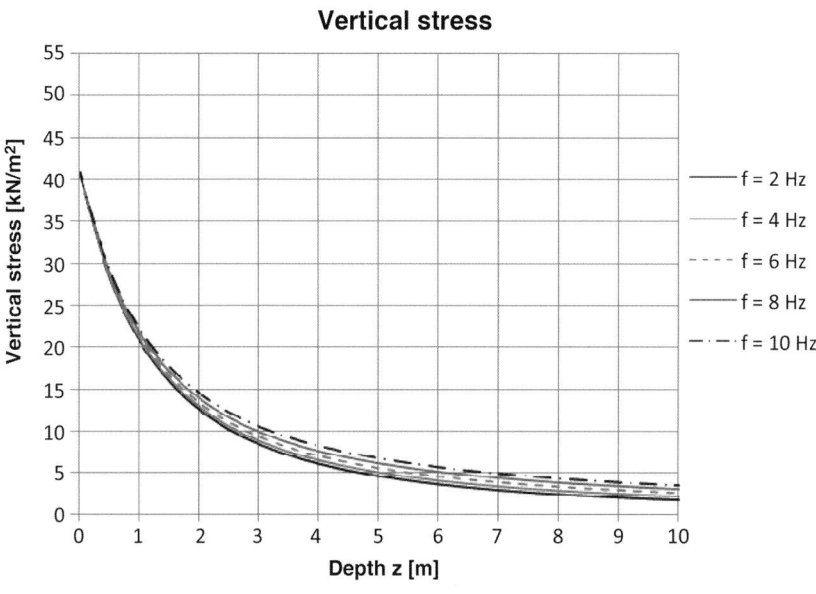

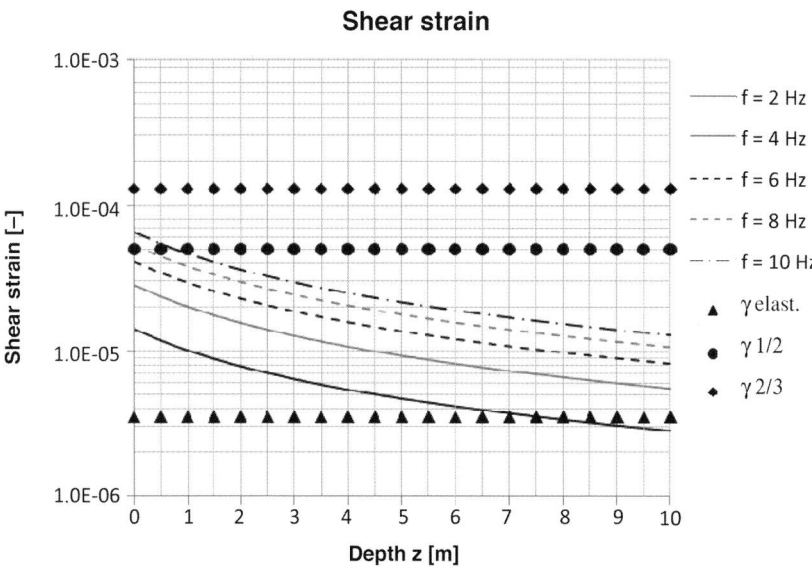

291

Input data:
Frequenz: f = 2 to 10 Hz   Surface area: Rectangle where:
                                                             B · L: 3.0 · 1.0 m
Load amplitude: 2 σ_{dyn} = 41.0 kN/m² Constrained modulus: $E_{s,k}$ = 75.0 MN/m²

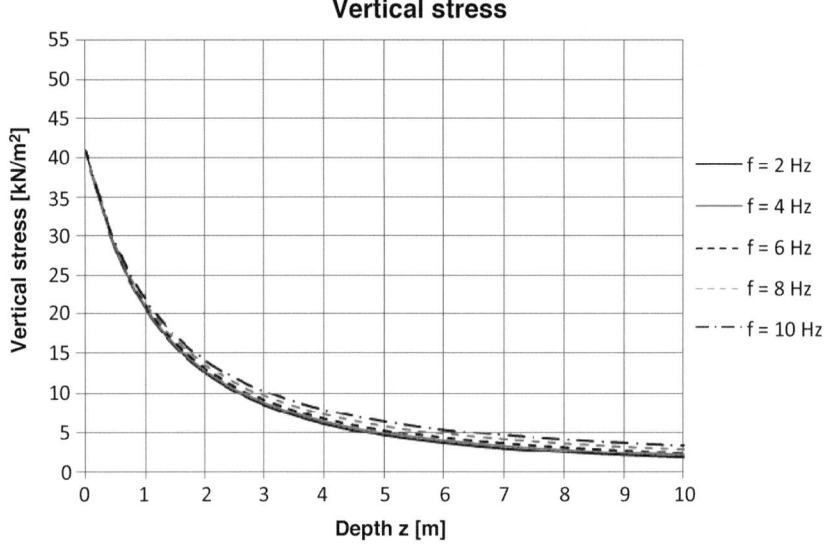

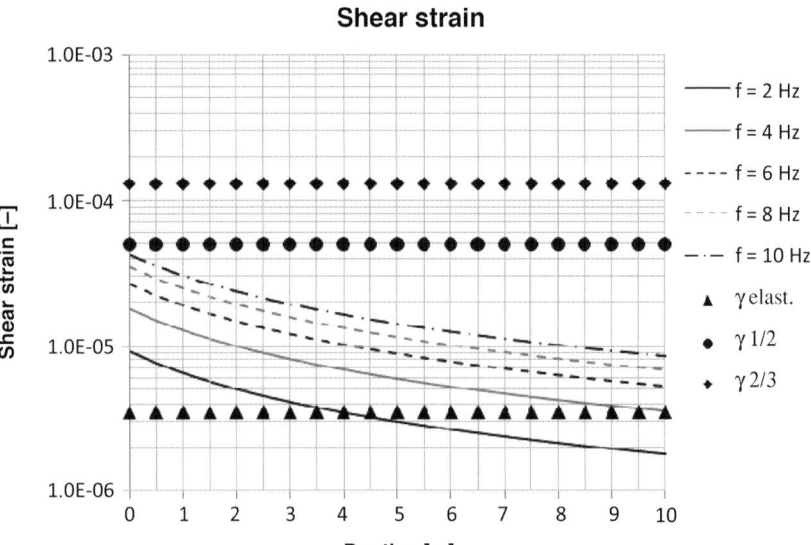

Input data:

Frequenz: f = 2 to 10 Hz   Surface area: Rectangle where:
                                                     B · L: 3.0 · 1.0 m
Load amplitude: 2 σ_{dyn} = 41.0 kN/m^2 Constrained modulus: $E_{s,k}$ = 150.0 MN/m^2

Vertical stress

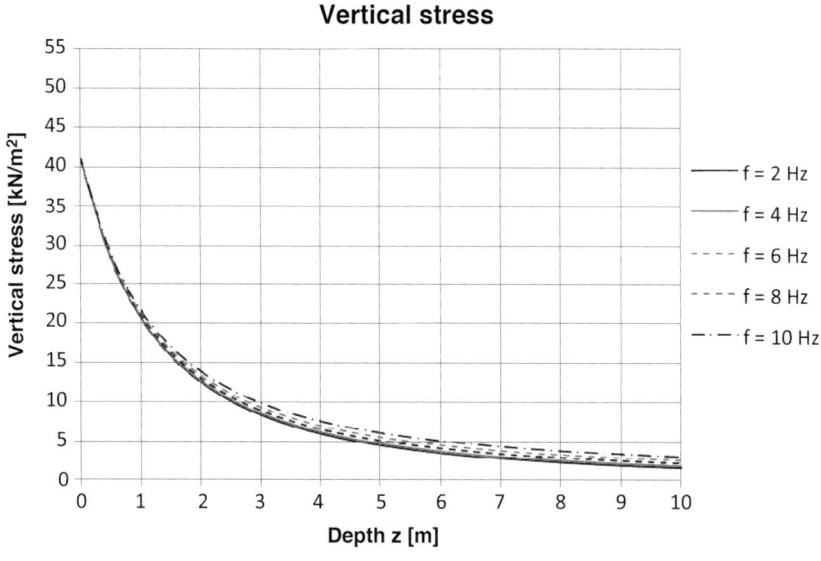

Shear strain

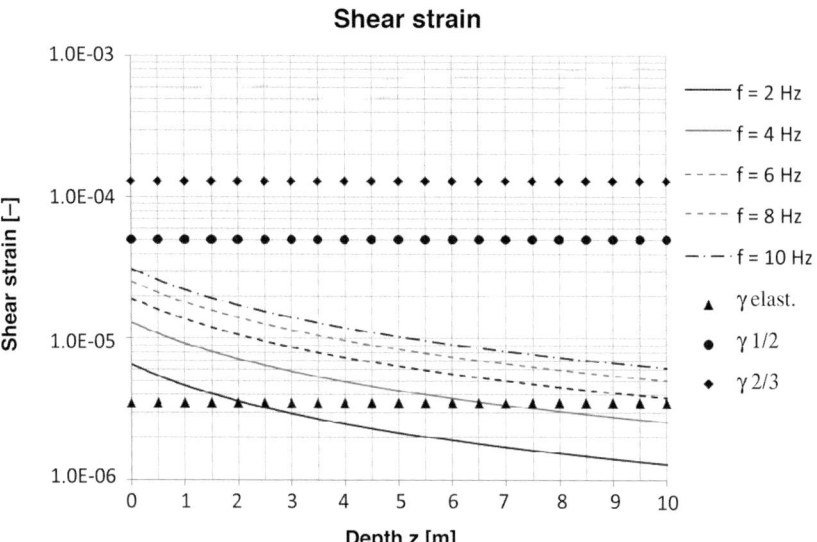

Input data:

Frequenz: f = 2 to 10 Hz   Surface area: Rectangle where:
                                             B · L: 3.0 · 1.0 m
Load amplitude: 2 σ_{dyn} = 23.8 kN/m²   Constrained modulus: $E_{s,k}$ = 50.0 MN/m²

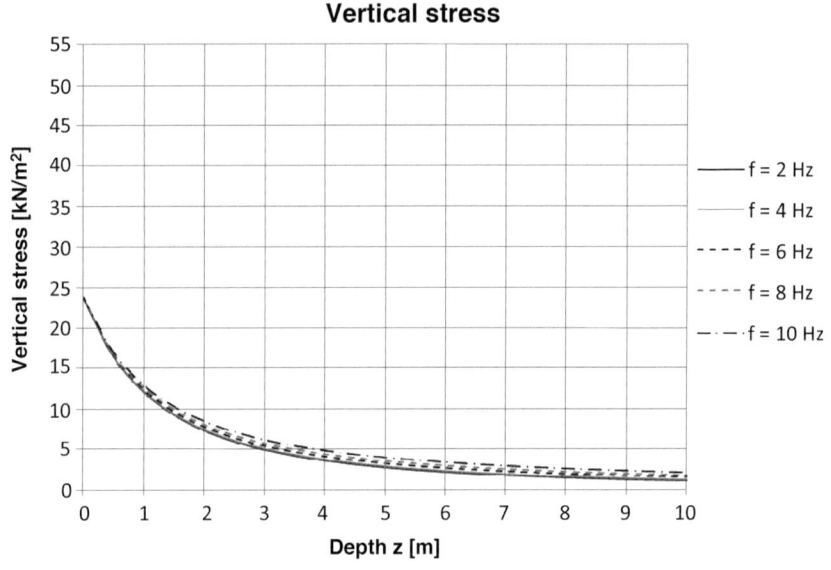

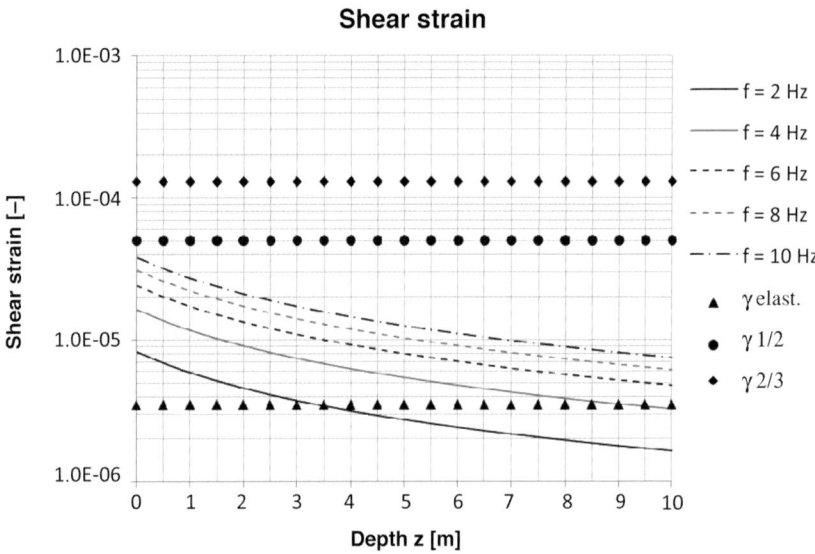

294

Input data:

Frequenz: f = 2 to 10 Hz   Surface area: Rectangle where:
                                             B · L: 3.0 · 1.0 m
Load amplitude: 2 σ_{dyn} = 23.8 kN/m² Constrained modulus: $E_{s,k}$ = 75.0 MN/m²

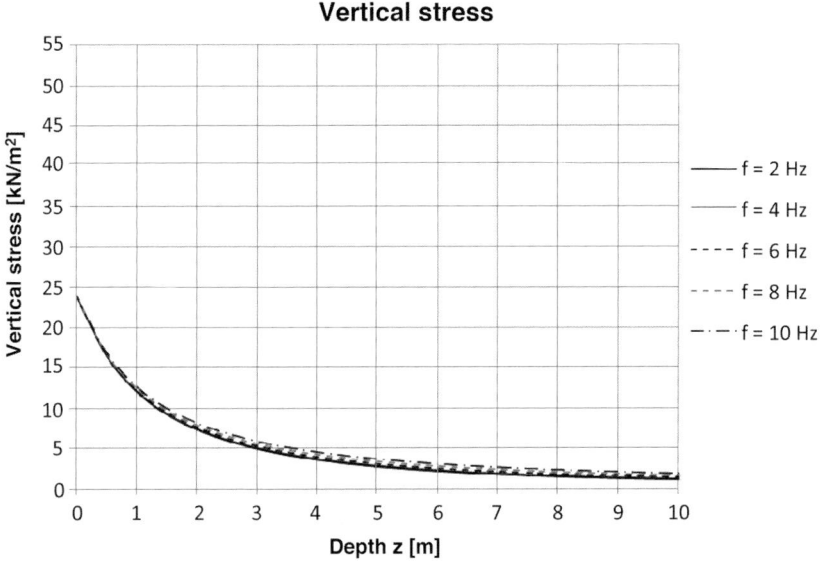

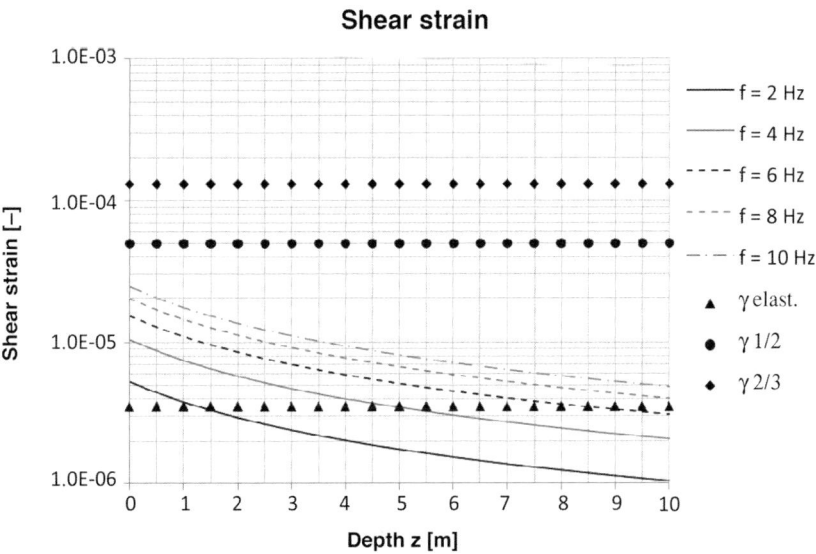

Input data:
Frequenz: f = 2 to 10 Hz   Surface area: Rectangle where:
B · L: 3.0 · 1.0 m
Load amplitude: 2 σ_{dyn} = 23.8 kN/m² Constrained modulus: $E_{s,k}$ = 150.0 MN/m²

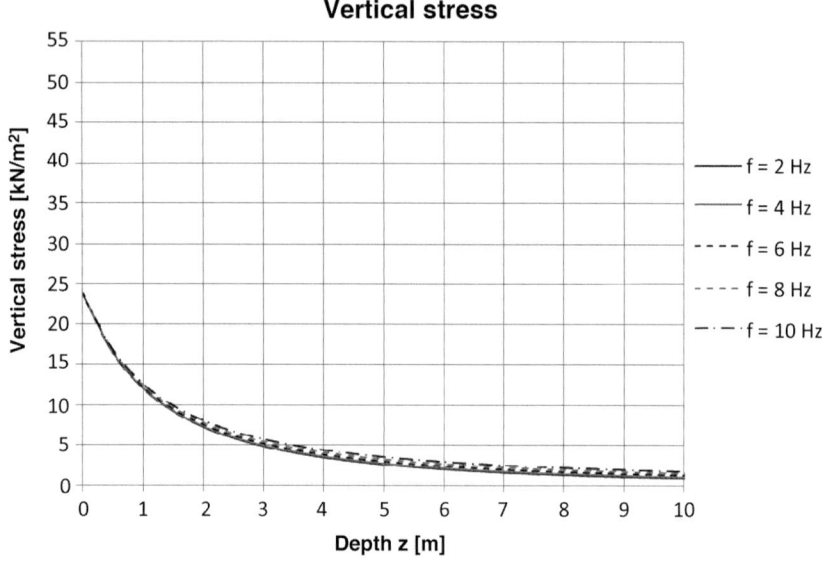

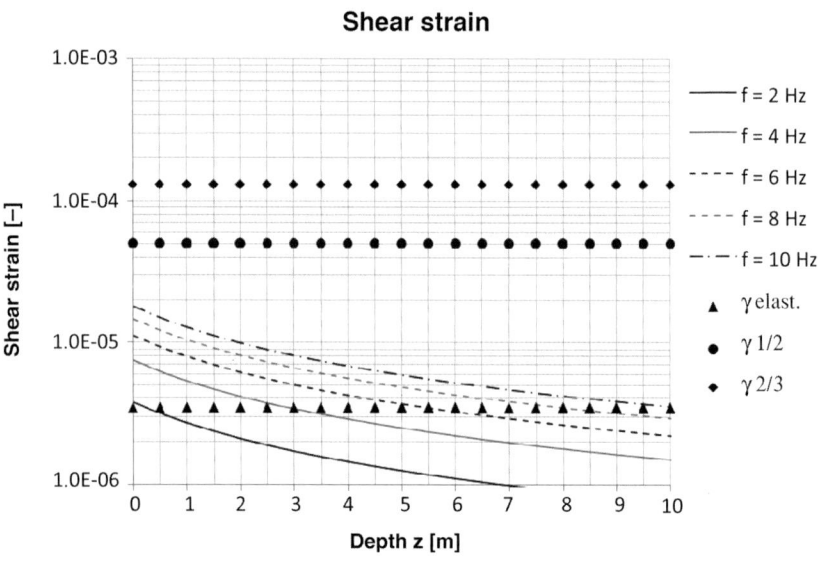

Input data:

Frequenz: f = 2 to 10 Hz Surface area: Rectangle where:

  B · L: 3.0 · 10.0 m

Load amplitude: 2 σ_{dyn} = 52.0 kN/m² Constrained modulus: $E_{s,k}$ = 50.0 MN/m²

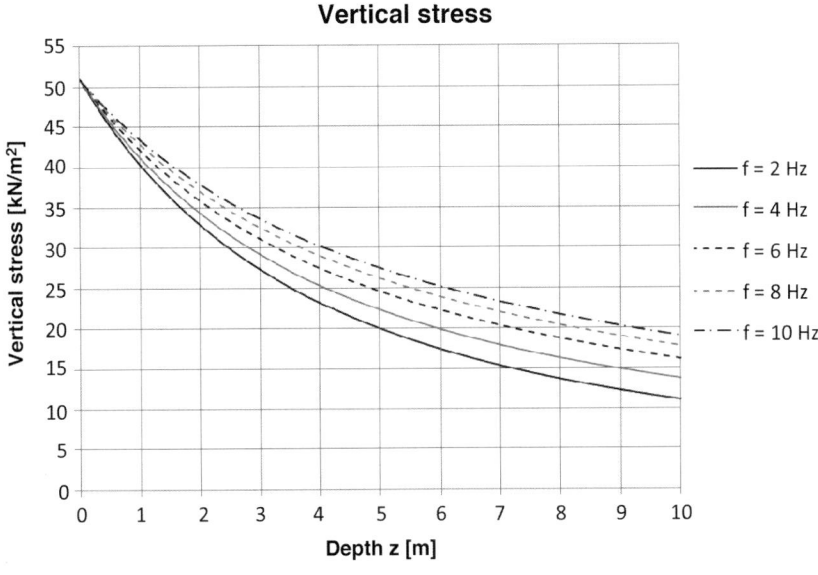

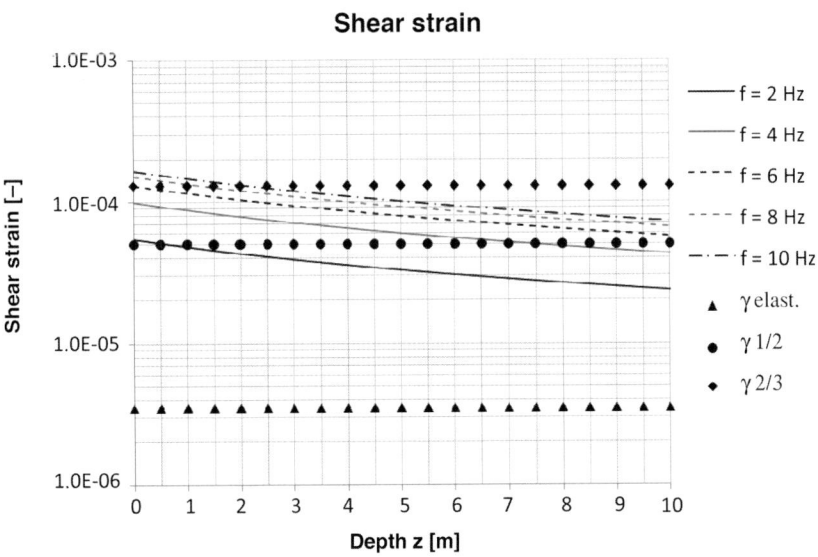

Input data:
Frequenz: $f = 2$ to 10 Hz   Surface area: Rectangle where:
                                                        B · L: 3.0 · 10.0 m
Load amplitude: $2\,\sigma_{dyn} = 52.0$ kN/m^2 Constrained modulus: $E_{s,k} = 75.0$ MN/m^2

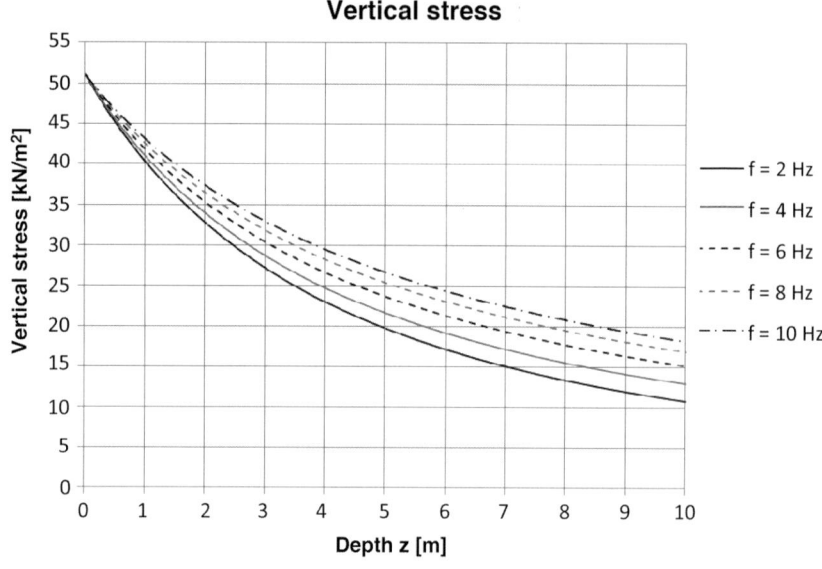

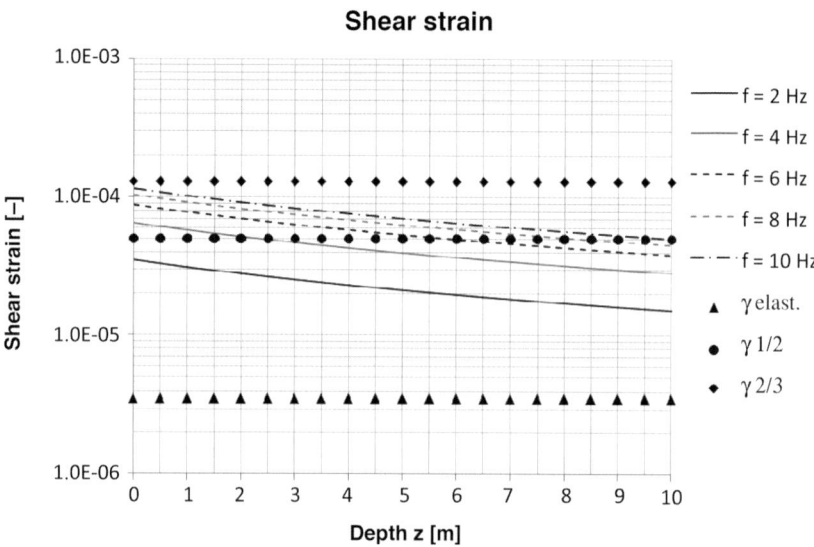

Input data:

Frequenz: f = 2 to 10 Hz   Surface area: Rectangle where:

                                                      B · L: 3.0 · 10.0 m

Load amplitude: 2 σ_{dyn} = 52.0 kN/m^2 Constrained modulus: $E_{s,k}$ = 150.0 MN/m^2

Vertical stress

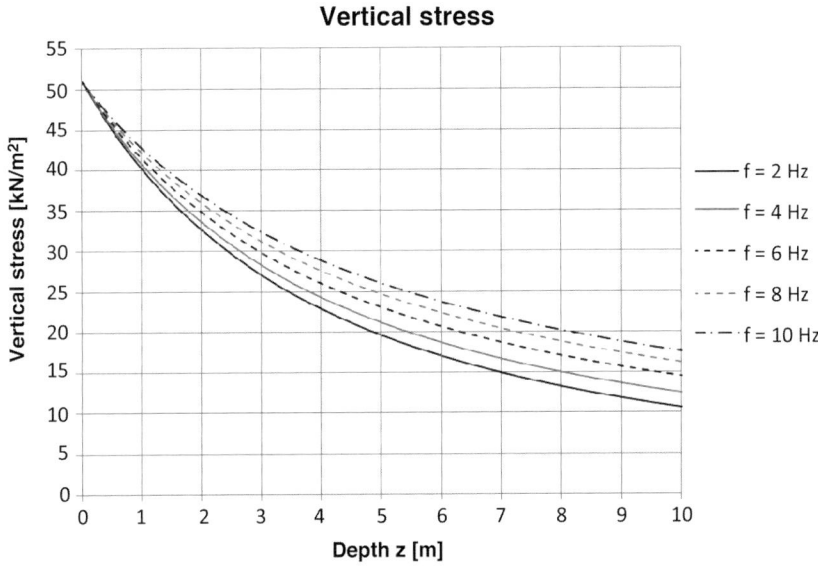

Shear strain

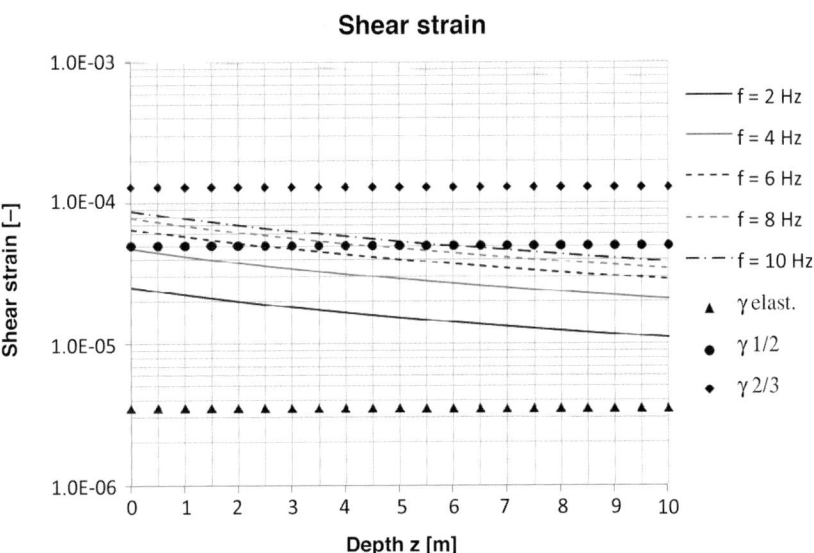

Input data:

Frequenz: $f = 2$ to 10 Hz   Surface area: Rectangle where:
                                                B · L: 3.0 · 10.0 m
Load amplitude: $2\,\sigma_{dyn} = 41.0$ kN/m^2 Constrained modulus: $E_{s,k} = 50.0$ MN/m^2

Vertical stress

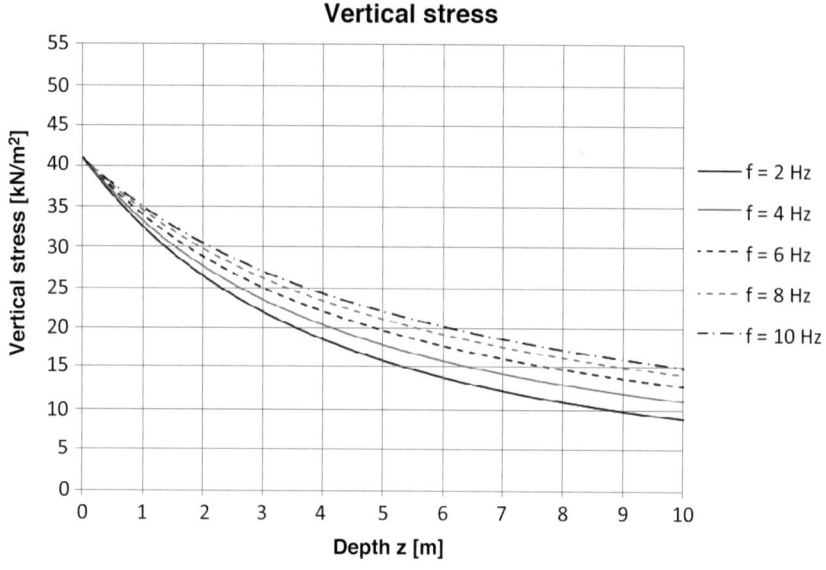

Shear strain

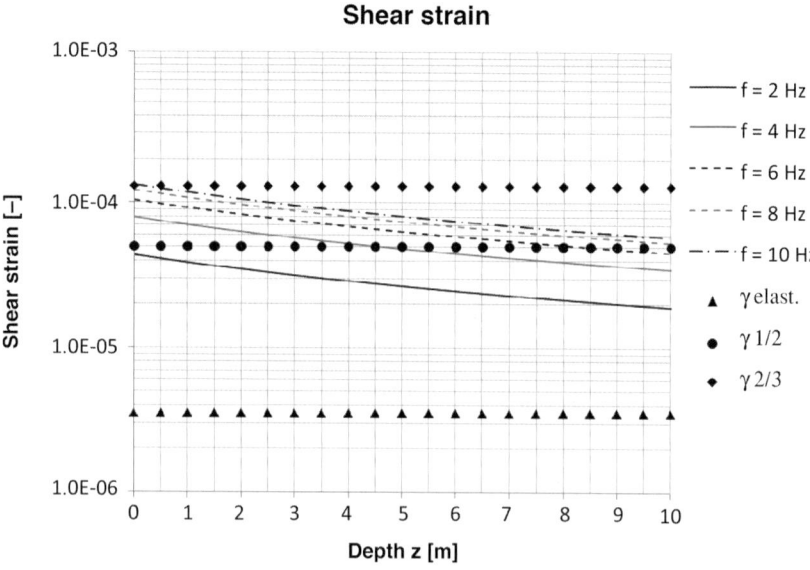

300

Input data:

Frequenz: $f = 2$ to 10 Hz Surface area: Rectangle where:
 B · L: 3.0 · 10.0 m
Load amplitude: $2\,\sigma_{dyn} = 41.0$ kN/m^2 Constrained modulus: $E_{s,k} = 75.0$ MN/m^2

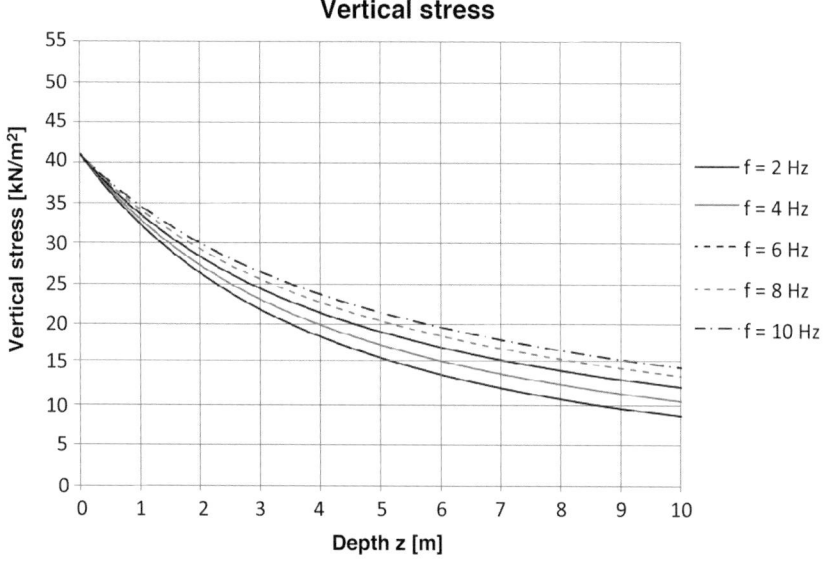

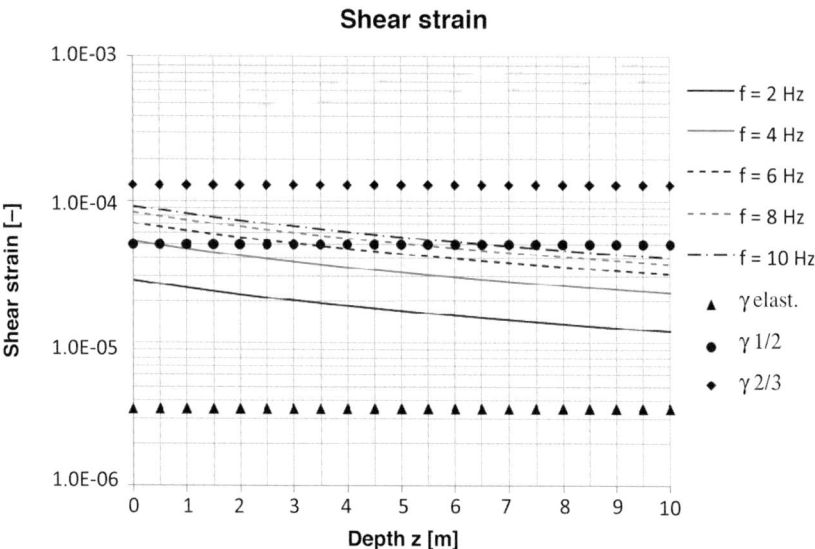

Input data:

Frequenz: $f = 2$ to 10 Hz   Surface area: Rectangle where:

                                                       $B \cdot L: 3.0 \cdot 10.0$ m

Load amplitude: $2\,\sigma_{dyn} = 41.0$ kN/m^2 Constrained modulus: $E_{s,k} = 150.0$ MN/m^2

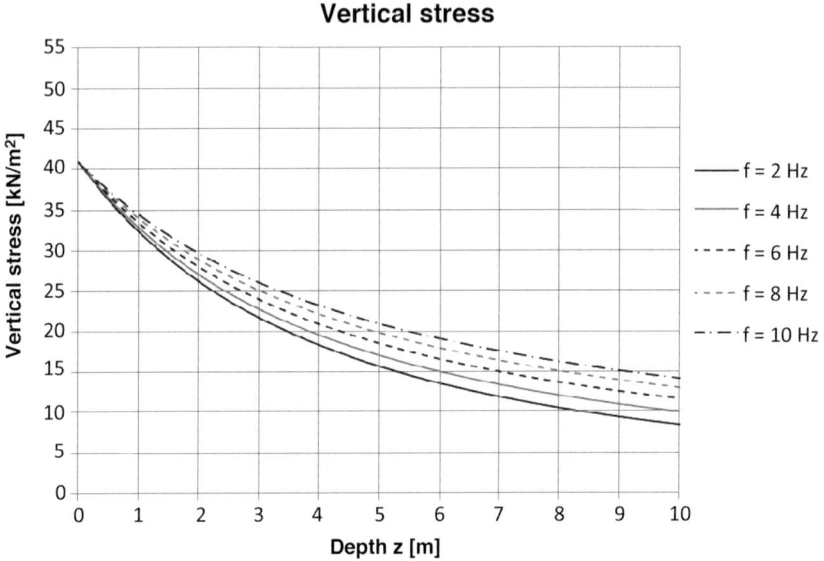

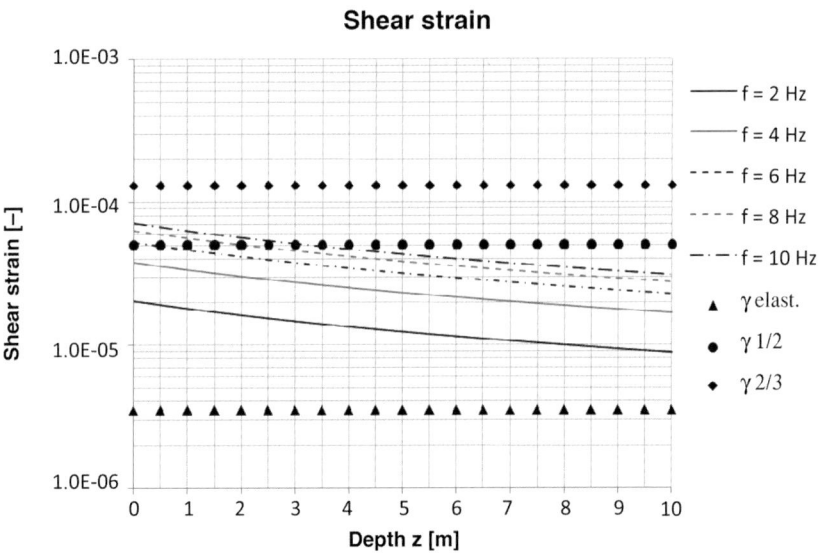

Input data:
Frequenz: f = 2 to 10 Hz   Surface area: Rectangle where:
                                                                 B · L: 3.0 · 10.0 m
Load amplitude: 2 σ_{dyn} = 23.8 kN/m^2 Constrained modulus: $E_{s,k}$ = 50.0 MN/m^2

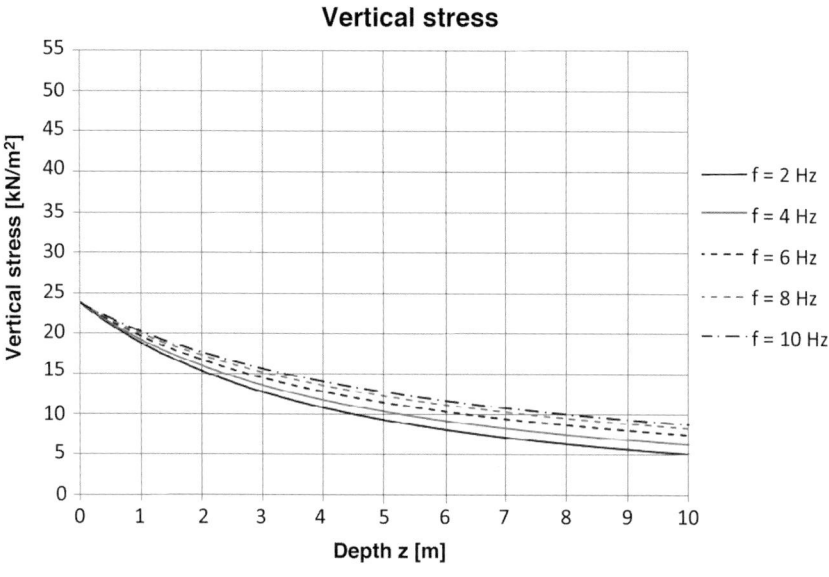

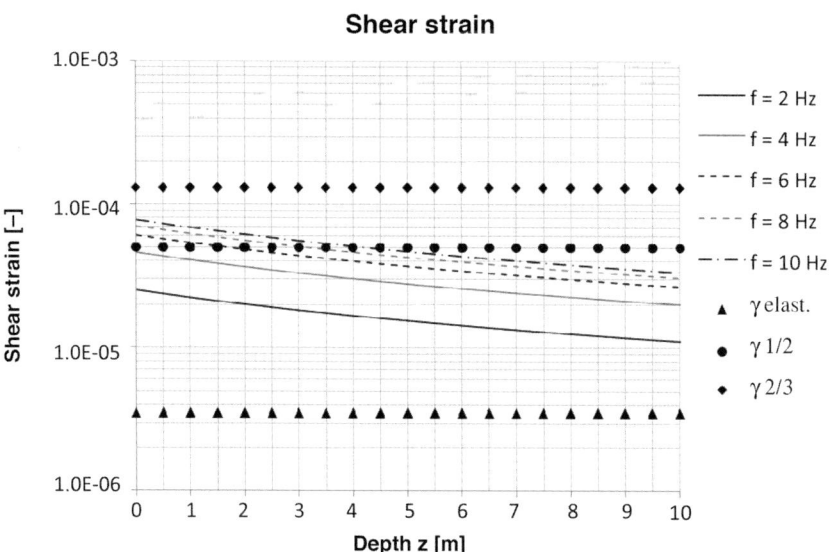

Input data:

Frequenz: $f = 2$ to 10 Hz   Surface area: Rectangle where:
$B \cdot L: 3.0 \cdot 10.0$ m

Load amplitude: $2\,\sigma_{dyn} = 23.8$ kN/m^2 Constrained modulus: $E_{s,k} = 75.0$ MN/m^2

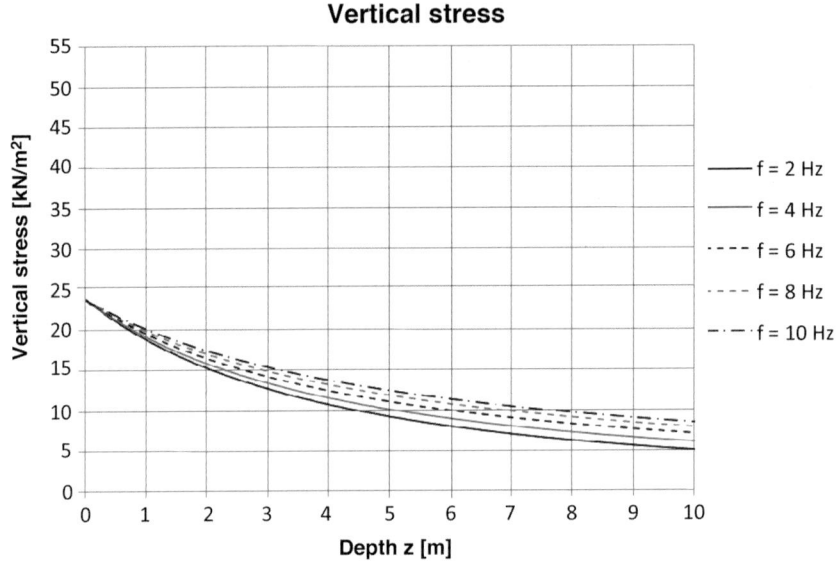

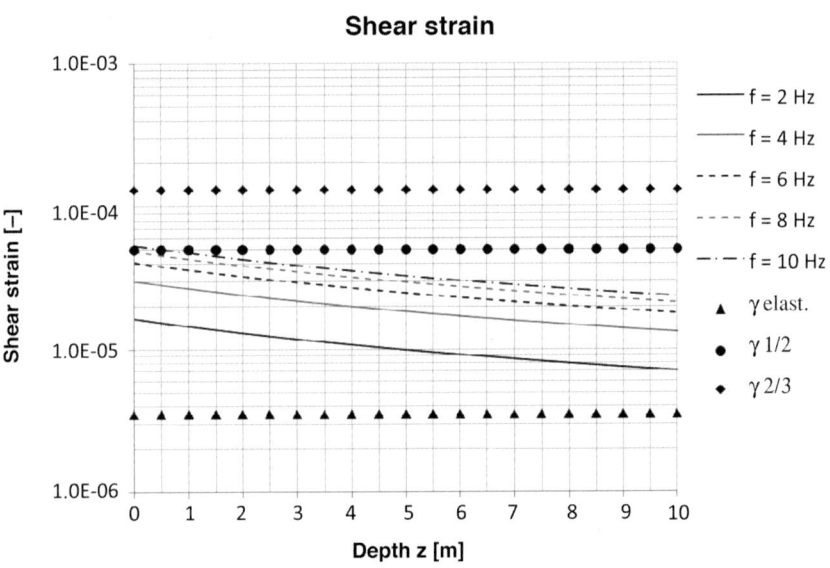

304

Input data:
Frequenz: f = 2 to 10 Hz   Surface area: Rectangle where:
                                                          B · L: 3.0 · 10.0 m
Load amplitude: 2 σ_{dyn} = 23.8 kN/m²   Constrained modulus: $E_{s,k}$ = 150.0 MN/m²

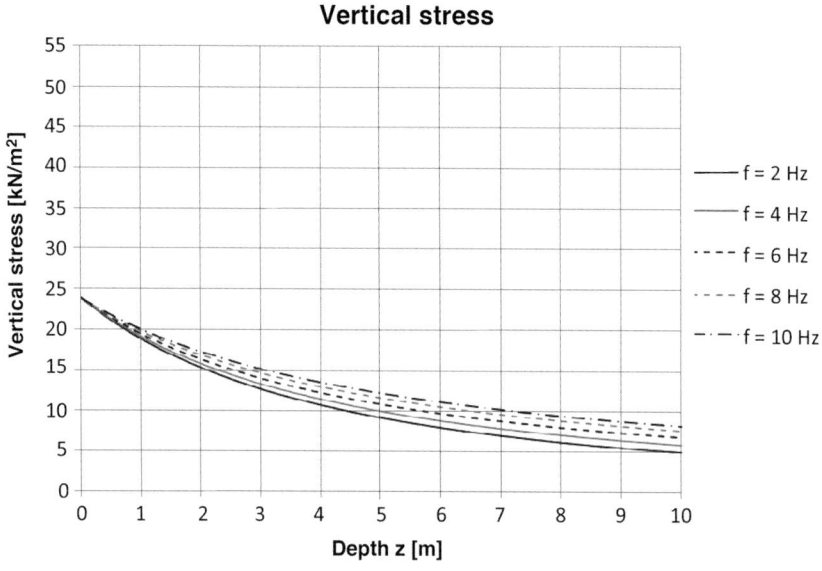

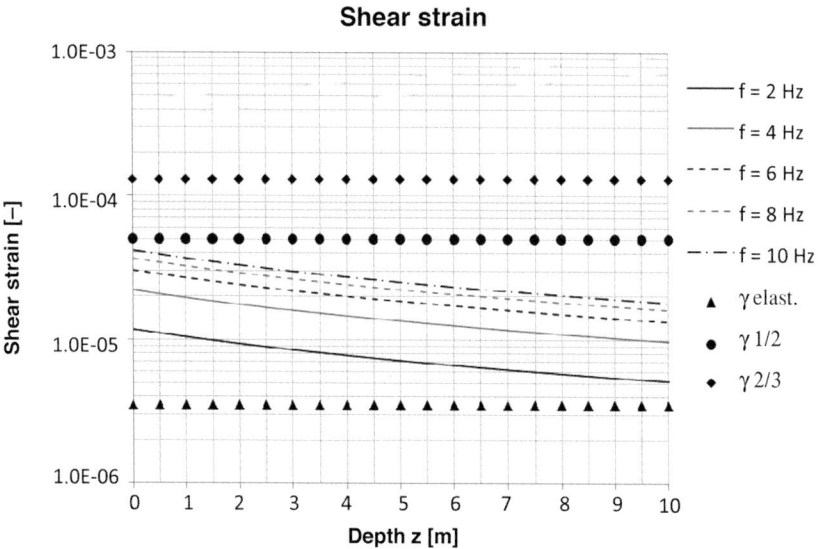

13 Figures

14 Tables

Advertising List